Chemical Engineering Series

Crushing and Grinding

THE SIZE REDUCTION OF SOLID MATERIALS

Crushing and Grinding

THE SIZE REDUCTION OF SOLID MATERIALS

George Charles Lowrison

LONDON
BUTTERWORTHS

THE BUTTERWORTH GROUP

ENGLAND
Butterworth & Co. (Publishers) Ltd
London: 88 Kingsway, WC2B 6AB

AUSTRALIA
Butterworths Pty Ltd
Sydney: 586 Pacific Highway, NSW 2067
Melbourne: 343 Little Collins Street, 3000
Brisbane: 240 Queen Street, 4000

CANADA
Butterworth & Co. (Canada) Ltd
Toronto: 14 Curity Avenue, 374

NEW ZEALAND
Butterworths of New Zealand Ltd
Wellington: 26–28 Waring Taylor Street, 1

SOUTH AFRICA
Butterworth & Co. (South Africa) (Pty) Ltd
Durban: 152–154 Gale Street

First published in 1974

ISBN 0 408 70574 4

Filmset by Keyspools Ltd, Golborne, Lancs.

Printed in England by Hazell Watson & Viney Ltd,
Aylesbury, Bucks.

To Eleanor, my wife,
who learnt to type in order to prepare this book

Preface

This book, so far as I am aware, is the first book devoted to both the theory and practice of the comminution of solid materials. It has been a very great pleasure to compile, though a severe exercise in eclecticism. In one bibliography (*Crushing and Grinding*, H.M.S.O., 1958) more than 2800 references are given and now, I understand, another bibliography containing 4000 additional references is in course of preparation by the Institution of Chemical Engineers. Many of these papers justify reading and re-reading. Yet the subject remains largely empirical. That the empiricism is useful, that the subject is fascinating, that most interesting techniques have been used towards its elucidation, it is the purpose of this book to show. Should it be so fortunate in its reception as to warrant another edition, I propose to make notes on additional papers as I read them and modify the text accordingly. I should be most obliged to readers if they would draw my attention to papers I have not had time to read, which elucidate or conflict with points made in the text or deal with an aspect that I have neglected; to authors of appropriate papers if they would send me copies; and to those who compare statements made in the book with their experiences and send me their comments.

I struggled to prepare a chapter on comminution costs, but the information I was able to gather on items other than power consumption was insufficient to support any generalisations. I should be particularly grateful, therefore, for data on capital, installation, maintenance and operating costs.

Not the least part of the pleasure from writing this book has come from the meetings which it has been possible to arrange with manufacturers of comminution equipment and their representatives. In particular, I should like to thank the following for the information they gave me with so much care, patience and enthusiasm: Mr. N. E. Croft (Allis Chalmers, Great Britain Ltd); Mr. L. W. Hudson (Pegson Ltd); Messrs. P. E. Kellington and N. McKenzie (Newell Dunford Engineering Ltd); Mr. A. J. Moyes (W. C. Holmes and Co. Ltd); Mr. R. North (International Combustion Ltd); Mr. C. L. Prasher (Babcock and Wilcox Ltd); Mr. H. Taylor (Bradley and Craven Ltd); and Mr. W. A. Westbrook (Kek Ltd). I should also like to thank the many authors of papers and manufacturers who have sent me copies of their publications. How I have used and interpreted their information is, however, entirely my responsibility.

4 Cranbourne Road
Bradford
West Yorkshire, BD9 6BH.

G. C. L.

Contents

USUAL MEANING OF SYMBOLS USED vii

1 INTRODUCTION 1

2 THE MEASUREMENT OF SIZE REDUCTION 3
Methods of Measuring Particle Size 4
Methods of Measuring Surface Area 5
Comparison of Methods 6
Relationship between Surface Area and Particle Size Distribution 6
Representation of Size Distribution 10
Appreciation of Particle Size 10
Reduction Ratio 10

3 THE NATURE OF BREAKING 15
Resistance to Breaking 16
The Transformation of Forces 19
Conditions for Breaking 23
Points of Weakness 23
Behaviour of Single Crystals 25
Behaviour of Granular Forms 29
Propagation of Cracks 30
Crack Healing 31
Mechanism of Compression 32
Temperature Effects 33
Effect of Rate of Application of Force 34
Hardness 35
Abrasion 41
Ductility 44
Stickiness 46

4 THE ENERGY CONSUMED IN THE COMMINUTION PROCESS 49
Energy Consumed by Material 50
Grindability 57
Energy Supply Required by Comminuting Equipment 60

5 THE SIZE AND SHAPE OF COMMINUTED PARTICLES 67
Size Distribution 67
Breakage and Selection Functions 75
Degree of Fineness 79
Natural Grain Size 79
Geometry of Broken Particles 81
Definition of Shape 85
Effect of Continued Comminution 87
Kinetics of Grinding 91
Size Control During Milling 92
Closed-circuit Milling 93
Closed-circuit Calculations 94

6 THE BEHAVIOUR OF MIXTURES DURING COMMINUTION 101
Mixtures of Two Solid Components 101
Wet and Dry Comminution 103
Additives to Aid Grinding 108
Concomitant Effects of Comminution 111

7 THE CLASSIFICATION, SELECTION AND LAYOUT OF COMMINUTION EQUIPMENT 115
Earliest Classification 115
Classification According to Purpose 115
Classification According to Comminution Process 117
Mechanics of Comminuting Equipment 117
Preliminary Considerations for Selection of Equipment 119
Plant Layout and Design 123

8 NIPPING MACHINES 129
Nipping 129
Jaw, Gyratory and Roll Crushers 132
Jaw Crushers 133
Gyratory Crushers 146
Roll Crushers 157
Comparison of Jaw, Gyratory and Roll Crushers 168
Disc Crushers 170
Pan Mills 173
Ring Roll and Ring Ball Mills 181
Raymond Mill 183
Lopulco or Loesche Mill 185
Babcock and Wilcox Ring Ball Mill 186
Capacity of Ring Roll and Ring Ball Mills 188
Power Consumption of Ring Roll and Ring Ball Mills 189

9 IMPACTING MACHINES 191
Mechanism of Impact 192
Rotary Hammer Machines 193
Mechanism in Rotary Impact Mill 196

Rotor Speed 197
Number of Impactors or Hammers and Their Shape 198
Nature and Design of Rotor 199
Impact Plates 200
Size of Product 201
Shape of Product Particle 201
Method of Feeding 202
Air Flow in Machines 203
Maintenance 203
Capacity 204
Power Consumption 204
Additional Processing with Rotary Impactor 204
Pin Mills 205
Power Consumption 208
Capacity 208
Stamp Mills 209
Pestles 210
Drop 210
Water Usage 210
Size of Feed 210
Size of Product 211
Capacity 211
Power 211
Vibration Mills 211
Advantages of Vibration Mill 212
Disadvantages of Vibration Mill 213
Torus Mill 213
Cylindrical Mill 216

10 TUMBLING MACHINES 225
Rod-loaded Mills 226
Ball-loaded Mills 234
Autogenous Mills 249

11 CUTTING MACHINES 255
Mechanism of Cutting 255
Materials of Construction 256
Types of Cutting Tool 257

12 ATTRITING MACHINES 261
Buhrstone 261
Colloid Mill 262
Petit Pulveriser 262
Fluid-energy Mill 263
Sand Grinder 266
Ultra-rotor 267

13 OTHER METHODS OF COMMINUTION 271
Weathering 271
Thermal Size Reduction 271
Pressure Alteration 272
Ultrasonics 272
Electrical 273
Chemical 274
Explosion 275

INDEX 281

Usual Meaning of Symbols Used

Å = Angstrom units (10^{-10} m)
a = interatomic or interionic distance *or* a constant
b = diameter of material particle
[1]C = capacity in units of weight or of volume per unit time
c = pore diameter
d = diameter of item of equipment
E = Young's modulus
e = exponential series
F = sieve size through which 80% of feed passes
f = function of
g = acceleration due to gravity
H = hardness
h = height
J = proportion of volume of mill occupied by the bulk of the media

$$= \frac{\text{volume of media} + \text{volume of space between media}}{\text{volume of mill}}$$

K = constant
L = length of mill
l = grain size *or* length
m = exponent, particularly in Gaudin's equation *or* meter
N = number of strokes or revolutions in unit time
n = number of items *or* exponent, particularly in Rosin–Rammler equation
P = percentage
p = a particular stress
q = open setting of equipment
R = reduction ratio *or* radius
r = radius
S = surface area per unit of volume
s = specific heat
t = temperature, time *or* discharge setting of equipment
U = ultimate strength
V = rate of application of strain *or* proportion of voids between media occupied by material
v = half total length of crack *or* velocity
W = load *or* power consumption

W_i = Bond's work index for material
W_e = the value equivalent to Bond's work index actually obtained on the particular machine
x = sieve size through which particle just passes
Z = sieve size through which 80% passes

α = amplitude *or* a constant
β = a constant *or* a component
Γ = gamma function
γ = surface energy per unit of surface *or* a component
Δ = modified density
ε = strain
η = efficiency
θ = a specific angle
μ = coefficient of friction
π = ratio between circumference and diameter of a circle
ρ = density, usually of media
σ = Poisson's ratio = lateral strain/longitudinal strain *or* ultimate tensile strength *or* density of material *or* surface tension
ϕ = function of
ψ = inclination of cone side to vertical
ω = angular velocity or frequency

1

Introduction

The purposes for which solids are comminuted are (a) to decrease the size of material; (b) to increase the surface area of material; and (c) to free material from its matrices.

The origins of comminution are almost coeval with man. As an intentional process, TYLOR[1] suggests it began with 'the rudest tribes of savages . . . for powdering charcoal and ochre to paint themselves with or the more useful work of bruising wild seeds gathered for food. The whole apparatus consists of a roundish stone held in the hand and a larger hollowed stone for a bed. . . . Now anyone using the pestle and mortar may notice that it works in two ways, the stuff being either pounded by striking or ground by rubbing against the side of the mortar. . . .' Thus were crushing and grinding discovered and distinguished by our oldest ancestors.

Today there are few industrial, agricultural or domestic processes which do not employ comminution in some form. RUMPF[2] has estimated that, of all the energy artificially generated in the world, about one-twentieth is consumed in the size reduction of solids. In some industries comminution is the major process and accounts for most of the cost. A modern road, with three lanes and 30 metres wide, uses for every kilometre of its length more than 30 000 tonnes of material, all of which has probably been comminuted. The comminution of domestic waste has much to commend it as a means of disposal and it can be done very effectively on a huge scale. The waste is much reduced in volume. By the thorough mixing which takes place as a concomitant of the pulverisation bacterial decomposition is accelerated, heat is produced which repels flies, and rats are repelled by the thorough admixture of such materials as glass and brick dust with the organic matter. The mixture easily composts and provides some fertiliser (albeit in very low concentration) for any land to which it is added. Furthermore, if incineration is necessary, the material is in a convenient form for it.

Comminution is employed on a vast scale for the separation of the useful constituents of minerals. Nearly every metal that is used by man owes its freedom to size reduction processes. Coal prepared as a fuel, whether for the hearth or the power station, is equally dependent upon them. Milling is such an important part of the manufacture of flour that from early times the man so employed was called a miller. In the confectionery industry materials are made tasty by being reduced to a small size. In the manufacture of components for the electronics industry, of paints and dyes, of paper, of medicines, of cleaning materials, of fertilisers, of prepared foods, of ceramics from which to eat or imbibe them, of bricks, of cement and concrete; in the preparation of food for farm animals; in the preparation of the soil in which to grow crops; in all these and much besides comminution plays an important part.

REFERENCES

1. TYLOR, E. B., *Anthropology*, 1930. London; Watts
2. RUMPF, H., *Dechema Zerkleinern Symposion, 1962*

2

The Measurement of Size Reduction

Fundamentally, size reduction is the separation of a particle into two parts and the creation of additional surface equal to twice the area over which the separation has taken place; one part of the new area is on what can be considered the mother particle and the second part on the offspring (*Figure 2.1*). Unless any surface due to cavities or pores is blocked in the process, the more times a particle is divided the greater the surface area of the material composed of the parts.

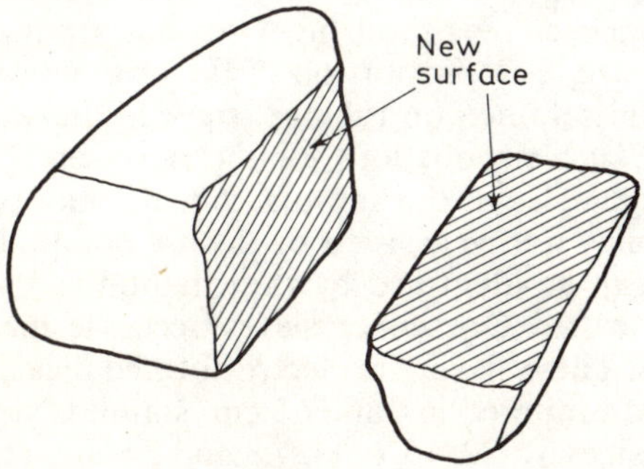

Figure 2.1. New surface produced when a particle breaks

The change in surface area of a solid is therefore a measurement of the change in size of the particles composing the solid, i.e. a measurement of size reduction. This is not to say that the change of surface area is exclusively due to the size reduction process; the size reduction process may just have been a means of exposing surface area which existed undetected before other parts of the particle were broken.

The surface areas of a series of geometrically similar figures are related as the squares of the lengths of any similar sides of those figures. There is an underlying assumption in most particle size determinations that the particles in a powder are geometrically similar (for a discussion on the evidence for such an assumption see p. 81) and therefore that surface area is calculable from the particle size determination.

METHODS OF MEASURING PARTICLE SIZE

The following are among the more usual methods of particle size determination. For an indication of their range of measurement see *Figure 2.5* and for definitions of standardised procedures for many of the methods see British Standard Specification number 3406.

Wedge gauge The wedge gauge is a method of determining the largest particle present in a fine paste[1]. It consists essentially of putting a film of the paste on the surface of an accurately ground stopper of small taper and pushing the stopper home into the neck of an accurately ground hole of the same taper; the extent to which the stopper is proud of its 'clean' position is a highly magnified (according to the angle of taper) measurement of the thickness of the film on the stopper and therefore of the size of the largest particle present in the film. The method was developed particularly for particle sizing of ground chocolate.

Sieving A sample of the powder is passed through a series of sieves each made of two sets of wires of uniform cross-section which are woven at right angles to each other, warp and weft; each wire in a set is fixed at a designated uniform distance from each neighbouring wire, thus producing a square mesh of holes of uniform size. The proportion of the original sample passing one sieve in the series and retained on the next sieve in the series is determined. The complete set of such proportions for the sieve series expresses the size distribution of the particles in the sample. Sieving may be on a dry or wet sample and may or may not be assisted by the use of liquid jets. The smallest particle size which can be identified by this method is about 50–75 μm, although by using so-called Daeschner sieves accurate measurements down to 5 μm are claimed[2]. These sieves are electro-formed nickel plates containing square apertures and mounted in 3 in (7.5 cm) stainless steel frames. The set consists of aperture sizes 31, 22, 15, 6, 11, 7.8 and 5.5 μm, i.e. in the $1/\sqrt{2}$ ratio common to British and American systems. MULLIN[2] describes a technique for locating these sieves in a bath of liquid and subjecting the whole to ultrasonic vibrations whereby separation is effected.

Sedimentation A sample of the powder is suspended in a liquid in which it is insoluble, and the resulting suspension is added to the top of a column of the same liquid. The heavy particles fall quickly through the liquid and the light particles fall slowly. ANDREASEN[3] pipetted off samples from the settling liquor at a fixed depth and weighed the suspension. A method of doing this automatically[12] is to place a weighpan near the bottom of the column of liquid, but in the liquid; the weight of solid reaching that position can be recorded at fixed time intervals. From a modified Stokes' equation the equivalent average sphere diameter in the time intervals can be calculated. A refinement of this method is to use an X-ray scanner over the height of the column to measure the light obscuration caused by the particles in various parts of the column after a certain time interval, and thus to measure the particle size distribution from this darkness density in various depths of the liquid. A modification is to increase the gravitational pull by centrifuging the material, which makes the sedimentation both faster and more effective for the smallest particles.

Elutriation A sample of the powder is subjected to a series of gas flows of increasing velocity; the lightest particles are entrained by the slowest flows and the heaviest by the swiftest flows. The solid in the series of gas flows is separated and weighed. The equivalent sphere diameters can be calculated by a modified Stokes' equation.

Microscopy A sample of the powder is placed on a microscope slide and the number of particles occupying grid areas of a range of sizes is counted. The proportion of the total in each size range is then calculated.

METHODS OF MEASURING SURFACE AREA

Among the methods used to determine surface area directly are the following. Standardised procedures for many are given in British Standard Specification number 4359.

Adsorption In certain circumstances the surface of solids can be covered by a molecular layer of another substance. The problem is to know when the surface is completely covered and when another layer is beginning to be built onto the monolayer. This problem for gaseous adsorption was solved by the investigations of Brunauer, Emett and Teller. By use of their formulations the area accessible to the gas used can be calculated from the quantity of gas adsorbed (the B.E.T. method). The area so determined includes all accessible cracks, crevices and pores down to about 20 Å (2×10^{-3} µm).

Permeability The frictional resistance to the flow of a fluid through a capillary, when the flow is streamline, is proportional to the area of the walls of the capillary. A closely packed bed of particles can be considered to be a complex of capillaries, the walls of which are the surface area of the particles. This principle is used to determine the surface area of a mass of particles by determining the pressure drop of a gas through a bed of the particles, particularly by the method of LEA and NURSE[13].

Rate of dissolution It is argued that the rate of dissolution, whether simple solubility or chemical reaction, of a solid in a liquid is proportional to the surface area of the solid exposed to the solvent. Thus by immersing a solid in a liquid solvent, with stirring to expose the whole particle surface to the solvent, for a fixed time and weighing the residue against standards, a weight of the dissolved solid can be related to a surface area. The method has been used with quartz and hydrofluoric acid by MARTIN[14]. There are dangers in this method (see p. 90) since some materials can undergo chemical change during size reduction which may enhance or inhibit reaction with a reagent. For example, GREGG[4] found that kaolin reacted much more vigorously with hydrochloric acid after a certain degree of reduction had been reached than could be explained simply as being due to increased surface area.

Other methods of determining surface area which could have particular application include the two following.

Magnetic coercivity A relationship has been established between the coercivity of magnetic materials and their surface area[5]. A direct measurement of coercivity is possible and, hence, a measurement of surface area.

Proton beam energy loss By measuring the average energy loss and spread of proton beams after they pass through a thin section of the material of the order of 1 mm thick, a value for the porosity of the material can be obtained. The advantage of the method (devised by U.K.A.E.A., Harwell) is that it can measure both open and closed pores and when combined with other methods could be used to determine the closed pores before and after comminution; thus a value could be obtained for the surface area actually created by the comminution process.

Figure 2.2 shows a classification of methods of particle size analysis[6].

COMPARISON OF METHODS

Most of the methods very briefly described above have been widely used and found to give reasonably reproducible results provided an exact procedure is followed such as is set down in the British Standard and A.S.T.M. Specifications[15]. What is disconcerting is that the results of one method rarely corroborate the results of another method. HEYWOOD[10] states that sieving, depending upon the method used, can give results which, if they are to include 95% of all results obtained, can vary by ±17% over the average of the results for any size range. PIRET and his co-workers[11] show that values obtained by the B.E.T. method for one kind of material (quartz) can be from 1.8 to 2.2 times as large as those obtained by the permeability method. Hydrofluoric acid dissolution on the same material gave areas three times as large as those obtained by permeability. The permeability method has been shown[7] to be susceptible to fatty or viscous additives, which decrease permeability and give a false measurement of particle size.

RELATIONSHIP BETWEEN SURFACE AREA AND PARTICLE SIZE DISTRIBUTION

While particle size distribution is, at least theoretically, convertible into surface area (see, for example, British Standard Specification 4359: Part 3: 1970, 'Calculation from particle size distribution'), surface area may not seem to have any obvious relationship to the way the various particles are

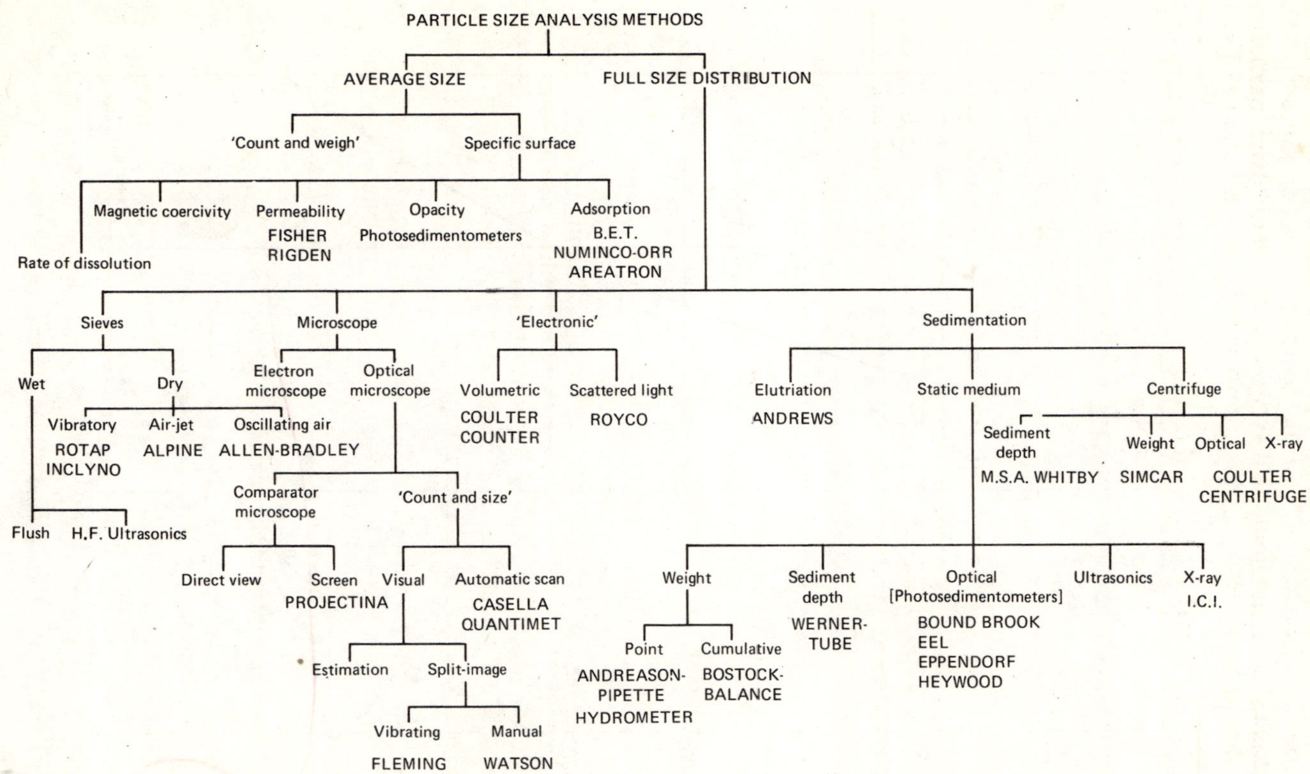

Figure 2.2. Classification of methods of particle size analysis. (After W. HURST *and* G. SPENCE[6]*)*

distributed into sizes. However, for a very long time an empirical index has been in common industrial use as a crude measure of surface area, viz. the screen size through which 80% of the material passes (Z). KIHLSTEDT[8] has examined this index and found that very often the surface area bears a direct relationship to it such that $S.\sqrt{Z}$ is approximately constant and has a value

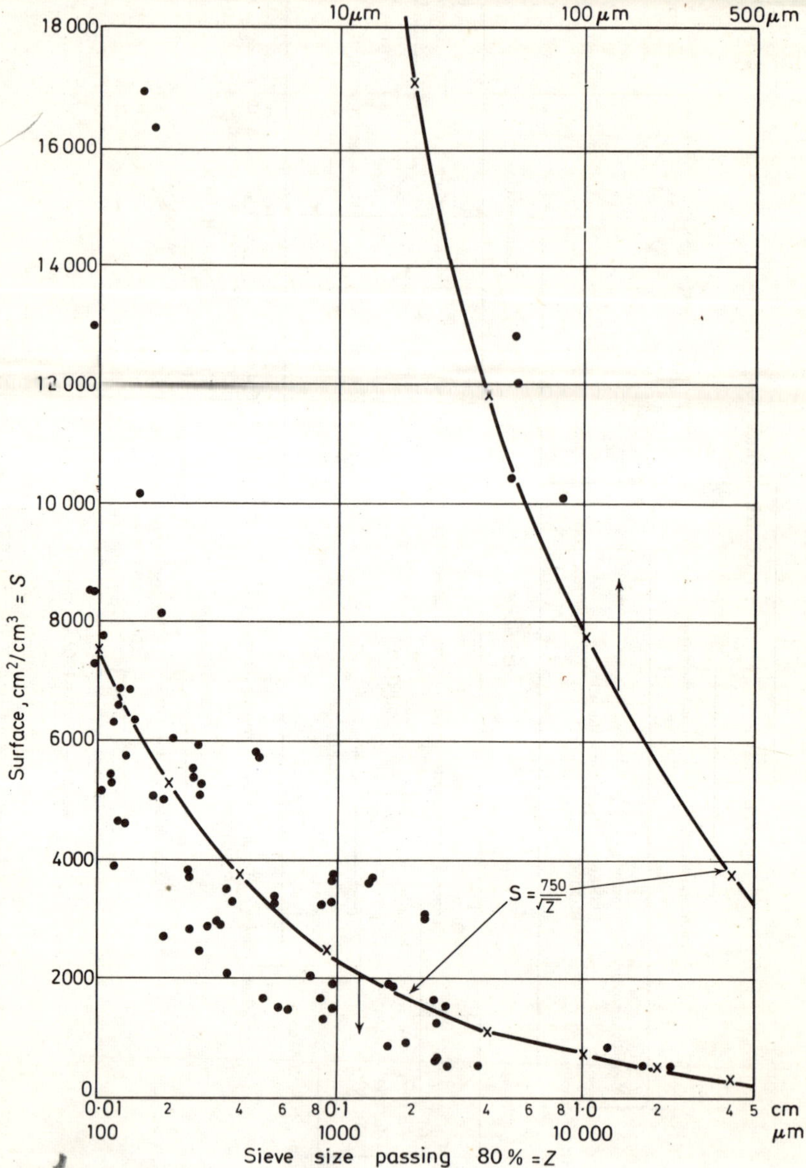

Figure 2·3. Kihlstedt's relationship between sieve size passing 80% and surface area

of around 750, where S = specific surface in square centimetres per cubic centimetre and Z = the screen size in centimetres through which 80% passes.

The Kihlstedt relationship is shown in *Figure 2.3* for a wide range of materials. The constant varies with the screen size, but for the normal sieving range, i.e. 75 µm (0.0075 cm)–50 000 µm (5 cm), is approximately 750.

Thus this figure—the screen size passing 80% of the material—crudely combines in one index both surface area and particle size distribution. The

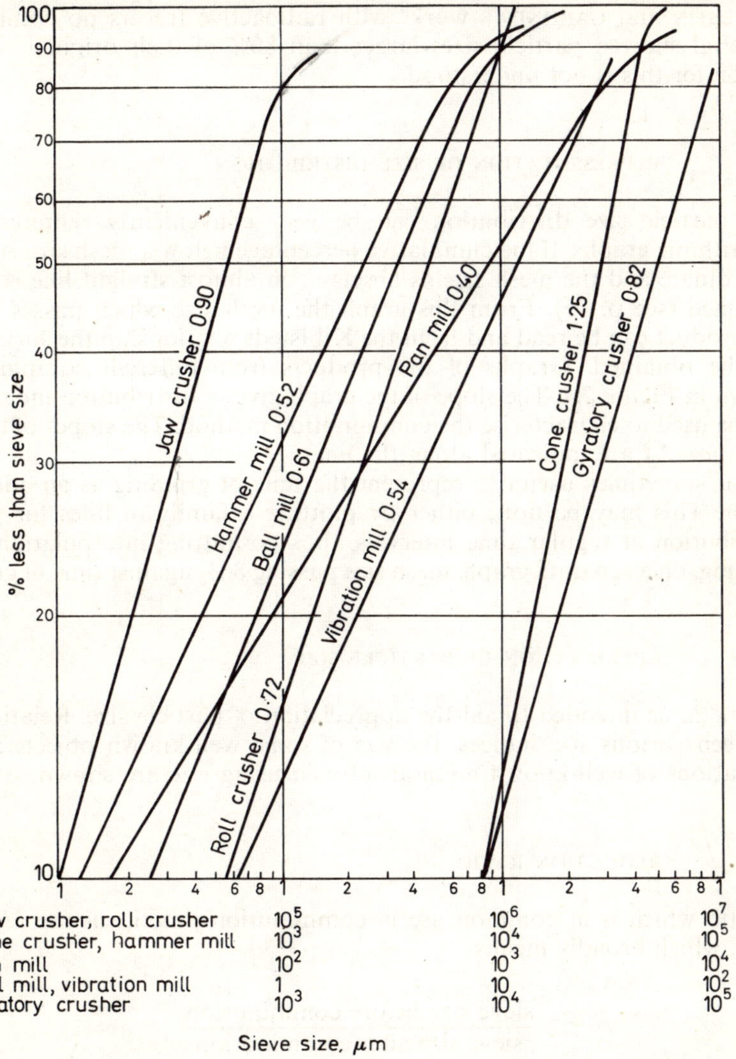

Figure 2.4. Method of representing size distribution and of distinguishing equipment by slope of graph

use of 80% passing rather than 100% passing to characterise a material avoids the undue influence of the topmost screen size, which may be determined by just an odd particle (see p. 12). To use less than 80% does not seem to be justified, because when percentage passing is plotted logarithmically against sieve size an approximately straight line is obtained up to about the 80% figure (see *Figure 2.4*); above this figure the line curves. To take a much smaller figure, say 50%, would in many cases be to remove characteristic sizes. It would be reasonable to assume that the unchanged (not size-reduced) material shows up more prominently at the largest sizes. In AUSTIN and GARDNER'S work[9] with radioactive tracers no comminuted material showed particle sizes larger than 80% of their original size. The reason for this is not understood.

REPRESENTATION OF SIZE DISTRIBUTION

The particle size distribution can be very conveniently represented on logarithmic graphs. If the cumulative percentage below a mesh size is plotted as ordinate and the mesh size as abscissa, an almost straight line is usually obtained (see p. 75). From this graph the mesh size which passes 80% of the product can be read and from the Kihlstedt relationship the surface area can be obtained. Graphs of the products from different equipment are shown in *Figure 2.4*. The slope of the graph gives a distribution index which can be used to characterise the comminution method. The slopes of the lines in *Figure 2.4* are indicated along the lines.

It is sometimes useful to represent the time of grinding as an additional factor. This may be done either by plotting a family of lines for the size distribution at regular time intervals, thus permitting interpolation, or by plotting, on a separate graph, mesh size passing 80% against time of grinding.

APPRECIATION OF PARTICLE SIZE

Figure 2.5 is intended to aid the appreciation of particle size. Relationships between various size indices, the size of some well-known objects and the limitations of well-known methods of measuring size are shown.

REDUCTION RATIO

A ratio which is in common use in comminution practice is the reduction ratio, which broadly means

$$\frac{\text{sieve size before comminution}}{\text{sieve size after comminution}}$$

Therefore it is always a number larger than unity. But, as will be readily appreciated, a particle does not break down into a number of particles of

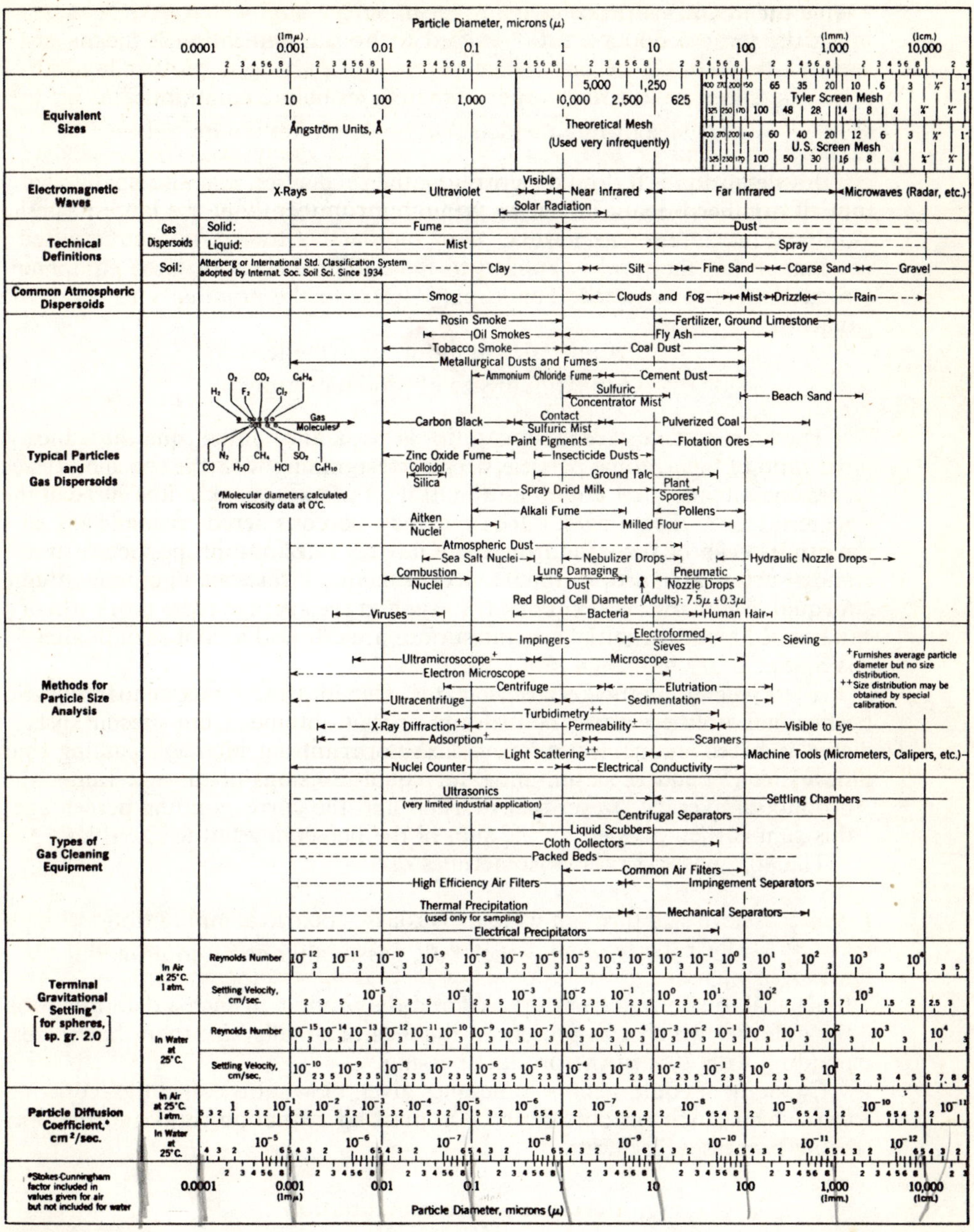

Figure 2.5. Characteristics of particles and particle dispersoids. (Reproduced by permission of The Bauer Bros. Co., Springfield, Ohio, USA)

equal sieve size; invariably the comminuted products are of different sizes. How then can this ratio have any meaning?

If the term 'reduction ratio' is used without qualification it means

$$\frac{\text{sieve size through which 100\% passes before comminution}}{\text{sieve size through which 100\% passes after comminution}}$$

If this definition is followed rigorously, then reduction ratio is usually a very small number, because there are almost invariably some particles near the original maximum size present.

Sometimes the term is applied to the equipment, so that the equipment reduction ratio is stated. This usually refers to the nominal settings of the machine:

$$\frac{\text{upper size limit for receiving feed}}{\text{maximum size of product setting}}$$

The term 'average reduction ratio' is used and, unless qualified, means the ratio of the average particle sizes corresponding with the specific surface areas before and after comminution. If the surface area of unit volume of the material is S, then this surface area may be considered as made up of n particles each of size x and therefore having a volume proportional to nx^3 and a surface area proportional to nx^2 or having surface area per unit volume S equal to $nx^2/nx^3 = 1/x$. Thus $1/S = x$ and the average reduction ratio of a material having an initial specific surface area S_1 and a final surface area S_2 is S_2/S_1.

A less elegant method of arriving at a value for the average reduction ratio which will not necessarily be the same as that obtained from specific surface area is based upon plotting sieve size (x) against percentage passing that sieve size (P) and then summing, for equal divisions of the size range, the product of average size in that division and the corresponding percentage; this sum is then divided by the sum of the percentages, thus: $\Sigma xP/\Sigma P$.

The 80% passing ratio is sometimes used:

$$\frac{\text{sieve size through which 80\% passes before comminution}}{\text{sieve size through which 80\% passes after comminution}}$$

It is usual to subtract from the feed and the product in such a determination the quantity which was present in the feed of smaller size than that corresponding with 80% of the product passing.

Capacity of equipment is sometimes given by what is called the reduction ton, which is the product of the 80% passing reduction ratio and the tons being processed per hour.

REFERENCES

1. LOCKWOOD, H. C., *Chemy Ind.* (1958), 1506
2. MULLIN, J. W., *Chemy Ind.* (1971), 1435; COLON, F. J., *Chemy Ind.* (1965), 263
3. ANDREASEN, A. H. M., *Trans. Ceram. Soc.*, **29** (1930), 239

4. GREGG, S. J., *Chemy Ind.* (1968), 611
5. GOTTSCHALK, V. H. and DAVIS, C. W., *Rep. Invest. U.S.*, No. 3268 (1935)
6. HURST, W. and SPENCE, G., *J. Br. Ceram. Soc.*, **6** (1969), 60
7. HIGI, G. G. and RABOTTINO, L., *Dechema-Monogr.*, **57** (1967), 427
8. KIHLSTEDT, P. G., *Dechema Zerkleinern Symposion, 1962*, 205
9. GARDNER, R. G. and AUSTIN, L. G., *Dechema Zerkleinern Symposion, 1962*, 218
10. HEYWOOD, H., *Symposium 'Particle Size Analysis', 1947*, Institution of Chemical Engineers and Society of Chemical Industry
11. JOHNSON, J. F., AXELSON, J. and PIRET, E. L., *Chem. Eng. Prog.*, **45** (1949), 708
12. BOSTOCK, W., *J. Sci. Instr.*, **29** (1952), 209
13. LEA, F. M. and NURSE, R. W., *Symposium 'Particle Size Analysis', 1947*, Institution of Chemical Engineers and Society of Chemical Industry
14. MARTIN, G., BOWES, E. A. and CRISTELOW, J. W., *Trans. Ceram. Soc.*, **25** (1926), 51
15. BRITISH STANDARD SPECIFICATIONS, 1796, 3406 and 4359; AMERICAN SOCIETY FOR TESTING MATERIALS: see for Specific Materials

3

The Nature of Breaking

There is no universal method of size reduction of solids because all solids are not the same in their behaviour towards size reduction methods. The term 'solid' as used in everyday language characterises anything which, without support other than a floor on which to rest, apparently retains its shape during storage. A solid can be a lump of marble, a piece of lead, a block of glass, a lump of pitch, crystals of salt, grass fibres or seaweed.

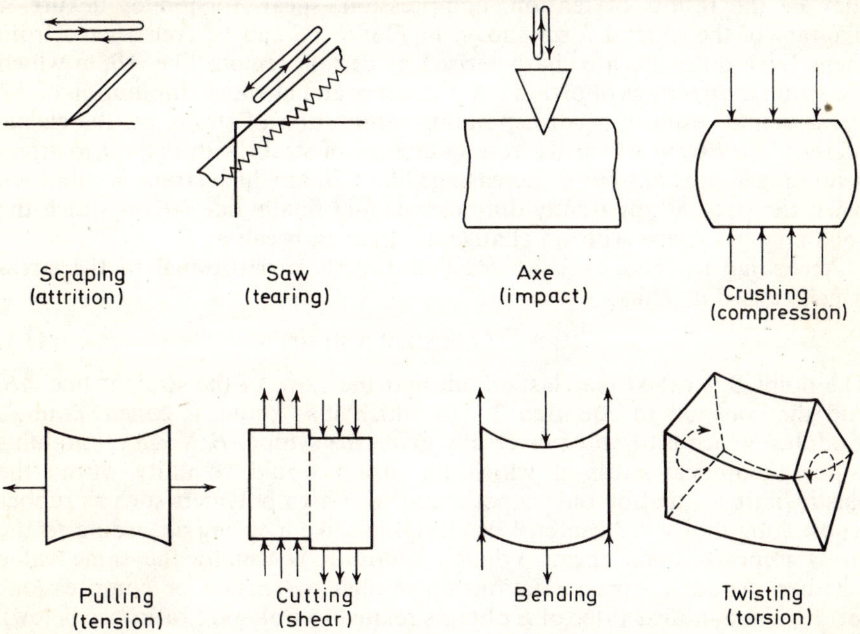

Figure 3.1. Methods of size reduction

In general, all that is necessary to break a solid is to change its shape beyond certain limits; pull it apart (subject it to tension); push it in (subject it to compression); twist it (subject it to torsion); bend it (subject it to flexure); cut it with another solid (subject it to shear); wear it away with another solid (subject it to attrition): subject a solid to any of these processes and, to a degree, the solid will break. *Figure 3.1* illustrates these processes.

RESISTANCE TO BREAKING

The resistance of a material to breaking is usually measured on a sample of the material shaped so as to include a length of uniform cross-section. The testing machine includes two clamps which are arranged to grasp the sample on either side of some of the length of uniform cross-section. The machine is operated so that the clamps subject the sample to known forces of tension, compression, shear, torsion or flexure. The changes in dimensions of the sample are measured. The ratio of the change in a given dimension to the original value is called the strain of that dimension. For tension and compression, strain is the proportion of the change in length. For shear and torsion, the angular displacement is measured. For flexure, curvature is used (see *Figure 3.2*). The force involved is calculated in terms of the force on the area over which it acts divided by that area, i.e. the force per unit area. This is called the stress.

If measurements of strain are recorded for a range of stresses, whether they be the results of tension, compression, shear, torsion or flexure, a diagram of the general form shown in *Figure 3.3* can be constructed from them. Such diagrams are characterised by certain regions like *AB*, in which the strain is directly proportional to the stress and in which diminution of the stress would result in a corresponding diminution of strain, i.e. the elastic region; like *BC*, in which the rate of change of strain with respect to stress is no longer constant but is increasing; like *CD*, in which strain occurs even when the stress is apparently diminished; and finally like *DE*, in which the solid changes shape with no change in stress to break at *E*.

According to Hooke's law, strain is directly proportional to the stress which caused it. Thus

$$\text{stress} = \text{constant} \times \text{strain} \qquad (3.1)$$

The point *B* is called the elastic limit and the slope of the straight line *AB*, and the constant in equation 3.1 to which it is equal, is called Young's modulus, which is almost universally given the symbol *E*. Young's modulus is the theoretical stress at which the strain would be unity within the elastic limit, a situation only experienced with high polymers such as rubber whose long molecules can curl or straighten like a spring according to the stress imposed upon them. Young's modulus is usually the same value whether the stress–strain relationship applies to tension or compression: for shear, a modified value of *E* obtains related to Poisson's ratio (see below). Thus for shear the value is $E/2(1+\sigma)$. As will be shown later, torsion is a form of shear and flexure is a combination of tension and compression.

The point *C* is called the yield point and the stress corresponding to the point *E* is called the ultimate stress. The region *CE* in which flow occurs with little or no increase in stress shows the ductility of the material. The larger the stress at the elastic limit, the stronger the material. The less the difference between the strain at the elastic limit and that at the yield point, the more brittle the material is said to be. Sometimes material is said to be brittle only when it is weak and not ductile or when it is weak, has a high Young's modulus (i.e. is 'rigid') and is not ductile. Sometimes it is said to be

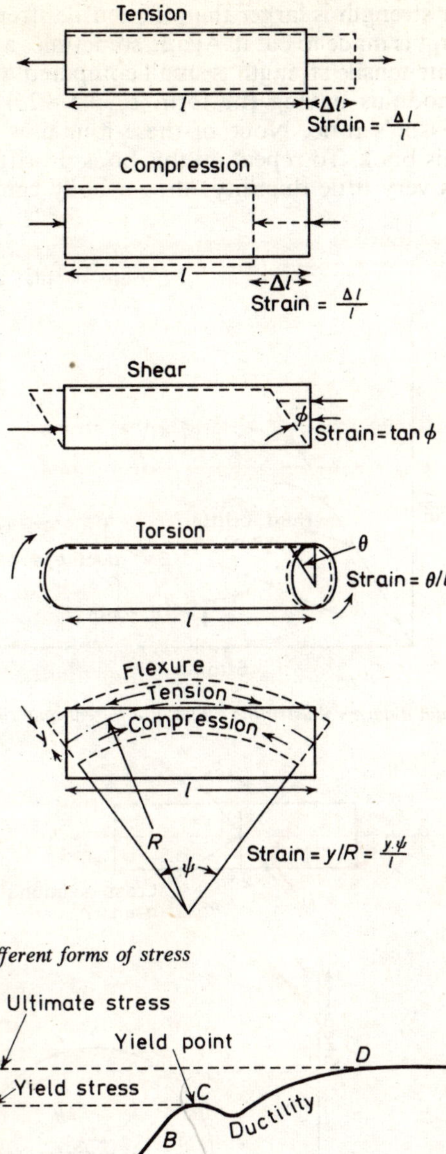

Figure 3.2. Strain from different forms of stress

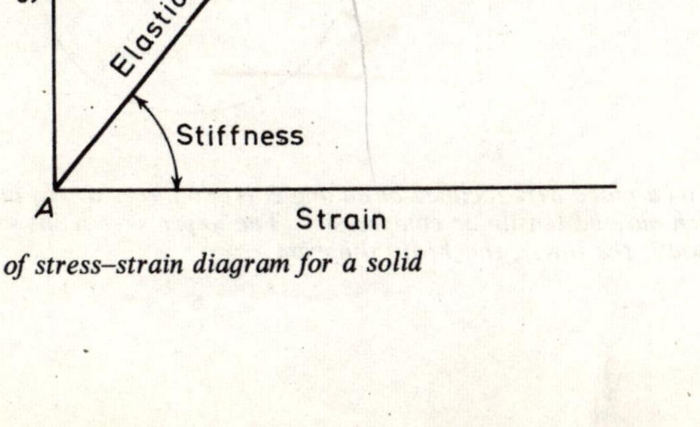

Figure 3.3. Features of stress–strain diagram for a solid

brittle if the shear strength is larger than the tensile strength, i.e. it will break up when an attempt is made to cut it. Again, sometimes a material is said to be brittle if its ultimate tensile strength is small compared with its so-called bulk modulus, bulk modulus having the form $E/[3(1-2\sigma)]$, in which σ is the reciprocal of Poisson's ratio. None of these four uses of the word 'brittle' is employed in this book. To repeat, in this book a brittle body is defined as one which shows very little ductility; thus a body can be both brittle and strong.

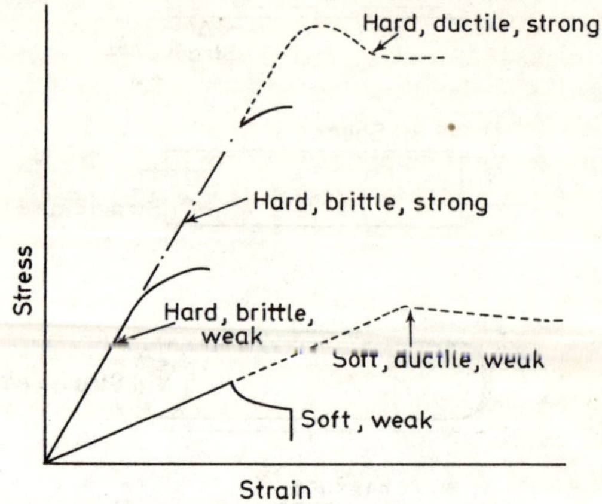

Figure 3.4. Stress–strain diagram illustrating effect of different qualities

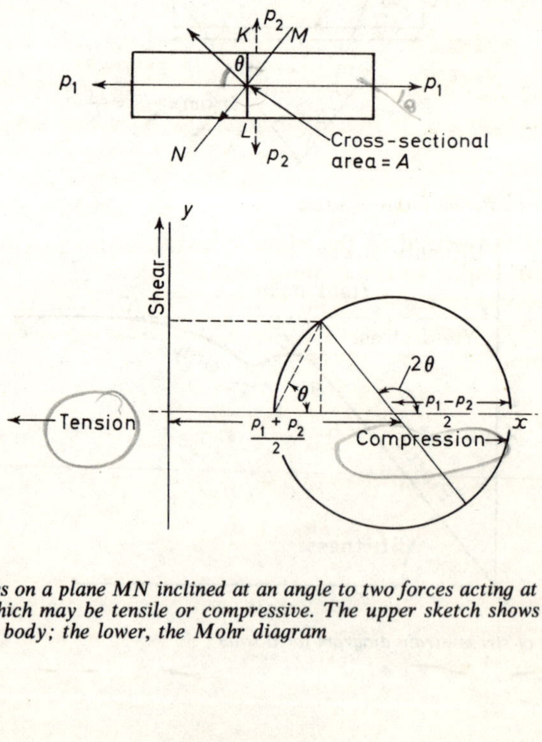

Figure 3.5. Forces on a plane MN inclined at an angle to two forces acting at right angles to each other either of which may be tensile or compressive. The upper sketch shows the arrangement of the forces on the body; the lower, the Mohr diagram

Figure 3.4 shows examples of substances having mechanical properties in different degrees.

THE TRANSFORMATION OF FORCES

Is there any relationship between the effects of these different forms of force—tension, compression, shear, torsion and flexure—or does a body react to any one independently of the existence of any of the others? Consideration of the forms (see *Figure 3.2*) reveals that torsion is a form of shear force, i.e. it manifests itself in the sliding of real or imaginary sections on each other, parallel to the direction of the force, and that flexure might be considered to bend the piece so that it forms part of the circumference of a circle in which those parts of largest radius are being stretched in tension and those of smallest radius about a neutral axis are in compression. Thus the elementary forms to be investigated are tensile, compressive and shear strengths.

If a body, taken to be rectangular for simplicity and having cross-sectional area A perpendicular to its principal axis, is subjected to a tensile force parallel to its principal axis (such as p_1 in the upper part of *Figure 3.5*), then the force can be resolved into components on any plane MN—say at an angle θ to the cross-section KL (see *Figure 3.5*). The component perpendicular to the plane can be seen to be

$$\text{total force} = \text{stress} \times \text{area over which it acts} = p_1 A \cos\theta$$

and, since it acts over an area MN,

$$\frac{p_1 A \cos\theta}{A \sec\theta} = p_1 \cos^2\theta = \frac{p_1}{2}(1+\cos 2\theta) \qquad (3.2)$$

The component tangential to the plane is acting over the same area and is

$$\frac{p_1 A \sin\theta}{A \sec\theta} = p_1 \sin\theta . \cos\theta = \frac{p_1}{2}\sin 2\theta \qquad (3.3)$$

The component tangential to the plane constitutes a shear force and the component perpendicular to the plane a tensile force.

If two tensile forces at right angles to each other act on the body (as p_1 and p_2 in *Figure 3.5*), then it can easily be shown that the shear force along plane MN becomes

$$\frac{(p_1-p_2)}{2}\sin 2\theta \qquad (3.4)$$

and the tensile force over that plane:

$$\frac{p_1+p_2}{2}+\frac{(p_1-p_2)}{2}\cos 2\theta \qquad (3.5)$$

In fact, when bodies are subjected to tension or compression they cause, respectively, compression or tension at right angles to them, as is manifested by the drawing out, i.e. decreased cross-section, of a body subjected to tension and the broadening out, i.e. increased cross-section, of a body subjected to compression (see *Figure 3.2*). The relationship between the longitudinal strain and the concomitant lateral strain is found to be constant for a given body, so that

$$\frac{\text{lateral strain}}{\text{longitudinal strain}} = \text{constant} = \sigma$$

This constant is known as Poisson's ratio, σ. Thus σ is always a fraction and may vary from close to zero up to 0.5 for a material such as rubber.

Within the elastic limit, the stresses corresponding with these strains are also in this same proportion, so that a tensile stress p_1 will cause a compressive stress $-\sigma p_1$ at right angles to it.

Thus any stress p_1 will have components in a plane at an angle θ perpendicular to the direction of the stress which can be represented by, parallel to the plane,

$$\frac{p_1 + \sigma p_1}{2} \sin 2\theta \tag{3.6}$$

(the sign is changed because subsidiary stress is compressive if principal is tensile) and, perpendicular to the plane,

$$\frac{p_1 - \sigma p_1}{2} + \frac{p_1 + \sigma p_1}{2} \cos 2\theta \tag{3.7}$$

Mohr constructed a circle from relationships 3.4 and 3.5 as shown in *Figure 3.5*. He represented the compressive forces as positive values on the x-axis and the tensile ones as negative on this axis, while shear forces were represented on the y-axis. The circle was constructed with centre on the x-axis and coordinate $(p_1 + p_2)/2$; it has radius equal to $(p_1 - p_2)/2$. Then if a radius is struck at an angle equal to 2θ, the component forces induced in a plane at an angle θ to the direction of the force p_1 may be read off as coordinates of the point at which the radius meets the circle, the shear force on the y-axis and the compressive force (tensile appearing as negative) on the x-axis.

From a study of this diagram certain facts emerge.

(1) The shear force disappears if $p_1 = p_2$.

(2) If a single force p_1 is applied together with its concomitant Poisson effect, shear force is

$$\frac{p_1(1 + \sigma) \sin 2\theta}{2}$$

i.e. the larger is σ, the larger is the shear force.

(3) If p_1 and p_2 have opposite signs, the radius of the circle is greater than if they have the same sign and therefore the shear force is greater.

(4) The value of the shear force is a maximum so far as the angle of the plane (θ) is concerned when $2\theta = 90°$, i.e. $\theta = 45°$. The value of the compressive or tensile force is a maximum when $\theta = 0$. Thus, if a body fails through tension or compression, it will fail along a surface perpendicular to the force represented by the largest coordinates on the x-axis, i.e. the break will be flat across the section of a rectangular piece. If the body fails in shear, it will be at an angle of about 45° to the minimum cross-section of a rectangular piece.

In practice, bodies do not shear at an angle of 45° but at a bigger angle, and Coulomb explained this by introducing the idea of internal friction. Resistance to shear, he said, was made up of two components, a static value q and another component which was proportional to the normal force being imposed on the plane of shear concerned.

Thus the effectiveness of two such forces as p_1 and p_2 to cause shear is

$$\frac{p_1-p_2}{2}\sin 2\theta-\mu\left[\frac{p_1+p_2}{2}+\frac{p_1-p_2}{2}\cos 2\theta\right] \tag{3.8}$$

The value of θ for which this expression is a maximum may be obtained by differentiating it with respect to θ and equating the derivative to zero. Thus

$$\frac{p_1-p_2}{2}\cos 2\theta+\mu\frac{p_1-p_2}{2}\sin 2\theta = 0$$

or $\mu = -\cos 2\theta$ and

$$\tan\theta = \mu\pm\sqrt{(\mu^2+1)} \tag{3.9}$$

which means, because of the negative sign, that 2θ for any practical value of μ will always exceed 90°, i.e. that θ will always exceed 45°. Thus, if a rectangular body is going to shear through the forces applied perpendicularly to its face (considering the matter two-dimensionally), it will fail at an angle $\theta = \tan^{-1}[\mu\pm\sqrt{(\mu^2+1)}]$.

The expression for shear with regard to the relationship between such forces as p_1 and p_2 is contained in equation 3.8; the body will fail in shear if the ultimate shear strength is less than this value while the ultimate tensile or compressive strength, as appropriate (depending upon the nature of forces p_1 and p_2), is more than either of the forces applied. Suppose that a compressive force p_1 is applied to the body which is just equal to U_C, the ultimate compressive strength of the brittle body. Then the body will break in shear if the effects of this force and its concomitant Poisson tensile force exceeds U_S, the ultimate strength of the body in shear, i.e. if

$$\frac{U_C(1+\sigma)\sin\theta}{2}-\mu U_C\left[\frac{1-\sigma}{2}+\frac{1+\sigma}{2}\cos 2\theta\right]$$

exceeds U_S. Now at the maximum value of shear $2\theta = \cot^{-1}\mu$, i.e. $\sin 2\theta = 1/\pm\sqrt{(1+\mu^2)}$ and $\cos 2\theta = -\mu/\pm\sqrt{(1+\mu^2)}$, and the required relationship is that:

$$\frac{1+\sigma}{2\sqrt{(1+\mu^2)}} - \mu\left[\frac{1-\sigma}{2} - \frac{\mu(1+\sigma)}{2\sqrt{(1+\mu^2)}}\right] > \frac{U_S}{U_C}$$

or

$$(1+\sigma)\sqrt{(1+\mu^2)} - (1-\sigma)\mu > \frac{2U_S}{U_C} \tag{3.10}$$

and, if typical values of σ and μ be taken of 0.25 and 1, means that $2U_S/U_C$ exceeds unity, or

$$U_S \text{ is less than } \tfrac{1}{2}U_C \tag{3.11}$$

If the principal force p_2 is tensile, then equation 3.7 becomes

$$\frac{p_2(\sigma+1)\sin 2\theta}{2} - \mu\left[\frac{p_2(\sigma-1)}{2} + \frac{p_2(\sigma+1)}{2}\cos 2\theta\right]$$

or, in terms of μ (equation 3.9),

$$p_2\left[\frac{(1+\sigma)(1+\mu^2)}{2\sqrt{(1+\mu^2)}} + \mu(1-\sigma)\right]$$

The condition for breaking taking place as a result of shearing rather than tension is

$$(1+\sigma)\sqrt{(1+\mu^2)} + (1-\sigma)\mu > \frac{2U_S}{U_T}$$

in which U_T is the ultimate tensile strength of the body. If, again, typical values are used of $\mu = 1$ and $\sigma = 0.25$, the condition is that $2U_S/U_T$ exceeds 2.5 or that U_S is less than $1.25U_T$.

The condition that a body subjected to a compressive force p will fail owing to tension is that the concomitant Poisson tensile force σp is greater than U_T when $p = U_C$, i.e. σU_C is greater than U_T, or, again taking a typical value of $\sigma = 0.25$, U_C is greater than $4U_T$.

Thus are derived the typical orders of precedence. A typical rectangular homogeneous body subjected to a compressive force on two of its opposite faces will break as a result of compression if U_C is greater than $2U_S$ and is also greater than $4U_T$. It will fail because of shear rather than tension if U_S is less than $1.25U_T$.

These inter-relationships will, of course, alter if μ differs from unity and σ from 0.25. For example, some quartz had the values $\mu = 1$ and $\sigma = 0.14$. The figures then mean (provided the quartz is homogeneous) that the body will break as a result of compression if U_C is less than $2.8U_S$ and less than $7U_T$, and that failure will be due to shear rather than tension if U_S is less than $1.22U_T$. Typical figures for quartz are $U_C > 110$ kg/mm², $U_T = 6.3$ kg/mm². Therefore quartz will not fail due to compression.

JOISEL[1] points out that few materials subjected to comminution on an industrial scale are homogeneous; that, when a body is compressed, there is almost always a part of the body which is more deformable than another; and that, therefore, parts on either side of that weak part are being subject to forces of shear. Materials met in practice in comminution processes usually fail in shear and it is the ultimate shear strength which should be used in calculations of material strength in such circumstances. As the ultimate shear strength is usually of about the same value as the tensile strength of mineral materials, it is the ultimate tensile strength which may be used.

CONDITIONS FOR BREAKING

It may be deduced from these considerations that, unless a material has undergone the whole series of changes from its original state to that in which it is at an ultimate stress whether it is of tension, shear or compression, it will not break. If a stress within the elastic limit is imposed, the material will be deformed, but it will not break; furthermore, it will return to its original shape when the stress is removed. If the material is ductile, it may distort permanently under the appropriate stress, but it will not break until the ultimate stress is reached. The importance of sustained forces for breaking is therefore apparent; that there is a considerable probability that most of the pieces, in a mass of material subjected randomly to a force, will escape breakage is very apparent.

Solids can exhibit a very wide range of behaviour when under stress. This behaviour may alter not only according to the chemical and physical constitution of the body but also with the degree and rate of application of the applied force. Broad generalisations are difficult to make accurately. Even such a generalisation as that metals will sustain a wide range of loads, after they have ceased to be in the elastic range, while siliceous minerals, such as quartz and granite, break when the load only slightly exceeds the point of departure from Hooke's law, will have to be qualified later.

POINTS OF WEAKNESS

If a gradually increasing breaking force is imposed upon a body, it can be expected to break first at such obvious points of weakness as visible cracks and weak joints of agglomeration. This is because stresses concentrate in regions where there are changes of area; *Figure 3.6* illustrates this. The maximum concentration occurs in areas where there are maximum curvatures, i.e. where the radii of curvature are the smallest (conversely, if the radius of curvature is increased by some means, the stress concentration is decreased). This principle is used when a sheet of metal is broken by bending it in two and hammering along the bend. The effect was quantified by

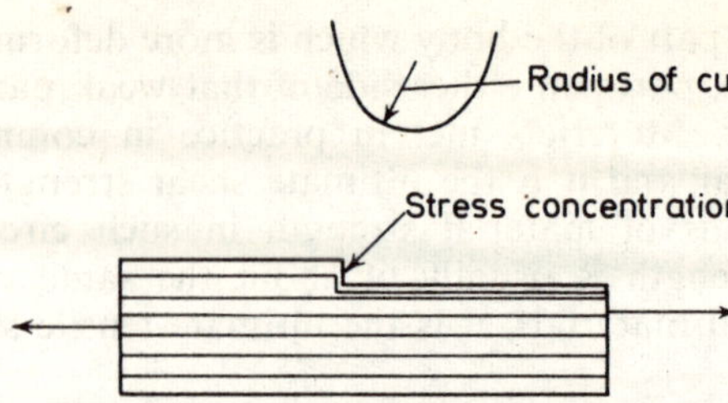

Showing how stresses concentrate at change of section

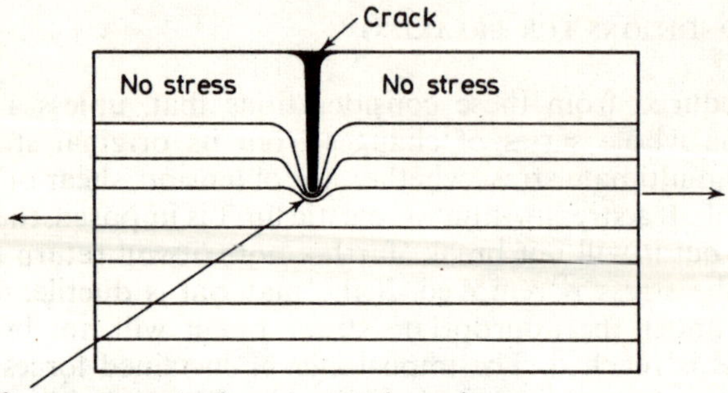

Stress concentration at end of crack

Figure 3.6. Nature and effect of radius of curvature and change of section to concentrate stress

INGLIS[2], who treated a crack as being elliptical in form and showed that

$$\text{stress} = \text{average stress} \times K$$

where

$$K = 1+2\sqrt{\left(\frac{\text{length of major axis of crack}}{2\times\text{radius of curvature at tip of crack}}\right)}$$

from which it follows that the average stress in a body is intensified as the length of the crack increases and as the radius of curvature of the crack decreases (i.e. as the crack becomes narrower). GRIFFITH[3] postulated that most bodies are a mass of cracks which vary in size from visible ones to those of the order of atomic dimensions and that, usually, breaking occurred by way of these cracks through the stress concentration effect explained by Inglis.

The initiation of a crack can be considered to be possible when the strain energy to be released in the vicinity of the potential crack at least equals the surface energy which would be created if the crack were formed. This can be calculated to take place when the concentration of stress is equal to $\sqrt{(E\gamma/v)}$, where E is Young's modulus, γ is surface energy per unit surface and v is half the total length of the crack (or, treating crack as an ellipse, length of major axis).

In the application and removal of forces such as occurs in a process of comminution, cracks can be expected to be enlarged and created even though fracture has not taken place. Thus not all energy applied which does not result in breaking is wasted provided material has more than one chance to meet the comminuting action of the equipment.

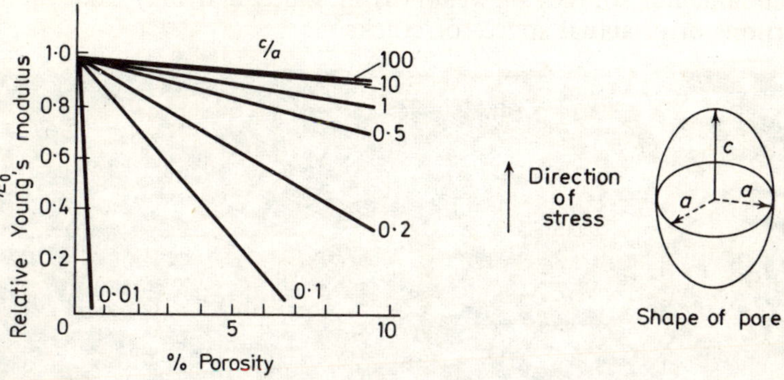

Figure 3.7. Effect of pore shape on Young's modulus

The presence of pores in a material can be expected to have the same effect as cracks[4]. For evenly dispersed pores of ellipsoid shape having axis *c* in the direction of the stress and equi-axes *a* perpendicular to the direction of stress, the effect on Young's modulus is shown in *Figure 3.7*.

BEHAVIOUR OF SINGLE CRYSTALS

Where there are no cracks, breaking can only take place between atoms. The true large single crystal consisting entirely of a precise regular array of ions or atoms (see *Figure 3.8*) is not encountered in ordinary industrial practice unless very special care has been taken to manufacture it, as, for example,

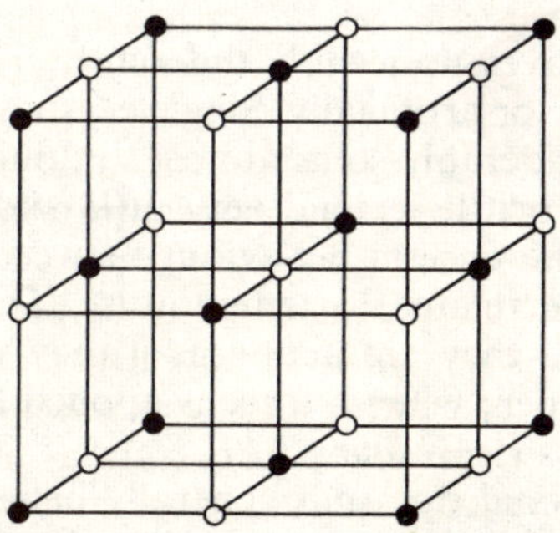

Figure 3.8. Array of ions or atoms in a single crystal. The example is magnesium oxide. Black circles = Mg^{2+}*; white circles =* O^{2-}

in the case of very special aircraft nickel-alloy turbine blades recently produced. Such single crystals are enormously strong structures. Most so-called crystals consist of growths arrested and recommenced (see *Figure 3.9*), so that the regular multiple cage-like structure is rarely exactly repeated for more than very small volumes (see p. 82). The changes in growth habit are the fundamental sources of weakness in solids and may take the form of distortions or of actual spaces or cracks.

Figure 3.9. The surface of a diamond magnified thousands of times, showing the complex structure of the apparently smoothest of surfaces. (Reproduced by courtesy of Professor S. Tolansky)

The single crystal behaves noticeably differently from crystals produced in bulk, whether naturally or artificially. Single crystals will either have great ductility and extend considerably so as almost to double their length before breaking or will be very brittle, extend very little and break abruptly. The precise reason for this difference in behaviour between single crystals is not known but appears to be related to temperature. The more ductile forms appear to extend because they are acting as if they were laminated like a block of sheets of notepaper: when a force is applied to them, they glide on some of these laminations (see *Figure 3.10*).

The bulk-produced crystal does not generally behave like this, although there are exceptions and glide lines can sometimes be seen on bulk specimens which have been extended (see *Figure 3.11*). In the bulk-produced crystal the regular array of atoms is interrupted, and these interruptions obstruct laminar movement or gliding. Such obstruction would be expected to strengthen the body, but in fact single crystals are much stronger. The explanation given is that the discontinuities—the cracks, pores and voids—in the bulk crystal counteract the advantages of the interruptions by concentrating forces.

As particle sizes are reduced, so plastic behaviour becomes more manifest than brittle and at sizes of the order of 1 μm (10^4 Å) plastic behaviour begins

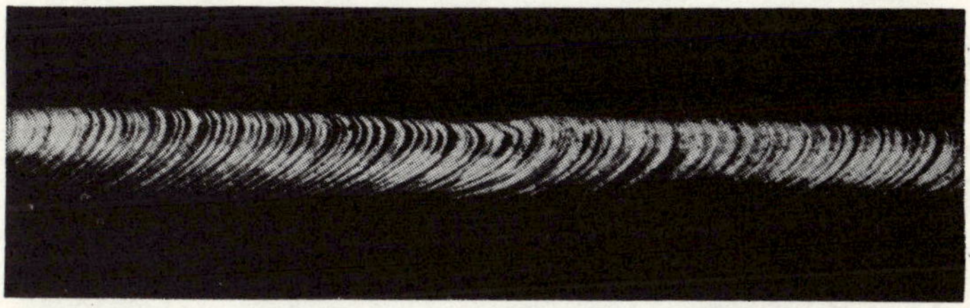

Figure 3.10. An extended single crystal of copper–aluminium alloy. (After ELAM[35]*)*

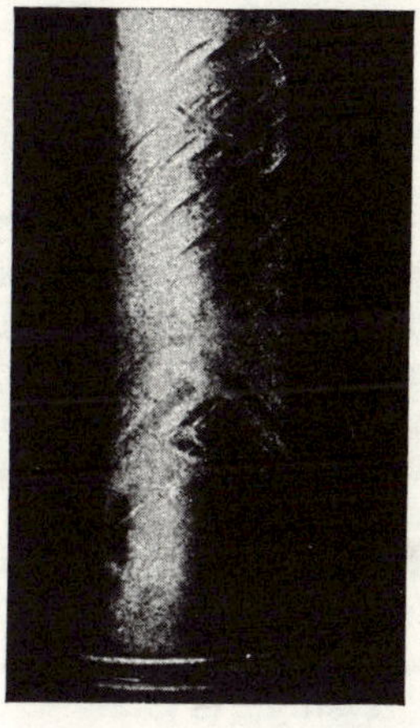

Figure 3.11. Stretched steel rod exhibiting glide lines. (After EWALD[36]*)*

to dominate. This could be because the residual particles approach more and more the wholly regular crystal form, i.e. single crystals. In this more plastic condition it is possible that welding is easier and therefore that agglomerations can more easily take place. There is some evidence for this (see p. 87).

In the true single crystal it appears that the cage-like structure will suffer a certain amount of change in shape without breaking: for example, when being heated or cooled and during tension or compression. But if these changes are sufficiently rapid or large or of long enough duration, some breakdown in the structure does occur, i.e. breaks and cracks are produced or the solid is transformed into a liquid or gas. For many substances the linear mechanical extension (strain) suffered before breaking is of the same order as the thermal linear expansion from the same temperature to the melting point (see *Table 3.1*), which indicates that the cause of breakdown in each case arises from excessive strain between neighbouring elementary particles, i.e. atoms or ions as the case may be.

Table 3.1. STRAIN AT ULTIMATE STRESS AND LINEAR EXPANSION FROM ROOM TEMPERATURE (15°C) TO MELTING POINT

Material	*Strain at ultimate stress*	*Theoretical max. strain, single crystal*	*Melting point,* °C	*Linear expansion,* 15°C *to m.p.*
Mild steel	0.002		1 050	0.01
Cast copper	0.02	0.2–0.4	1 084	0.02
Zinc	0.01	0.1	418	0.01
Cast aluminium	0.02		660	0.014
Lead	0.001		325	0.009
Silver	0.002	0.2	962	0.02
Silica	0.001		1 710	0.001
Aluminium oxide	0.001		2 037	0.016
Magnesium oxide	0.001	0.2	2 800	0.03
Carbon (graphite)		0.1	3 500	0.02
Carbon (diamond)		0.2		0.003
Glass	0.001		800	0.004
Polystyrene	0.02		125	0.008

For the single crystal the tensile force necessary to separate the ions or atoms in the lattice is very great. The Griffith relationship (p. 24) can be transposed for atomic conditions so as to give the maximum tensile strength of a single crystal without flaws to be $\sqrt{(E\gamma/a)}$, where a is the interatomic or interionic distance. From this expression the calculated value for the sodium chloride single crystal is 200 kg/mm^2, and for a normally grown crystal the value for the tensile strength is 0.4 kg/mm^2. Thus, comparing tensile strengths,

$$\frac{\text{single crystal}}{\text{bulk-grown crystal}} = 500 = \sqrt{\frac{E\gamma}{a}} \cdot \sqrt{\frac{v}{E\gamma}},$$

or $v/a = 2.5 \times 10^5$, or the effective total crack length in a cross-section of a bulk-grown crystal is $2.5 \times 10^5 \times$ interatomic distance for sodium chloride crystal, which is about $2.5 \times 10^5 \times 3$ Å; thus effective crack length is about 7.5×10^{-2} mm or about 0.1 mm. Such a crack length, if occurring continuously, would be visible to the human eye, which can see $2\text{–}3 \times 10^{-2}$ mm, and because the cracks are not usually visible, the calculated crack length may represent a family of cracks of that total length.

There is a minimum length of crack, below which van der Waals' forces act, of about 1 μm. Grinding would be expected to proceed down to this size more easily than below it. This is supported by experience and is discussed in more detail below.

BEHAVIOUR OF GRANULAR FORMS

Most solid materials from which objects are fabricated consist of agglomerations of such crystals as have been described above and are known as polycrystals. Additional sources of weakness exist in such structures at crystal interfaces (see *Figure 3.12*). Such interfacial weaknesses may be increased by substances which are either deliberately added or are impurities that gather at the boundaries and weaken the adherence of the grains. Sometimes such boundary substances link themselves into skeletons around the grains and strengthen the structure thereby. The larger that grains grow, the larger are the continuous surfaces over which shearing forces can act and therefore the weaker such polycrystals become. The smaller grains obstruct directional forces and therefore the finer-grained material is usually stronger. CLARKE[4] has shown that ultimate strength is inversely proportional to the square root of the grain size. Grain growth is therefore inimical to strength. Grain growth is normally favoured by increases in temperature, a usual concomitant of a size-reducing process.

Figure 3.12. Weakness at grain boundaries: fatigue crack

Grain growth is also inimical to ductility, and some substances can be made superplastic by being produced and maintained in a small-grain form. This fact would appear to conflict with the theory of gliding as a source of ductility since smaller grain size would appear to increase the obstruction to gliding. One explanation is that large grains are more full of distortions than small grains and the small grains provide a multiplicity of surface for the transfer of mobility of ions and thus the mobility of distortions.

The existence of weakness along grain boundaries is of considerable importance in comminution and can be used to advantage if the grain size happens to suit the size particle required. Particularly in impact mills or tumbling mills, size reduction can be effected down to the natural grain size. It is also possible to exploit this effect to remove constituents which increase wear in mills. For example, silica, which is particularly abrasive, can be separated from other constituents, if its natural grain size is sufficiently different from that of the other desirable constituents. This separation may be achieved sometimes by breaking down the mixture to the size range of these silica grains, when it is found that the natural grains of silica are unbroken while the material that was originally the matrix has been broken to considerably smaller sizes: the silica can then be sieved off. If these size differences are not large enough to enable sieving to be used, other methods such as those which depend upon differences in density of the constituents can sometimes be used, e.g. elutriation or cyclone separation.

PROPAGATION OF CRACKS

From what has been stated above it may be deduced that to scratch a surface is seriously to weaken the solid. Cracks are rarely initiated on the surface, however, during the comminution processes, which usually take effect through the propagation of existing cracks. The speeds of propagation of cracks in material are enormous and can approach the speed of sound in the material (N. F. Mott calculated that the maximum is about 40% of the speed of sound). Such speeds are usually of the order of thousands of kilometres per hour. The propagation of cracks is obviously related to the ability of the materials to propagate strain. Materials which are amorphous obstruct propagation of strain more efficiently than do crystalline materials; thus they tend to break locally at tips, where the strain is contained, whereas crystalline breaking tends to be more diffuse. Amorphous material comminutes to a mixture of coarse and fine particles, i.e. the extremes of size; crystalline material breaks into a narrower range of sizes.

Similarly, heterogeneities obstruct the propagation of cracks and can dissipate energy, so that rupture is less likely. The use of fibres which provide a broad front to the propagating waves of strain, when set in a matrix, is an exploitation of this effect.

Cracks when once formed can be self-propagating provided there is enough strain energy in the material. The law of conservation of energy holds just as much in these instances as in any other, and breaking can only take place so long as there is energy remaining after the various concomitant processes in the creation of new surface have been provided with energy: processes such as the creation of sound; the use of potential energy in changing the position of pieces of material; and the evolution of light, frictional heat and electricity. Thus material which has high surface energy will not propagate cracks as readily as material of low surface energy. CLARKE and his co-workers at Harwell[4,6] have shown that cracking will be self-propagating if

$$\gamma = \frac{\varepsilon^2 E(l-c)}{24(1-\sigma^2)}$$

where γ is the surface energy at the grain boundary, ε is the strain, l is the grain size, σ is Poisson's ratio and c is the pore diameter at which the fracture initiates. CLARKE[4] has suggested a pore at the grain boundary as a common origin for fracture.

CRACK HEALING

KING and TABOR[7] showed that when pressures were imposed hydrostatically on rock salt, lead sulphide or ice, their yield stress was increased considerably and plastic deformation occurred, i.e. brittle fracture did not occur. This is a phenomenon which could be related to the closure or sealing up of the cracks by the pressure as in the case of a metal sheet on to the largest surface of which another piece is welded: it is weakened by the cracks caused in the welding process at the weld toe to the extent of 40% of its original tensile strength and can be restored to nearly its original strength by peening or hammering the cracks together.

Some materials, particularly metals, have the ability to self-seal cracks formed in them. This ability is due, it is thought, to the electron cloud or mobile electrons which characterise such substances; they are usually ductile for this reason. In a similar way if a substance which is soluble in a certain liquid and does not contain cracks above a certain size is extended under the liquid, its ultimate strength is enhanced. For example, sodium chloride crystals under water are enhanced from 450 to 5000 g/mm^2, owing, it is thought, to the cracks being filled with the viscous saturated liquor, produced by solution and acting as a solid[3], thereby allowing of the transmission of forces across the cracks and preventing stress concentration.

Perhaps the effect is of surface cracks on the crystal being dissolved to leave a much stronger body even if of reduced size. This viewpoint is supported by some work of JOFFE[12], who showed that glass behaved in a similar way when extended under hydrofluoric acid solution. For a similar reason some makers of shafts required for very heavy duty, e.g. the main shaft in a large gyratory crusher, bore out the centre of the shaft to remove material which might have cracked during the forging and whose cracks might be propagated during use of the shaft. However, WILLIAMS[8] has shown that polyethylene sheet will show a ductile tear in air, while under methyl alcohol, in which polyethylene is very insoluble, the tear is brittle and accomplished by less load. Other investigators[9] have shown that glass in water, ethanol or turpentine, chrome steel in caustic soda and zinc in mercury all showed less strength than in air.

In a similar way, gases may have an influence on the strength of solids. MACHLIN and MURRAY[10] suggest that ozone improves the strength of some materials. RUMPF[11] suggests that the adsorption of gases onto nascent surfaces during crushing can be a significant user of energy during crushing. The effect could be due to annealing by the heat produced in the adsorption process. Cracks could be caused to disappear by absorption by crystal growth. The converse could presumably apply, of incompatible substances entering cracks and preventing healing. This might be an explanation of the decreased power consumption when certain 'additives' are introduced in a grinding process (see p. 108).

An interesting slant on this problem of the weakening of solids by cracks is provided in glass manufacture. Soda-lime glass is suspended in a bath of molten salt which has anions of larger ionic radius than sodium, e.g. potassium:

Ionic radius	$\mu m \times 10^{-4}$
Sodium	1.86
Potassium	2.23

Exchange of ions takes place near the surface (a penetration greater than 50 μm is achieved) and the entry of the larger ion strains the glass surface and tightens up the cracks. The effect is stronger glass which resists crack formation.

MECHANISM OF COMPRESSION

Most substances accumulate cracks in accordance with the history of their distortion since they last froze from the liquid state. The existence of these cracks is made manifest in the difference between measurements of strains under tension and under compression; strain is usually less under compression because cracks are pushed together, whereas under tension they are pulled apart. Under compression the material is made more homogeneous; under tension, more porous. Solids which have had a lot of distortion break into smaller pieces than those which have not, e.g. quartz which has been under pressure in the earth crushes much finer than quartz which has not. Thus a lump with many cracks in it, when submitted to additional strain, enough to break it, will form smaller pieces than if it had not so many cracks in it.

It follows from these considerations that small particles are likely to have shorter lengths of cracks in them than do large particles and that therefore they are likely to be stronger. This is further discussed in Chapter 5 (p. 88). Smaller particles are more difficult to break down; there may be other reasons for this but it is usually considered that there is a lower limit to the size of particle producible by a grinding process.

As has been previously observed, the movements of parts of a solid under strain can be seen in certain materials (see *Figures 3.10* and *3.11*); the action is a sliding one. It is difficult to conceive of a compressive action being anything but a sliding one. To compress is to compact and it seems impossible for compression to lead to breaking unless there is complete collapse of the structure like a shell or, in the extreme, of the penetration of the atomic nucleus by the electrons as is supposed to occur at hitherto unachievable pressures when atoms would lose their identity. Rocks at enormous pressures in the earth and at the depths of oceans are not pulverised. What appears to take place in a solid under compression is that the atomic layers of the substance, in numbers which appear to vary from a few hundred to a few hundred thousand, slide on one another as would jam on the bread of a compressed jam sandwich, i.e. there is a pushing out at right angles to the direction of the compression and it is this action which is responsible for the breaking. Twisting (or torsion) is similar in its action in that it makes layers slide over one another and pull away from one another. The effective action in breaking appears to be always tensile or shear.

A method was used by JOFFE[12] to measure the strength of a body wholly under tensile forces. Spheres of rock salt were cooled in liquid air and then immersed in a bath of molten tin. From the fact that the spheres were not shattered a minimum tensile strength was calculable and found to be one-third of the theoretical strength for a single crystal.

However, no machine has been devised which works on large quantities of material so as to put it directly into a state of tension or shear. Nobody has devised a method of holding one part of a lump and pulling another part of it away and of doing this at the rate of tons per hour. The development of a giant pair of scissors seems equally impractical. The use of compressive forces in grinding machines is yet another source of inefficiency in the use of energy in the comminution process.

TEMPERATURE EFFECTS

Much of the energy used in the comminution process appears as heat in the broken solid, a temperature rise of more than 100°C being quite common in some mills. As solids become hot, so they become more ready to adhere to one another and to the surface of the machinery. A problem is therefore to impart not so great a quantity of energy to the solid that it becomes plastic, but sufficient to break it. Some equipment, such as air-swept ring and roll and ball mills, pin mills which induce their own very high air flow and fluid-energy mills using compressed air, coincidentally cools material by its air flow. Progressive increases in temperature must, if necessary, be avoided by removal of heat. Temperature is sometimes decreased by adding solid carbon dioxide or liquid nitrogen to solid prior to or during grinding. Some materials such as nylon and certain foodstuffs have to be size-reduced in this way. An additional advantage is that volatile materials which aid flavour or aroma are preserved as well as heat-sensitive materials such as vitamins, colouring matter and those which can explode or burn. In these cases care must be taken with the material of construction of the equipment: cast-iron, for example, is very brittle at low temperatures and it may be necessary to use a cold-resisting alloy steel. With some mixtures which contain low-melting-point constituents, it may be more convenient deliberately to soften the mixture by heating so that the solid suspension may be ground in a low-viscosity liquid rather than a viscous mass; examples of such materials are asphalt and wax-based products. In the extreme, it may be necessary to fuse the solid and to spray it as a liquid and then to cool the spray, to get it in the degree of subdivision required.

It has been argued by DJINGHEUSIAN[13] that there is an absorption of energy in the comminution process, viz. by the creation of more surface energy, and that, in accordance with thermodynamic principles, an increase in temperature should aid comminution. Djingheusian states that he improved the rate of production of new surface by 6–19% by raising the temperature of the feed by about 11°C. SCHULZ[14] could find no effect from increased temperature in the range 25–400°C on quartz, taconite and magnetite.

EFFECT OF RATE OF APPLICATION OF FORCE

Silicone rubber will break if subjected to a swift violent blow, but will flow like a liquid if a slow continuous force is applied. Careful observation shows, however, that this behaviour of silicone rubber is merely a particularly obvious example of two general phenomena. (1) All solids flow when subjected to continuous pressure for a long enough time: rocks at great depths in the earth show severe permanent deformation; salt domes formed from evaporated sea-water as it receded have evidently flowed from under dense sedimentary rocks deposited long after the sea-water had disappeared. (2) Apparently all solids increase in strength as the velocity of application of a

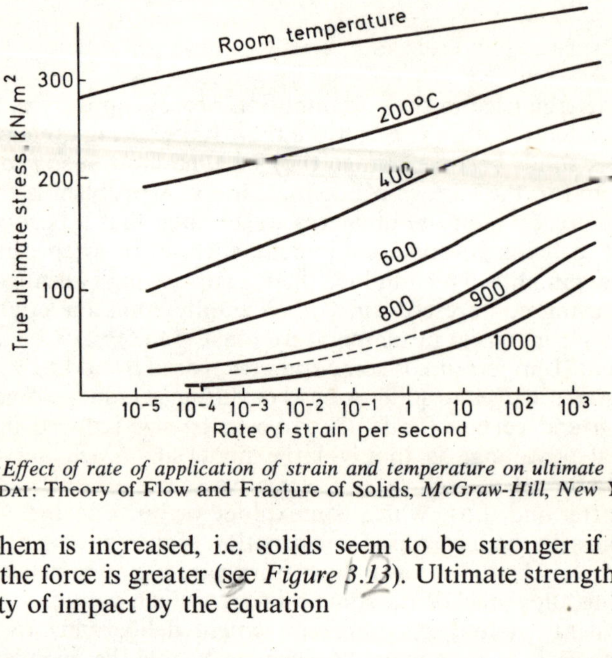

Figure 3.13. Effect of rate of application of strain and temperature on ultimate stress of copper. (After A. NADAI: Theory of Flow and Fracture of Solids, *McGraw-Hill, New York, 1950)*

force to them is increased, i.e. solids seem to be stronger if the speed of impact of the force is greater (see *Figure 3.13*). Ultimate strength is related to the velocity of impact by the equation

$$\text{ultimate strength (stress)} = \text{constant} + \text{another constant} \times \log \frac{V}{V_0}$$

where V is the rate of application of the strain and V_0 is some comparative rate of application of strain.

An extension of these ideas is that a small force, if it is maintained for a very long time, will eventually break a solid. Thus a so-called relaxation time can be derived beyond which a sustained stress, of any value, will cause the material to flow (*Table 3.2*). The meaning of this value for fluids is that below this relaxation time the fluid behaves as an incompressible or inextensible solid.

From this concept follows also the idea of a 'life' for material under a sustained load before breaking. MCCANCE[15] has shown that, for lead, this life varies as shown in *Table 3.3.* PRESTON[16] has derived a relationship for glass,

$$\text{breaking stress in kg/cm}^2 \times \log_{10} \frac{\text{(time in ms)}}{6} = 4600$$

which can be interpreted in its extreme form as stating that for an infinite time the breaking stress is zero, while if the time is less than 6 ms, no force is great enough to break glass.

The importance of these ideas to comminution is that a very small sustained load is more likely to break a material than a large impact of short duration; this has been confirmed by HEYWOOD[17] and many others, but some results reported by FAHRENWALD, NEWTON and HERKENHOPF[18] on work with light high-velocity impact compared with a slow one on Ottawa sand and by WILLIAMS[8] on work with polyethylene appeared to show the opposite. Three specimens of the polymer sheet were made to have the same length of crack in the three ways: (1) slow steady entry of a razor blade, (2) striking the side of the specimen with a razor blade, and (3) slow steady entry of a razor blade for a short distance, the specimen then being vibrated at 2000 c/min until a crack of the same length as the others appeared. The three specimens were then submitted to a tensile load and it was found that: Specimen 1 showed considerable crack branching and sustained the highest tensile load. Specimen 2 had one crack but it was augmented by many hackles. Specimen 3 broke with a clean straight crack at the lowest tensile load. Thus the more slowly induced crack evoked the strongest resistance to breaking.

Table 3.2. RELAXATION TIMES OF SELECTED MATERIALS

Material	*Relaxation time*, s
Concrete	10^6
Nylon	10^3
Steel	10^{-2}–10
Waxes, asphalt, putty	10^{-4}–10^4
Air	10^{-10}
Water	10^{-13}

Table 3.3. LIFE OF LEAD UNDER DIFFERENT STRESSES

Stress, kg/cm²	*'Life' before breaking*
107	1 s
95	1 min
85	1 h
70	1 day

HARDNESS

Hardness is a property which most users of comminution equipment consider to be of primary importance mainly in its effect on output and on cost of maintenance. Hardness can be considered to be resistance to being scratched and, on first thought, seems to be (a) quite distinct from strength, though perhaps not from resistance to crushing and grinding, and (b) not relevant to resistance to impact, because a very hard material such as diamond is relatively easy to shatter by impact.

Mohs' table of hardness places materials in an order determined by their ability to scratch the preceding material:

1. Talc
2. Gypsum
3. Calcspar
4. Fluorspar
5. Apatite
6. Telspar
7. Quartz
8. Topaz
9. Corundum
10. Diamond

Thus a substance which would scratch apatite and be scratched by quartz would be said to have a Mohs hardness between 5 and 7.

HEDDLE[19] has given some practical tests by which hardness on the Mohs scale can be assessed in the field. Talc will mark paper; 2.5 is the hardness of the finger nail; 4.0 is the hardness of a copper coin; 4.5 is the hardness of the common quality of mild steel; 5–5.5 is the hardness of window glass; 6.5 is the hardness of a good pocket knife blade.

More refined methods of measurement than that of Mohs have been developed, but they all depend upon the measurement of the ability of one material to scratch or to penetrate into another.

Methods of measuring hardness can be classified as follows: (1) Comparative scratchability. (2) Penetrability of a loaded hard definitely shaped object expressed either as a function of depth of penetration or number of scratches to reach a certain depth. (3) Degree of reaction or bounce to a falling body used as a pendulum and expressed as a function of rebound or the number of swings to break the specimen.

The chief methods are:

The Brinell method, in which a steel ball of 10 mm diameter under a load of 3000 kg is pressed into the specimen for 15 s and the diameter of the depression (d mm) is measured. The Brinell hardness number is given by

$$\frac{191}{10-\sqrt{(100-d^2)}}$$

The method is criticised because progressive indentations are not geometrically similar to one another.

The Vickers method, in which the indentations of a diamond pyramid having sides with apices of angle θ under a load of p kg are measured across the maximum diagonal d mm. The Vickers hardness number is

$$\frac{2p \sin(\theta/2)}{d^2}$$

The Rockwell methods provide for different hardness ranges by using various combinations of penetrator design and pressures. The penetrator may be a steel ball or a diamond cone and a feature of the instrument is a direct reading dial. First a small standard is added to the penetrator and the dial adjusted to read zero with that weight upon the penetrator. Then a larger weight is added and maintained in place for a fixed time, and then removed. The distance the penetrator has moved is read from the dial as a measure of hardness. A list of the various combinations is shown in *Table 3.4*. The combinations used are distinguished by letters from A to G.

Table 3.4. ROCKWELL HARDNESS TESTS: WEIGHT AND PENETRATOR COMBINATIONS

Rockwell letter	*Minor load*, kg	*Major load*, kg	*Penetrator*
A	10	60	diamond 120° cone
B	10	100	$\frac{1}{16}$ in diam. steel ball
C	10	150	diamond 120° cone
D	10	100	diamond 120° cone
E	10	100	$\frac{1}{8}$ in diam. steel ball
F	10	60	$\frac{1}{16}$ in diam. steel ball
G	10	150	$\frac{1}{16}$ in diam. steel ball

The Shore method uses a small pointed hammer which is dropped from a fixed height; hardness is measured by the rebound of the hammer on a fixed scale.

A comparison between the Mohs figures and the Vickers number in the units kgf/mm^2 is:

Mohs No.	1	2	3	4	5	6	7	8	9	10
Vickers No.	40	70	100	150	300	600	800	1 100	2 100	8 000

If the logarithm of the Vickers number is plotted against the Mohs number, an approximate straight line is obtained. A comparison between various hardness numbers is given in *Table 3.5*.

Table 3.5. HARDNESS INDICES COMPARED

Mohs	*Brinell*	*Vickers*	*Rockwell C*	*Shore*	*Knoop*
	100	100			
	197	201	11		
5	285	300	28.5	42	470
	375	398	40.0	53	
	461	497	49.0	63	
6	534	609	56.0	73	
	578	704	60.0	80	
7	620	800	63.3	86	960
	725	900	66.8	96	
	760	1 000	69.9	98	
8		1 150			1 840
9					
		2 500			8 500

A scale of hardness with more equal intervals of Vickers hardness might be (Vickers hardness indicated in brackets):

1. Talc (40)
2. Opal (350)
3. Apatite (600)
4. Quartz (800)
5. Hard steel (1000)
6. Zircon oxide (1150)
7. Garnet (1400)
8. Topaz (1575)
9. Tungsten carbide (1750)
10. Zircon carbide (2100)
11. Tungsten titanium carbide (2250)
12. Titanium carbide (2400)
13. Vanadium carbide (2500)
14.
15. Silicon carbide (3000)
16. Manganese (3300)
17.
18. Tungsten diboride (3500)

KURTH[20] related hardness directly with yield stress. MARSH[21] has confirmed this for the Vickers hardness number, which he has shown is approximately equal to three times the yield stress. *Table 3.6* illustrates this

Table 3.6. RELATIONSHIP BETWEEN HARDNESS AND YIELD POINT AND BETWEEN HARDNESS AND MELTING POINT

Material	*Vickers hardness,* kg/mm^2	*Yield point,* kg/mm^2	$\frac{\textit{Hardness}}{\textit{Yield point}}$	*Melting point,* °K	$\frac{\textit{m.p.}}{\textit{Hardness}}$
Steel	220	70	3.1	1 830	8.3
Cast-iron	210	12	17.5	1 450	6.9
Corronel B	245	44	5.6		
Copper	88	27	3.2	1 350	15.3
Zinc	50	10	5.0	690	13.8
Aluminium	45	14	3.2	930	20.6
Vulcanised fibre	12	2	6.0		
Lead	6.7	2.1	3.2	590	88
Diamond	8 000	5 400	1.5	4 670	0.6

relationship. FRIEDERICH[22] showed that hardness was a function of the quotient of the valency of the element with the highest valency in the compound and the two-thirds power of the molecular volume (i.e. the molecular surface):

$$\text{hardness} = f\left(\frac{\text{valency of element}}{\text{molecular volume}^{2/3}}\right)$$

Thus materials are hardest which contain elements of the highest valency and possess the smallest molecular volume. PIRET[23] shows that, for brittle materials in a controlled grinding process, the new surface produced per unit volume per unit input of energy is approximately linearly related to Mohs hardness multiplied by density of the material.

Hardness in a crystal varies with the direction of the crystal face, i.e. crystals are usually anisotropic with respect to hardness. This is discussed later (p. 82). MCCANCE[15] showed that hardness was a function of the degree of dislocation in a material. SCHELLINGER[24] showed that hardness was related to the surface energy of materials. *Table 3.7* illustrates this.

Table 3.7. RELATIONSHIP BETWEEN HARDNESS AND SURFACE ENERGY

Material	*Surface energy**, erg/cm^2 (γ)	*Vickers hardness*, kg/mm^2 (H)	γ/H
Calcite	29 000	96	300
Pyrites	60 000	250	240
Quartz	107 000	300	360
Rock salt	26 100	14	1 860

It has been claimed that hardness is related to the absolute melting point (*Table 3.6*). Some materials having high melting points are very soft. Abrasive properties seem to be related more to high melting point than to hardness (see p. 41). However, it must not be overlooked that hardness can vary considerably over an area. When hardness tests are made, areas of uninterrupted surface are selected. If attempts are made to comminute very hard materials which are also free from cracks, very large forces are needed for success. Most comminuting equipment fails with, for example, chromium metal and titanium diboride because these materials are usually free from cracks. Most hard materials such as carborundum and corundum, which yield to impact very easily, are really composed of polycrystals; they yield along cleavage planes.

Some materials develop a hard surface when moved against foreign surfaces. For example, lime in rotary kilns develops this glaze, whereas in vertical kilns it does not. Hammer mills induce this glaze on some materials. BEILBY[25] suggested that polishing produced an amorphous layer of distinct physical characteristics and others have suggested that all crystals have such a layer (others claim to have proved its non-existence). Some metals decrease in hardness after being immersed in certain substances, e.g. oleic acid and molten paraffin wax. Possibly in a similar way glass can be drilled more easily if turpentine is used. Again, this can be an explanation of the effects on rate of grinding when certain substances are added to the feed (see p. 108).

A soft material grinds finer than a hard material in most machines for the same time of grinding (see *Figure 3.14*), and if a hard material is ground to pass a certain sieve size, it will have less fine material in it than a soft material ground to this same limit. When a mixture of hard and soft materials is ground, the proportion in which energy is imparted to the components is rather complicated. Provided the ratio of hardness of the harder material to hardness of the soft material is less than a certain figure, the proportion of the energy absorbed in the grinding process which is imparted to the harder material is smaller than the proportion of the mass of the harder material in the mixture. If the hardness ratio exceeds this figure, then the relationships of energy to mass invert and the harder material obtains more than its share of energy compared with its mass proportion. However, the larger the proportion of mass of the harder material in the mixture the larger this hardness ratio has to be for the harder material to have imparted to it a

* Schellinger's figures for surface energy were of the order of 100 times the usually accepted figures. RABINOWICZ[33] found the relationship $\gamma = KH^{1/3}$ close for a large number of materials.

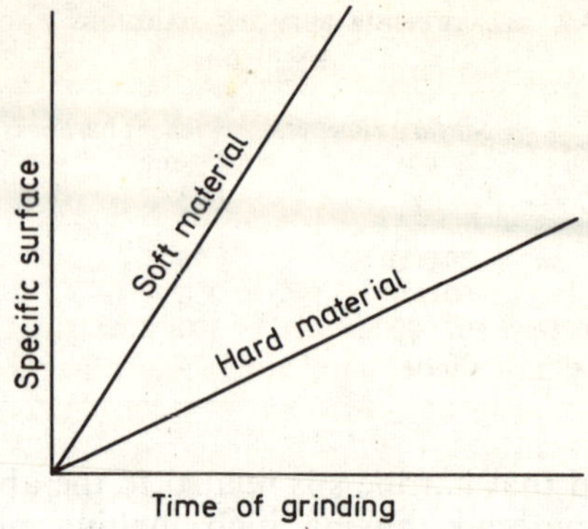

Figure 3.14. Relative rates of grinding: hard and soft materials

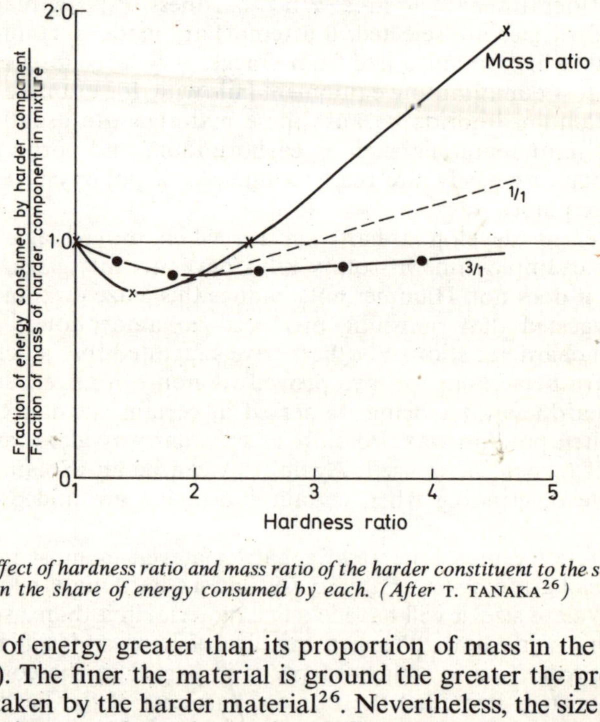

Figure 3.15. Effect of hardness ratio and mass ratio of the harder constituent to the softer constituent of a mixture on the share of energy consumed by each. (After T. TANAKA[26]*)*

proportion of energy greater than its proportion of mass in the mixture (see *Figure 3.15*). The finer the material is ground the greater the proportions of the energy taken by the harder material[26]. Nevertheless, the size distribution of the constituents of a mixture is unaffected by the proportion of the constituents of a mixture. This aspect is discussed further in Chapter 6 (p. 102).

The effect of temperature on hardness is peculiar to the material and a change in the position in the scale of hardness can occur during the comminuting process. Relationships have been found for metals of the kind

$$H = Ke^{-at}$$

where K and a are constants and t is the temperature in deg C. For non-metallic materials the relationship is more complex and has not been stated generally, although that for sodium chloride has been investigated.

ABRASION

An inevitable concomitant of abrasion is the separation of material. Any process in which very small pieces of material separate from a larger piece is likely to be given the name of abrasion. It is probable that most actions which are called abrasion consist of a mixture of actions. What is undoubtedly common to them all is high surface stress. Hertz's study of the subject showed that if a spherical surface was pressed in contact with a flat plate, the spherical surface deformed and touched the plate in a circle of radius a such that

$$a \propto \left[\frac{Wr}{2}\left(\frac{1}{E_1}+\frac{1}{E_2}\right)\right]^{\frac{1}{3}}$$

where r is the original radius of the spherical surface, W is the load with which the surfaces are pressed together, and E_1 and E_2 are the Young's moduli of the materials forming the spherical and plane surfaces. Beneath this area of contact, there spreads out in the thickness of the flat plate a volume of compressive strain which is approximately a frustrum of a sphere; the stress then changes to a tensile one, the isobars of which curve up round the sphere to near to and parallel with the surface. A coordinate of one point of maximum tensile stress is found to be about half the radius of the circle of contact vertically beneath the centre of that circle and its magnitude is found to be about 0.3 times the average load on the circle of contact, i.e. $0.3 \times (W/\pi a^2)$. On either side of this point in a vertical direction, the tensile forces decrease, whereas all tensile forces concentrate at or near the circumference of the circle of contact on the surface.

The conditions in such a plate are therefore very complicated. Cracks tend to be formed so that they fan out from the circle of contact; with maximum stresses occurring a short distance beneath the circle of contact, the conditions are present for pieces to be separated from the plate.

In addition to Hertzian fracture, the effect of friction must also be considered. This subject has been studied intensively by BOWDEN and TABOR[27]. Even the apparently most smooth surface when magnified consists of very undulating country with mountains, hills and valleys. When two surfaces are brought together it is 'like Switzerland being inverted on Austria', as Bowden and Tabor have it. When these surfaces move relative to each other, their mountain peaks push into the other's valleys and deform and melt, or their mountains tear each other and break pieces off. Their temperature is increased in the process and sparks appear even from such surfaces as glass and quartz.

The abrasive materials are therefore often not hard but are distinguished by high melting points. Agalmatolite, an aluminium silicate, has a Mohs hardness of 1 and yet it is used to polish ceramic glazes; however, it has a melting point of 1700°C. Camphor (m.p., 179°C) can be used to polish Wood's alloy (Mohs hardness about 3; m.p., 61°C) but not tin (Mohs hardness 1.5; m.p., 232°C). Quartz (Mohs hardness 7; m.p., 1710°C) may be polished with zinc oxide (Mohs hardness 4; m.p., above 1800°C). That wear has very little to do with hardness is shown by *Table 3.8*, which gives the comparative wear of balls of different hardnesses[28] in a ball mill.

Table 3.8. WEAR AND HARDNESS OF BALLS IN A BALL MILL COMPARED

Material of ball	*Wear rate (W)*	*Brinell hardness (H)*	*H/W*
Cast steel	183	170	0.9
White cast-iron	154	364	2.4
Austenitic manganese steel	154	212	1.4
Pearlite Cr–Mo steel	121	387	3.2
Heat-treated forged steel	129	477	3.7
Climax 6–1 steel	131	200	1.7
Martensitic Cr–Mo steel	110	555	5.1
Ni-hard (sand-cast and stress-relieved)	100	600	6.0
28% Cr iron	95	560	5.9
16% Cr–39% Mo steel	90	512	5.7

If scratching or sliding or impact are combined with indentation acting on material with granular or other heterogeneities, the situation is quite complicated. Grain boundaries can be penetrated; grains can be gouged out or disintegrated; and homogeneous volumes can suffer Hertzian fracture. Coarseness of particle size also has a strong influence. The possibility of chemical action must not be overlooked, particularly in wet grinding. In fact, abrasion cannot be generally defined and it is necessary to use clearly defined tests which allow results on the same material to be reproduced; clearly, the closer these tests are to the circumstances in which abrasion is to be predicted, the better. Several testing procedures have been devised and a few of them will be described.

Morgan Marshall

The Morgan Marshall test is much used for ceramics. A known weight of alumina (corundum) particles is fired at fixed pressure of compressed air at a fixed area of surface of the specimen, and the weight lost by the specimen is determined. The abrasive index (A) is given by

$$\mathrm{A} = \frac{\text{loss of weight per unit surface}}{\text{bulk density of specimen}} = \frac{\text{volume removed}}{\text{surface area}}$$

The Morgan Marshall index has been shown[34] to be analysable into two functions, one having the form

$$A = \frac{\frac{1}{2}mv^2 \cos^2\theta \sin 3\theta}{\phi} + \frac{\frac{1}{2}mv(v\sin\theta - K)^2}{\chi}$$

and the other

$$A = \frac{\frac{1}{2}mv^2 \cos^2 \theta}{\phi} + \frac{\frac{1}{2}mv(v\sin\theta - K)^2}{\chi}$$

in which v, m, θ, ϕ and χ are respectively the velocity, total mass fired of the particles, the angle between particle direction and the abraded surface, the kinetic energy absorbed by the surface to release unit volume of material by respectively erosion and deformation. Thus the part of both equations with the denominator ϕ is abrasion due to erosive wear and the part with the denominator χ is due to deformation wear. The two functions meet at $\theta = 18.5°$ and the first equation applies below 18.5° and the second above. For small values of θ, erosion is seen to dominate while for high values deformation wear has more influence. A brittle material (see p. 16) will show little abrasion at high angles: comparatively, a ductile material will show abrasion at all angles because it tends to be softer (see p. 46).

Yancy, Geer and Price

The YANCY, GEER and PRICE test was originally devised for prediction of wear in equipment used for pulverising coal[29]. Test plates are rotated in a mortar containing a sample of the feed material. Four removable blades, 2 in (about 5 cm) long and 1.5 in (about 4 cm) wide made of specified mild steel, 16 gauge (about 1.6 mm) thick, are carried so as to form radial paddles on a shaft. The edges of the blades nearest the walls of the mortar are radiused at 0.625 in (about 1.6 cm) horizontally so as to present a nipping surface with the walls of mortar as the shaft rotates. There is a clearance of 0.2 in (about 5 mm) between the edge of the blades and the walls of the vessel and also the bottom of the vessel. The bottom of the shaft rests in a bearing and the top is directly driven. The mortar is 8 in (20 cm) internal diameter and 9 in (23 cm) deep and into it is placed a 4 kg sample of the material to be tested screened less than 4 mm. The shaft rotates at 1500 rev/min. A revolution counter is incorporated with which is connected a switch which stops the rotation of the shaft at a set number of revolutions. After the rotating period, the blades are removed, cleaned free from adhering material (e.g. by rubbing with a cloth or if necessary a fibre brush wheel) and weighed.

Factors which were found[29] to be important to reproducibility of results were (1) clearances between the blades and the mortar, (2) speed of rotation, (3) weight of sample in mortar, (4) total number of revolutions and (5) particle size of material tested. Multiplication of number of revolutions did not increase abrasion in the same proportion, owing, it seems, to the comminution of the material particles during the process decreasing the abrasiveness of the material. Thus factors (4) and (5) are manifestations of the same cause. A material with poor grindability (see p. 57) will thus maintain its abrasiveness over a longer period than one which is easily comminuted. Moisture increases the abrasiveness of some materials and wet grinding is notoriously more wearing than dry; the reason is not clear.

The investigators after whom the test is named suggested that, with coal, 25–30 tests could be made with the same set of blades, but, clearly, this is a matter of the extent of wear over each test; coal abrades away not more than 0.2% and 30 tests would only lose 6%. If the geometry of the test piece is seriously changed, then the tests must be decreased accordingly.

Bond

The BOND method[30] might at first appear to measure the same effect as the previous method. It is true that the method consists of a paddle being rotated in a drum which is itself turning and lifting and tumbling material over the paddle, but some thought will show that impact is involved in the Bond process, whereas it is not involved to any great extent in the Yancy, Geer and Price method.

A flat paddle 3 in (about 7.5 cm) by 1 in (about 2.5 cm) and $\frac{1}{4}$ in (about 6 mm) thick, made of SAE 4325 steel and hardened to 500 Brinell, is inserted about 1 in into a rotor 4.5 in (11.5 cm) in diameter so that it sticks out like a limb. The rotor rotates at 632 rev/min and the drum rotates at 70 rev/min in the same direction as the rotor. Material screened between 0.75 in (19 mm) and 0.5 in (12.5 mm) and weighing 400 g is added to the drum. The whole is then rotated for 15 min; the drum is emptied, a further 400 g is added and rotation is recommenced. In all, four lots of 400 g are added, so that the paddle is abraded for 60 min. The loss in weight of the paddle in grams is the abrasion index. The paddle may be used again by turning it round and inserting the worn end in the rotor; a paddle is used only twice.

DUCTILITY

Ductility is the ability of a material under stress to flow rather than break. Toughness is the property of resistance to the propagation of cracks and is the effect of a material having stress-relieving capacity. A material has stress-relieving capacity if it can increase the radius of curvature of a crack during its propagation and thus decrease the effect of stress concentration. A material can increase the radius of curvature of a crack if it has the ability to flow considerably under stress and thus widen the curve of the crack, i.e. if it is ductile. The opposite of toughness is fragility, not brittleness, though many brittle materials are fragile.

Ductility and elasticity cannot be exhibited at the same time, though a material can be ductile when it is not elastic. Great extensibility can represent the accumulated effects of the ability to sustain great elastic strain and to be ductile beyond the elastic limit, but a material can show ductility when the stress is imposed gradually and elasticity if the stress is imposed impulsively. Thus some of the lighter silicones are quite plastic when moulded in the hands, but when thrown onto a hard surface bounce back.

Ductility is related to the form of bonding of the constituent atoms or ions. Covalent and ionic bonds do not favour ductility but metallic electron cloud forms are favourable to ductility. Materials such as silicon nitride, silicon carbide and beryllia have metallic bonding to some degree.

Single crystals are very often ductile and this is explained by the lack of obstruction to the gliding of the crystalline planes of atoms or ions over one another. Ordinary crystals, being of a more disjointed structure, do not flow so easily because gliding is blocked. The phenomenon of work-hardening is the creation of blockages during processing. MCCANCE[15] attributes the hardening to the creation of surface energy in the increased surface of the new slip plane boundaries. Gliding can be blocked by the insertion of a foreign atom or ion in the lattice; or even on the macroscopic scale, by the addition of finely divided material to the melt before crystallisation (for example, finely divided oxides can be added to metal in this way); or by saturating a melt with a substance which is either precipitated or remains supersaturated when the solution is cooled (age-hardening). Some substances, of which carbon and hydrogen are examples in iron, form clouds which diffuse through the interstices of the matrix atoms and assemble at dislocations and thus inhibit their movement. The effect of this phenomenon, called strain-aging, is that the yield stress is higher than that of the uninhibited matrix, but rapidly drops as movement ensues when the inhibitors are being dispersed. (The higher yield point is restored by reheating and rediffusing the inhibitors to collect on the dislocations.) In fact, elasticity itself appears to depend upon the inclusion of foreign substances in the lattice, as was shown by MCCANCE[15].

However, foreign materials can also enhance ductility. Heterogeneous bodies can slow down the speed of propagation of cracks and can relieve stress concentration. FORWOOD and FORTY[31] diffused gold into crystalline sodium chloride by inducing an electrical potential difference of 100 V/cm at a temperature of 650°C across a sodium chloride crystal using a gold cathode and a platinum anode. The gold ions diffused into the salt crystal and displaced the sodium ions (atomic radii: Na, 1.86 Å; Au, 1.44 Å). The crystal was then held at 450°C for some time to cause the gold ions to mix. Crack propagation was shown to be decreased from 5×10^4 cm/s to less than 1×10^4 cm/s. There was some evidence of increased plasticity.

Pseudo-plasticity, which is really toughness, can be conferred on a material by embedding foreign materials of thread-like shape so that the interfacial strength between the fibrous material and the matrix is less than one-fifth of the tensile strength of the material. The factor of one-fifth results from an analysis of the forces in the vicinity of the crack tip. The effect of this construction is to allow crack tips to meet the fibre, for the forces to separate the fibre from the matrix and for the radius of curvature of the crack tip to increase. The increase in radius decreases the stress concentration (see p. 23) and the crack propagation may be arrested.

It appears to be necessary for a minimum stress to be present before planes can be observed to slip (see p. 27). Compressive strength is usually greater than tensile strength for reasons explained above (p. 32) and some materials are brittle under tension and ductile under compression. Marble behaves in this way: at pressures above 100 MN/m^2, marble is ductile.

Increase in temperature usually improves ductility, thus conforming with the idea that ductility is related to atomic mobility. Toughness and ductility, although usually associated, can also be dissociated from each other, because ductility can aid the flow of dislocations and cause the spaces created by them to coalesce into cracks or cavities. Substances are most extensible which are least atomically dense, i.e. which have the largest interatomic spacing (a) but have the highest Young's modulus (E). Ductility can be expressed[32] by $\frac{1}{8}a^3E$, which is proportional to $\frac{1}{8}$ molecular volume $\times E$, whereas

$$\text{hardness} = f\left(\frac{\text{valency of element}}{\text{molecular volume}^{2/3}}\right)$$

Thus hardness and ductility are incompatible, although compromises are possible: material containing elements of high valency but having intermediate molecular volume and high modulus of elasticity could be both hard and ductile. Iron is such a material, as is shown in *Table 3.9*.

Table 3.9. COMPROMISE BETWEEN HARDNESS AND DUCTILITY

Material	*Valency*	*Molecular volume*	*Young's modulus* $\times 10^6$ lb/in^2	*Brinell hardness*	*Ultimate strain*
Iron	3	7.1	29	106	0.12
Diamond	4	3.4	170	2 500	0.05
Lead	4	18.2	2.3	4	0.69

A rock containing a mixture of very hard constituents and very ductile ones set in a matrix can present comminuting problems. Both compression and impact results in a compaction, while abrasion will result in the harder material dominating. Thus both parts of the mixture display those of their properties which most oppose comminution.

Some materials, particularly metals, have the ability in certain conditions to undergo very large extensions, of the order of 1000%, without fracture. The effect is called superplasticity and is obtained by very low application of strain, not greater than 0.1% per minute, to material at a temperature not less than half that of its absolute melting point. The grain size of the material has to be about 1 μm, and the smaller the grain size below this, the faster may the strain be induced. The materials showing this property are usually alloys, because the grains of pure materials grow too quickly at the temperature necessary. An alloy of 80% zinc and 20% aluminium is superplastic between 200 and 250°C and a steel alloy of 26% chromium, 5.5% nickel is superplastic between 900 and 1000°C.

STICKINESS

Stickiness is a cause of considerable concern in some comminuting processes. Materials, at or near their melting points or sometimes containing a liquid, often in very small amounts, may stick together or upon solid surfaces of the equipment. Other substances, apparently quite dry and very much below their melting points, also show this disposition; talc, chalk and many pigments are examples. Such materials agglomerate and ball up in comminution equipment and block up openings and screens. Although they are often quite soft, they affect equipment in this way so that throughput is reduced often below that of quite hard materials.

The explanation for such behaviour, apart from closeness to melting point, appears to be moisture associated with very small particle size. As is shown in Chapter 6 (p. 103), the effect of pastiness increases considerably as the size of the solid particles decreases for the same quantity of liquid, and it would appear that some substances exhibit stickiness because they contain very fine particles which are dispersed and have associated with them the minute quantity of water which is all that is necessary to cause them to become pasty: they form a film on the larger particles and act like an adhesive.

In such instances comminution may be aided as follows: (a) Dry the material, if it will tolerate heat, either with hot air inside the comminuting equipment or by other means outside the equipment. (b) Cool the material if it will bear the usually high cost. (c) Add certain substances of high absorptive qualities such as kaolin and kieselguhr, which have the effect of drying the material even when added in small quantities. They are usually added to a rotating drum of the material and thus form a coating over the particle surfaces. This procedure is usually not as effective as the others, because as soon as comminution has taken place, fresh surfaces have been created with the original sticky properties. (d) Add materials whose function is *not clear* such as coal and a substance known as TDA (see p. 108), which have the effect of permitting comminution to proceed.

REFERENCES

1. JOISEL, A., *Dechema Zerkleinern Symposion, 1962*, 49
2. INGLIS, C. E., *Trans. Instn. Nav. Archit.*, **55** (1913), 219
3. GRIFFITH, A. A., *Phil. Trans. R. Soc.*, **221A** (1920), 163
4. CLARKE, F. J. P., *Acta Met.*, **12** (1964), 139
5. ZWICKY, F., *Phys. Z.*, **24** (1923), 131
6. DAVIDGE, R. W. and TAPPIN, G., *J. Mat. Sci.*, **3** (1968), 165 and 297
7. KING, R. F. and TABOR, D., *Proc. R. Soc.*, **A223** (1954), 225
8. WILLIAMS, J. G., *Chem. Br.*, **7** (1971), 196
9. BENDICKS, C., *Chim. Ind.*, **30** (1948), 103
10. MACHLIN, E. S. and MURRAY, G. T., *J. Appl. Phys.*, **30** (1959), 1731
11. RUMPF, H., *Dechema Zerkleinern Symposion, 1962*, 1
12. JOFFE, A., *The Physics of Crystals*, 1928. New York; McGraw-Hill
13. DJINGHEUSIAN, L. E., *Trans. Can. Min. Inst.*, **57** (1954), 157
14. SCHULZ, N. F., *Univ. Minn. Publ.* No. 3421 (1951)
15. MCCANCE, A., *Engineering, Lond.*, **172** (1949), 570
16. PRESTON, F. W., *Nature, Lond.*, **156** (1955), 55
17. HEYWOOD, H., *J. Imp. Coll. Chem. Eng. Soc.*, **6** (1950/2), 26
18. FAHRENWALD, A. W., NEWTON, J. and HERKENHOPF, E., *Eng. Min. J.*, **138** (1937), 45
19. HEDDLE, M. F., *Encyclopaedia Britannica*, 9th Edn (1883), Vol. 16, 378
20. KURTH, A., *Z. Ver. Dt. Ing.*, **52** (1908), 1560; **53** (1909), 85
21. MARSH, D. W., *Proc. R. Soc.*, **A279** (1964), 420
22. FRIEDERICH, E., *Fortschr. Phys. Chem.*, **18**(12), (1926)
23. PIRET, E. L., *Chem. Eng. Prog.*, **49** (1953), 56
24. SCHELLINGER, A. K., *Min. Eng.*, **4** (1952), 369
25. BEILBY, G., *Aggregation and Flow In Solids*, 1921. London; Macmillan
26. TATSUO TANAKA, *Dechema Zerkleinern Symposion, 1962*, 360

27. BOWDEN, F. P. and TABOR, D., *Friction and Lubrication of Solids*, 1950. Oxford University Press
28. CROFT, P. E., Private communication
29. YANCEY, H. F., GEER, M. R. and PRICE, J. D., *Trans. AIME*, March (1951), 262
30. BOND, F. C., *54th Ann. Meeting Am. Inst. Chem. Eng., 1963*
31. FORWOOD, C. T. and FORTY, A. J., *Phil. Mag.*, **11** (1965), 1067
32. OROWAN, E., in *Dislocation in Metals* (ed. M. COHEN), 1954. New York; American Institute of Mining and Metallurgical Engineers, Institute of Metals Division
33. RABINOWICZ, E., *Friction and Wear of Materials* (1965). New York; J. Wiley and Sons
34. NIELSON, J. H. and GILCHRIST, A., *Wear*, **11** (1968), 111
35. ELAM, C. F., *Oxford Engineering Science Series* (1935). Oxford University Press
36. EWALD, P. P., PÖSCHL, Th. and PRANDTL, L., *The Physics of Solids and Fluids* (1936). London; Blackey

4

The Energy Consumed in the Comminution Process

The search for a formula for the energy consumed in the comminution process has been ardent and, to date, unsuccessful. The primary problem is that of the energy required to break the material. The factors of breaking have been sketched in Chapter 3. There is also the secondary problem of imparting energy to the material by machines. Not the least part of the solution to the problem is to distinguish the energy absorbed in the equipment from the energy absorbed by the material. Breaking can only be effected through equipment which can absorb energy in all the same ways as the material to be comminuted. The first part of this section will be concerned with the energy consumed in the comminution of material; the second part, with the energy which has to be generated in the equipment before it can impart sufficient energy to the material to fracture it.

It is to be expected that there is a relationship between the energy required to break material and the surface created in the breaking process (see Chapter 2) but this relationship can only be made manifest if the energy consumed in creating new surface can be separately measured. But the larger the intensity of energy input and the less directed it is to the sole activity of comminution, the greater the difficulty of this separate measurement. A material which is plastic will consume energy in changing shape, a shape which it will retain without creating significant new surface. Even in a simple experiment, such as will be described later, of releasing a ball to fall upon a piston set in a cylinder and thus confining material to be crushed, the absolute energy consumed by the material is confused with the performance of the equipment. For example, if a mild steel ball were used, it would probably have to drop four times the distance that a hard nickel steel ball would have to drop to have the same breaking effect, because a mild steel ball distorts more than the hard steel one and more strain energy is therefore absorbed by the steel. If the material were so confined in the cylinder and were of such a size that the plane of maximum stress passed through that part of the particle which was in contact with the cylinder, then the particle would be supported by the cylinder and would exhibit enhanced strength so that more energy would be required to break it than if the particle were not so situated.

The simplifying notion adopted in most of the considerations in this section will be of a brittle material which does not absorb energy in processes such as elongation or contraction or in heating which is not finally utilised in breakage; furthermore it will be assumed that the particles of the material are separated from one another without constraint from their environment or from each other. Needless to say, such a hypothesis is usually far removed from circumstances in large-scale industrial practice.

ENERGY CONSUMED BY MATERIAL

RITTINGER in 1867[1] suggested that the energy consumed in the size reduction of solids was proportional to the new surface produced. Algebraically, this can be expressed as:

$$\text{energy} = K\,(\text{new surface}) = K\,(\text{final surface area} - \text{initial surface area})$$

$$= K\,(\text{final number of particles} \times \text{surface area of representative final particle} - \text{initial number of particles} \times \text{surface area of representative initial particle})$$

$$= K\left(\frac{1}{D_2^3}\cdot D_2^2 - \frac{1}{D_1^3}\cdot D_1^2\right) = K\left(\frac{1}{D_2} - \frac{1}{D_1}\right)$$

in which D_1 and D_2 are the initial and final sizes, respectively, of the particle.

Various experimenters have shown that in certain carefully controlled circumstances this is approximately the case. MARTIN, BOWES and TURNER[2] showed by grinding a quantity of sand in a ball mill, for a range of times, that the work done bore a close relationship to the additional surface produced. Some of their figures are shown in *Table 4.1*. KWONG and co-

Table 4.1. RELATIONSHIP BETWEEN NEW SURFACE AREA AND WORK DONE. BALL MILLING SAND

Change in surface area, ft^2	*Work done*, ft-lb	*Work done per unit area*
3977	243 375	61.3
7852	470 250	59.9
11 170	699 190	62.6
14 941	892 346	59.7
17 899	1 097 300	61.3

workers[3] have shown a similar linear relationship for quartz, calcite, fluorite, labradorite, sphalerite, apatite, topaz and glass by measuring the effect of a ball dropped onto a piston in a cylinder containing the sample of material. The same team of investigators, joined by AXELSON[4,5], failed to show a linear relationship for sodium chloride crystals and for quartz when the range of energy input was extended. They also showed that the first set of minerals was brittle, i.e. showed little plasticity. In a third set of experiments[5] the team showed that over the extended range of energy input and using B.E.T. rather than permeability methods of measuring surface area (see p. 5), and slow hydraulic compression (see p. 35), a linear relationship was obtained. However, using the same surface area measurement methods but the dropped ball method of breaking, and over an even more extended range of energy input, they found that the relationship was not linear. FAIRS[6] found a linear relationship for limestone and anhydrite. There is at first sight a fundamental fallacy in Rittinger's law. The stress necessary to fracture a body in tension or compression does not depend upon the distance over which these forces act, i.e. the so-called length of a body is immaterial to the value of the ultimate stress, but the energy absorbed in the process does depend upon the change in length, which in turn depends upon the original length. Thus particles which are longer in proportion to their cross-sectional area would require more energy. This argument is invalidated by the fact that particles of material as met in practice from run-of-quarry and other size-reduction processes are geometrically very similar for a given feed.

KICK in 1885[7] postulated that the energy required for producing 'analogous changes in configuration in geometrically similar bodies of equal technological state varied as the volume or weight of those bodies'. Deformation energy of a body is proportional to its volume, and just prior to rupture of a brittle body the strain energy, as it is called, is still proportional to volume. Thus unit volume of particles which are in a state just short of rupture contains this deformation energy independently of the size of the particles. Every time that some group of particles of unit volume jointly reach a state just short of rupture, they contain this same deformation energy. Thus the deformation energy at maximum stress is the same for unit volume at each stage of comminution or

$$\text{total deformation energy} = K_1 \times \text{number of stages of size reduction}$$

and if the original average particle size is D_1 and the final size D_n in unit volume of material, and the number of stages of size reduction is n and the reduction ratio at each stage is r, then

$$\left(\frac{D_1}{D_n}\right)^3 = r^n$$

and

$$n = \frac{3(\log D_1 - \log D_n)}{\log r} = K_2 \log\left(\frac{D_1}{D_n}\right)$$

or

$$\text{total deformation energy} = K_1 K_2 \log\left(\frac{D_1}{D_n}\right) = \text{say, } K \cdot \log\left(\frac{D_1}{D_n}\right)$$

Some substantiation for this formula has been found in crushing cubes of stoneware[8] and the similarity of size distribution of products of successive crushing operations on the same equipment suggests that quanta of energy are used in the stages of comminution. The strain energy is only the energy of deformation up to the stage of rupture and does not include the energy actually required for fracture. TAPLIN[9] has suggested that Kick's law covers the energy required up to the point of fracture and that Rittinger's law covers the energy actually required for fracture after the particle has been brought to the state of deformation at which it will fracture, and that the ratio of these two quantities of energy is related to the size of the particle concerned. Thus if the energy states at the two levels are represented by α, preparatory to fracture (Kick), and β, energy of fracture (Rittinger), $\alpha/\beta = x$, where x is the particle size. Thus, when x is large, Kick's law dominates; and when x is small, Rittinger's law dominates.

But the energy of fracture should only be the same as the additional surface energy induced, and this can be shown to be usually less than 1% of the strain energy. For single crystals (see p. 28) the tensile strength is given by $(E\gamma/a)^{\frac{1}{2}}$ and the volume strain energy for a unit cube would be tensile strength × half the strain or $(E\gamma/a)^{\frac{1}{2}}.(E\gamma/a)^{\frac{1}{2}}.(1/2E) = (\gamma/2a)$. For a cube of side x this would be $\gamma x^3/2a^4$. The surface energy for a cube would be $6\gamma x^2$. Thus the ratio of strain energy to surface energy is $x/12a^4$. Now a in most crystals is of the order of 2 Å and for a particle of greater size than 1 μm (10^4 Å) $= x$, the ratio of strain energy to surface energy becomes more than, say, 50 or surface energy would be less than 2% of strain energy for single (i.e. uncracked) crystals. For real crystals a (atomic distance) is replaced by v (crack length), which is of course larger than a, but 1 μm is a very small particle in ordinary comminution terms.

A dimensional argument has been urged[10] that surface change cannot be a measure of energy or work since they are equivalent to force multiplied by distance, while surface can only be a measure of force; but surface energy arises from the potential energy of the surface atoms or ions which would be realised if these particles were able to enter the interior of the solid by the very small distance over which the strong attractive forces act and therefore has the same dimensions as work or energy (ML^2/T^2).

There is less broad evidence for Kick's law than there is for Rittinger's law.

BOND[11] developed an equation for the energy content of a material based upon a function, the reciprocal of the square root of the screen size through which 80% passes, which will be represented by $Z^{-\frac{1}{2}}$. Thus for a particle of infinite size this function would be zero. Bond suggests that any comminution process can be considered to be an intermediate stage in the breakdown of a particle of infinite size to an infinite number of particles of zero size, so that the energy possessed by a group of particles is proportional to $Z^{-\frac{1}{2}}$ and the energy acquired by a group of particles characterised by Z_1 in breaking to a group of particles characterised by Z_2 is proportional to the difference $Z_2^{-\frac{1}{2}} - Z_1^{-\frac{1}{2}}$. This is also proportional to the energy (W) required to break particles of size Z_1 to particles of size Z_2. Thus

$$W = K(Z_2^{-\frac{1}{2}} - Z_1^{-\frac{1}{2}})$$

where K is a constant for the material concerned. In the form of this equation in which K is replaced by $10W_i$, W_i is generally called the Bond index so that

$$W = 10W_i(Z_2^{-\frac{1}{2}} - Z_1^{-\frac{1}{2}})$$

It should be pointed out at this stage that there is nothing in the argument that energy is associated with a degree of subdivision between the infinitely large and the infinitely small which is peculiar to the use of the square root in Bond's formula; such an argument can equally well be applied to the Rittinger or Kick formulas. What is unique about Bond's formula is that for a vast number of materials a work index can be derived in practice with a relatively constant value over a range of sizes. Some are listed in *Table 4.2.* They are for materials ground dry and can be converted to indices for wet grinding by dividing them by 1.3.

Table 4.2. SELECTION OF BOND WORK INDICES AND HARDGROVE GRINDABILITIES

Material	*Specific gravity*	*Hardgrove grindability*	*Bond work index*
Basalt	2.91		19
Bauxite	2.20	39–76	10
Cement clinker	3.15	33	15
Coke	1.31	33	17
Corundum			30–35
Dolomite	2.74	112	13
Feldspar	2.59	30–75	12
Ferrosilicon	4.41		11
Flint	2.65		29
Fluorspar	3.01	74–92	10
Glass	2.53		14
Granite	2.66		11
Gypsum rock	2.69		7
Hematite	3.55	35–96	14
Lead ore	3.45	51–81	13
Limestone	2.65		14
Magnesite	3.06	44–64	12
Magnetite	3.88	40	11
Malartic			9–13
Marble			4–12
Morenci			9
Phosphate rock	2.74	47	11
Potash ore	2.40		9
Pyrites	4.06		10
Quartz	2.65		15
Quartzite	2.68		11
Rutile ore	2.80	26	14
Sandstone			11
Silica sand	2.67	31	16
Silicon carbide	2.75		29
Slag	2.83	17	11
Slate	2.57	72	16
Zinc ore	3.64	136	12
Zircon sand			20–30

Bond explains the factor $Z^{-\frac{1}{2}}$ as being a measure of crack length to break the particle of size Z; since the surface area of unit volume of material is proportional to Z^{-1}, the crack length in unit volume is considered to be proportional to one side of that area and therefore proportional to $Z^{-\frac{1}{2}}$.

HOLMES[12] has proposed a modification to Bond's energy equation, substituting an exponent r in place of Bond's 0.5, so that the equation reads:

$$W = 10W_i(Z_2^{-r} - Z_1^{-r})$$

Values of r which Holmes determined for various ores are given in *Table 4.3*.

Various attempts have been made to show that the relationships of Rittinger, Kick and Bond are interpretations of single general equations:

$$\frac{dW}{dd} = -Cd^n \qquad \text{(WALKER et al.[13])}$$

$$W_R = a + \frac{b}{d_0^m} \qquad \text{(DOBIE[14])}$$

$$W = \frac{Kd_0^{3-r}\cdot(R^r - 1)}{2^r - 1} \qquad \text{(HOLMES[12])}$$

where W is energy consumed (or work done), d is size of particle, d_0 is initial size of particle, W_R is energy consumed for size reduction ratio R, and r and m are exponents. *Table 4.4* shows the values to be attributed to these symbols in order to arrive at a particular law.

Table 4.3

Ore	*Holmes exponent* (r)
Amygdaloid	0.25
Malartic	0.40
Springs	0.53
Sandstone	0.66
Morenci	0.73
East Malartic	0.42
Chino Nevarda Consolidated	0.65
Real del Monte	0.57
La Luz	0.34
Kelowna Exploratory	0.39
Utah Copper	0.50

Table 4.4

Energy law	*Values of symbols*				
	n	a	b	m	r
Rittinger	−2	0	b	1	1
Kick	−1	a	0	—	0
Bond	−1.5	0	b	0.5	0.5

HUKKI[15] suggests that the relationship between energy and particle size is a composite form of the Rittinger, Bond and Kick laws. If lines representing these laws are drawn on logarithmic-scaled paper over the range of particle sizes represented by the limits of about 1 μm, 5000 μm (5 mm) and

10 cm, respectively, a composite curve is produced (see *Figure 4.1*) which bears some relationship to a series of curves obtained in practice. Rittinger's law seems to apply fairly well over the range 10–1000 μm. For sizes smaller than this the effect of prolonged milling is not to change surface area and even to decrease it (see *Figure 5.7*). This effect is discussed in Chapter 5 (p. 87).

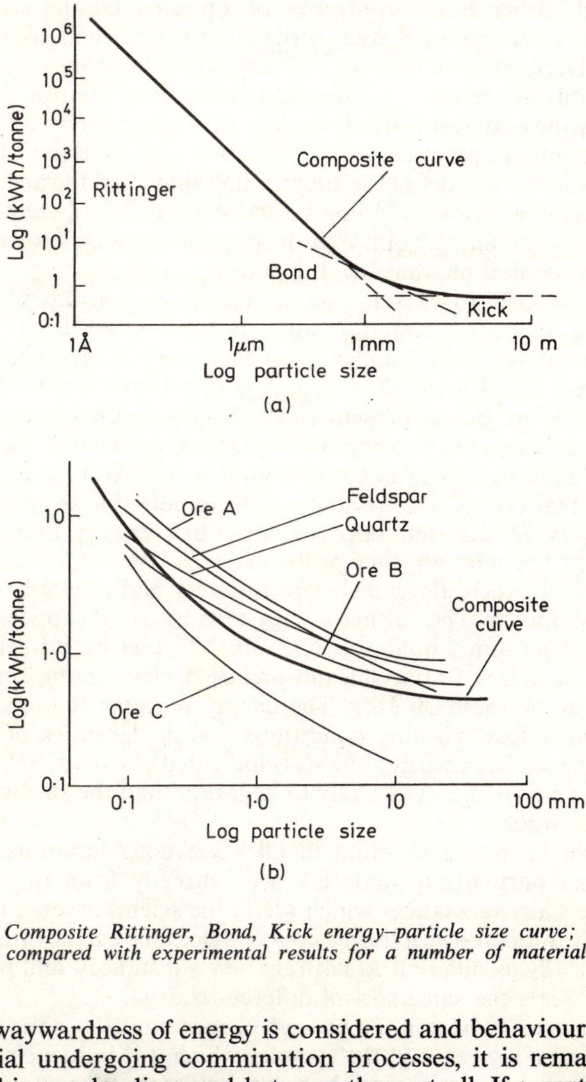

Figure 4.1. (a) Composite Rittinger, Bond, Kick energy–particle size curve; (b) portion of composite curve compared with experimental results for a number of materials. (After R. T. HUKKI[15]*)*

When the waywardness of energy is considered and behavioural possibilities of material undergoing comminution processes, it is remarkable that any relationship can be discerned between them at all. If a particle is compressed within its elastic limits, it will merely deform for the duration of the stress and then return to its original shape. Particles which rub together or collide exchange kinetic energy for heat, sound, vibration, electricity or light. Particles which are shattered fly off with kinetic energy. Then there are the differences in macro- and micro-structures of the particles themselves; the areas to which the forces are applied; and the question whether the environment is healing or aggravating to cracks. Above all these factors, so far as the scientist or engineer is concerned, is that of difficulty of measurement—measurement of change of surface area from the point of view of method and instruments and from that of what surface area has actually been created rather than discovered or revealed by the size reduction; measurement of energy imparted to the solid rather than consumed by the equipment. Difficulties of control are very formidable indeed.

It is probably not possible to suggest a more useful relationship for energy consumed by the material which is broken than that proposed in principle by Rittinger which is so practical. The consequence of comminution is increase in surface area. The physics of the subject may show that the absolute increase in surface energy is a relatively small fraction of the energy required to bring a particle to the breaking point and over it; nevertheless increase in surface area is an expression of what has happened.

Bond's relationship is a very useful vehicle for conveniently comparing energy requirements of materials and, as has already been stated (p. 8), Kihlstedt has shown that Bond is in effect determining changes in surface area. Kihlstedt has shown that $Z^{-\frac{1}{2}}$ is proportional to surface area. The reasons for this are not at present clear. What is probably misleading is to express energy consumed compared with surface energy acquired as an efficiency. If a material will not break until it has acquired a certain strain energy, then that energy has inevitably to be imparted to the material and the useful recovery of the difference between that energy and the acquired surface energy becomes another problem.

Quantitatively, the difference between theory and practice is not so great as might first appear. The surface energy of many solids appears to be of the order of 10^{-1} μJ/mm^2, both theoretically[16, 17] and experimentally[18]. This means that particles of about 10 μm and of density about 2 mg/mm^3 have surface energy of about 30 J/kg. The energy required for breaking of such materials under free crushing conditions[19] is of the order of 30 kJ/kg, i.e. 1000 times the surface energy; but, as is indicated above (p. 52), for a particle of 10 μm, the strain energy is likely to be larger than the surface energy by a factor of this order.

What must be borne in mind in all these considerations is that most materials and particularly materials won directly from the earth do not begin to approach substances which are in the scientific sense homogeneous mechanically. A breakdown from a coarse sandstone can be into grains which are mechanically as different as entirely new substances, and no doubt on a microscopic scale the same sort of differences exist.

The appearance of heat in the process of comminution is the manifestation of the energy having reached the stage in which it is influencing the velocity of the atoms or ions composing the solid. Before that the energy was mainly used in processes of distortion. Equilibrium corresponds with energy being utilised by the atoms themselves. The mystery is not the manifestation of heat, but what is happening prior to the speeding up of the atoms.

It is astounding, too, that at the level of industrial equipment power usage is of the order of 100 times the usage at the experimental free crushing level. The large machine seems to be so prodigal with power. This aspect is discussed below (p. 60).

GRINDABILITY

In consequence of the theoretical considerations outlined in the previous section, various investigators have suggested methods of comparing the ease with which materials can be comminuted. In the methods suggested the general concept has been to start with the material in a fixed size range and either to determine the power consumed in reducing the material to another fixed size range in a specified piece of equipment or after submitting the material to a defined comminuting process in such equipment to determine the size range of the product. The translation of such results into a grindability index has taken many forms, some of which will be the subject of this section. The reciprocal of such quantities is sometimes used and is called the crushing resistance.

Gradually more precise indices have evolved based upon closely defined methods of measurement. STADTLER[10] suggested that the energy contents of the various size fractions of a material can be assessed according to Kick's law, particularly if the sieve ranges are arranged so that for ease of calculation they conform with a geometric relationship. Thus, if each sieve size in the screen series is related to the preceding and larger screen by, say, the square root of 2, then a fixed size reduction ratio exists between all screens in the series and, according to Kick's law, the energy associated with the *n*th screen size is *n* times that associated with the first screen size in the series. Thus, if such a series of screens constitutes a fixed set and if to each sieve is allocated a consecutive ordinal number which is largest for the smallest screen and unity for the largest screen, an assessment of energy content before and after comminution can be made by multiplying by the appropriate ordinal numbers each of the fractions passing one screen in the series and retained by the next. These values can be summed and a comparison made between the sum before comminution and that afterwards; by relating the difference to the energy consumed, a figure can be calculated which represents energy associated with unit ordinal number by unit weight of material for that particular comminuting process. Comparisons can then be made for different materials processed in the same equipment and for the same material processed in different equipment.

CAREY and STAIRMAND[19] have developed this idea by actually determining the energy associated with particular size ranges of a material in a simple hydraulic press under free crushing conditions (see p. 49). They construct a graph for the material from which the energy may be read for any mean particle size and, using the same summing technique as Stadtler, they arrive at the energy associated with a particular size distribution of a material. They emphasise that the figures they give as examples are tentative, but it may be interesting to compare the figures they give for coal and quartz as shown in *Table 4.5*. The Bond index is itself an expression of grindability, since the index is a measure of power consumption by a material comminuted between fixed size ranges. Bond has devised several methods, presently to be described, of determining this index.

Grindability has more recently been based upon performance in a carefully defined piece of equipment according to a strict procedure. Among the various methods are the following.

Table 4.5. ENERGY UNITS ASSOCIATED WITH PARTICLE SIZE

Material	*Particle size,* μm			
	100	1 000	10 000	100 000
Coal	1.43	0.53	0.28	0.13
Quartz	3.1	2.1	1.45	0.7
Quartz/coal	2.2	4.0	5.2	5.4

Hardgrove test. Hardgrove devised his machine with the ring ball mill particularly in mind. The results of this test have therefore particular applicability to that kind of machine without of course necessarily restricting them in that way. The machine[26] consists of a top rotating ring with eight one inch diameter balls which roll on a bottom stationary ring cavity (see *Figure 4.2*). The top rotating ring is loaded to 64 lb. Into the mill is introduced 50 g of material screened between 16 and 30 mesh ASTM and then revolved 60 times. The weight (W) passing 200 ASTM is determined and the Hardgrove index is $13+6.93W$.

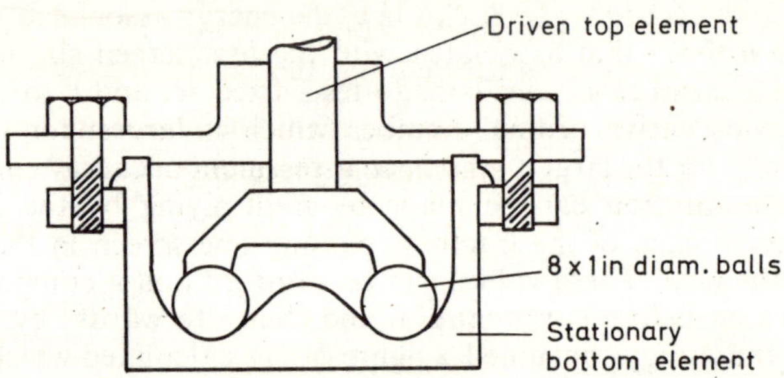

Figure 4.2. Hardgrove grindability machine

Bond ball mill test[20]. A standard ball mill, 12 in internal diameter and 12 in internal length, charged with 285 balls weighing in total 20 125 g between $\frac{3}{5}$ in and $1\frac{1}{2}$ in diameter, of which 120 are over 1 in diameter, is filled with 700 cm^3 bulk volume of dry minus 6 mesh (3330 μm) material. The mill is rotated 100 times at 70 rev/min and then stopped and the contents are screened at a test mesh size which is part of a series. Fresh feed is added to the oversize to bring total volume to 700 cm^3 again and the mixture is added to the mill. The mill is rotated for a number of times calculated to produce a circulating load of 250%, i.e. containing 28.6% at the end of the cycle which will pass the mesh size. The process is continued until the net weight of undersize produced per revolution is constant: let this be G g. Then the grindability is said to be G. If F is the sieve size through which 80% of the feed will pass and P is the sieve size through which G passes, then the Bond index for wet grinding (see p. 53) is

$$W_i = \frac{44.5}{P^{0.23}} \times G^{0.82} \times \left[\frac{10}{(P)^{\frac{1}{2}}} - \frac{10}{(F)^{\frac{1}{2}}}\right]$$

from which the index for dry grinding may be obtained by multiplying by 1.3.

Bond rod mill test[27]. A standard rod mill (12 in internal diameter and 24 in internal length), charged with six $1\frac{1}{4}$ in diameter and two $1\frac{3}{4}$ in diameter steel rods, all 21 in long and weighing 33 380 g, is fed with 1250 cm^3 of $<\frac{1}{2}$ in material and rotated at 46 rev/min as for the Bond ball mill test except that 100% circulating load is used. Then G is determined as in the ball mill test, and the Bond work index for wet grinding is given by

$$W_i = \frac{62}{P^{0.23}} \times G^{0.625} \times \left[\frac{10}{(P)^{\frac{1}{2}}} - \frac{10}{(F)^{\frac{1}{2}}}\right]$$

Bond impact test[27]. Two equal hammers weighing 30 lb are arranged so that they can strike simultaneous blows on opposite sides of a test piece. The pieces used are of minus 3 in and plus 2 in. During the test, a piece is located with its smallest dimension (d) between the hammers. The hammers are raised to a known height and dropped so that they hit the piece. The hammer is raised to a greater height and a succession of tests made until the piece breaks. Ten samples are tested and the results averaged. Then

$$I = \frac{2 \times \text{mass of one hammer} \times \text{final height of hammer}}{\text{thickness of piece } (d)}$$

where I is impact crushing strength, and the Bond work index for wet grinding is

$$W_i = \frac{2.59\, I}{\text{density of piece}}$$

The strength index. The strength index is used particularly for coal[21]. The sample is closely graded to be not less than a certain size (f), and a second size (p) less than f is chosen. The sample is then passed through a standard piece of equipment and screened. Then

$$\text{strength index} = \frac{1}{(\%\text{ passing } f)^{4/3}\,(\%\text{ passing } p)}$$

HOLMES[12] has noticed a relationship between P and G, where P is the sieve size through which passes G, the value determined by the Bond ball or rod test, viz. $G = KP^b$, so that a logarithmic plot gives $\log G = b \log P + \log K$, a straight line. Values of b which Holmes determined are: Par Harbour quartz, 0.77; Nigerian granite, 0.78; Lake Shore Ontario gold ore, 0.62; Derbyshire Carboniferous Limestone, 0.45.

All of the above tests are open to criticism because they select a size range which can exclude important elements in a heterogeneous sample. However, not to select size ranges would risk departure from free grinding conditions.

Bond proposed a relationship between his work index and that of Hardgrove:

$$W_i = \frac{435}{(\text{Hardgrove index})^{0.91}}$$

but tests made at Warren Spring, on a wide selection of minerals, indicate that this relationship is only very approximately true. A comparison between the two indices is contained in *Table 4.2.*

ENERGY SUPPLY REQUIRED BY COMMINUTING EQUIPMENT

The second aspect of the problem of energy of comminution is the transfer of sufficient energy from a prime mover to the piece of material to cause it to fragment.

The efficiency with which equipment imparts energy to the material is idiosyncratic to the equipment. Clearly some equipment is constructed so that much energy is absorbed in its accessories and ancillary parts which is not necessarily available to the material. In a ball mill, for example, an assessment was made as follows:

	%
Bolt friction	4.3
Gear losses	8.0
Heat losses from drum	6.4
Heat absorbed by air circulation	31.0
Heat absorbed by product	47.6
Theoretical energy for size reduction	0.6
	97.6

In spite of this, approximate figures have been found to apply[19] as power consumption for various ranges of size reduction. Thus

		kWh/t
Coarse crushing	down to 10 cm	3–4
Intermediate crushing	between 1 and 10 cm	5–6
Fine grinding	between 125 μm and 1 cm	20–30
Superfine grinding	below 125 μm	100–1000

One approach to the problem has been[22] to regard the energy consumed by the equipment as constant and only the energy consumed by the material to be variable, this latter energy being regarded as directly proportional to the mass of the material comminuted, so that a straight line relationship exists between the total energy consumed and the weight of material processed. Thus, if u is the proportion of energy required by the material at optimum throughput and i is the proportion of energy required to drive the equipment when empty, so that in optimum conditions $u+i = 1$, and if q is the proportion of optimum quantity of material being processed, then the total energy required is $qu+i$; if

$$\text{relative energy} = \frac{\text{total energy required}}{\text{optimum energy}} = \frac{qu+i}{u+i}$$

$$= qu+i$$

then

$$\text{relative energy per ton} = \frac{qu+i}{q}$$

HEYWOOD[22] evaluated u and i for some equipment (see *Table 4.6*). However, the relationship is found to be very approximate and i varies quite appreciably with size for some kinds of equipment, e.g. for roll crushers

Table 4.6. RELATIONSHIP BETWEEN ENERGY REQUIRED TO DRIVE EQUIPMENT WHEN EMPTY (i) AND ENERGY CONSUMED BY OPTIMUM LOAD OF MATERIAL (u)

	i	u
Ball-loaded mill	0.9	0.1
Ring roll mill	0.4	0.6
Impact mill	0.25	0.75

values ranging from 0.2 to 0.7 are obtained. Other investigators have determined the proportions of energy consumed in breaking the material in certain specified conditions. Examples are given in *Table 4.7*.

A relationship developed by BOND and WANG[23] suggested that

$$\text{horse power consumed per ton} = \left(\frac{n^{\frac{1}{2}}}{P}\right)^{\frac{1}{2}}$$

where n is the quotient of sieve size in inches through which 80% of feed will pass divided by size through which 80% of product will pass, i.e. F/P. Their graph is shown in *Figure 4.3*. They assign values for various pieces of equipment.

Another approach is to calculate the Bond work index from the power being actually consumed. *Table 4.8* shows the factors so determined for a number of different sizes of several types of equipment. The results obtained with limestone have been selected where they are available. It is appreciated that limestone has notoriously irregular physical properties, but it happens to be a material which many manufacturers use for testing their equipment. Thus, if W is the power being consumed, measured in kilowatt-hours per

Table 4.7. ENERGY CONSUMED BY MATERIAL IN VARIOUS EQUIPMENT

	*Percentage of total energy consumed**									kWh/ton*
	2	6	19	22	22	24	*White*	*Wilson*	*Bennet*	23
Jaw crushers										0.25–3.0
Gyratory crushers	0.5								33	1.6–2.1
Crushing rolls										1–8
Ring roll				60	30					
Stamp mill							10.5		35	
Swing hammer, coarse		20/30								
Swing hammer, fine		5/10	36	75	20					2–160
Ball-loaded	0.02–0.06	10/13	6/15.5	10	10	45	185	20/40	23	200/5 000
Ball-loaded Hardinge			6.3							50/600
Air-energy mill		4								

* Numbers and names below these headings are of references and investigators, respectively. The data from reference 2 are based on absolute surface requirements; those from reference 6 are based on kW readings, empty and loaded; those from reference 19 are calculated from energy associated with state of sub-division (see p. 57) and compared with energy consumed by equipment.

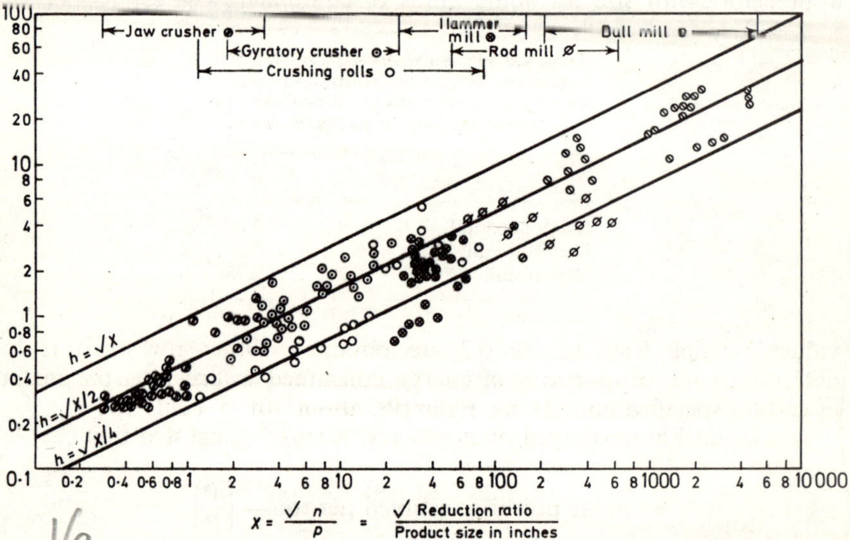

Figure 4.3. Bond and Wang's relationship between energy consumption by equipment type and reduction ratio

ton, W_e is the calculated Bond index, and the screen sizes in microns through which 80% of the feed and product passes are, respectively, F and P, then

$$W = 10W_e(P^{-\frac{1}{2}} - F^{-\frac{1}{2}})$$

Although the values obtained for different sizes of the same type of equipment have a wide range, a broad pattern can be seen. The biggest consumers are jaw and gyratory crushers and pan and tumbling mills (W_e about 30); intermediate consumers are rotary impactors (hammer) and

Table 4.8. CALCULATION OF ACHIEVED BOND INDEX (W_e) IN VARIOUS TYPES OF EQUIPMENT

Type of equipment	*Feed size,* F (μm)	*Product size,* P (μm)	$\frac{1}{P^{\frac{1}{2}}}-\frac{1}{F^{\frac{1}{2}}}$ (S)	*Energy consumed, E* (kWh/tonne)	$\frac{E}{10S}$ (W_e)	*Material (where known)*
Jaw crusher	76×10^4	127×10^3	1.6×10^{-3}	0.94	58	
	122×10^4	305×10^3	0.9×10^{-3}	0.27	30	
Gyratory crusher	3×10^4	3×10^3	12.5×10^{-3}	7.46	60	
	3×10^5	4×10^4	3.2×10^{-3}	0.45	14	
	12×10^4	12×10^3	6.2×10^{-3}	1.28	21	
Roll crusher	508×10^2	64×10^2	8.0×10^{-3}	0.75	9	Limestone
	5×10^3	5×10^2	30.7×10^{-3}	1.46	5	
Pan mill	15×10^4	14×10^3	5.8×10^{-3}	1.95	34	
Ring roll, Raymond	3 170	59	112.2×10^{-3}	10.25	9	
Lupulco	19×10^3	76	107.7×10^{-3}	10.10	9	
Ring ball, E type		72	118×10^{-3}	7.5	6	
Impact rotary	103×10^4	25×10^4	1.0×10^{-3}	0.22	22	
	610×10^3	$1\,525\times10^2$	1.3×10^{-3}	0.40	31	
Vibration mill	4 760	74	101×10^{-3}	22.4	22	Limestone
	5 000	130	73×10^{-3}	13.3	18	Sea salt
	4 760	74	101×10^{-3}	28.3	28	Limestone
Pin mill	3 000	50	123×10^{-3}	13.3	11	
Ball mill	125×10^3	295	49.4×10^{-3}	13.5	27	Limestone
	9 550	147	72.2×10^{-3}	17.9	25	
	6 350	74	104×10^{-3}	21.8	21	
	38 100	1 651	19.5×10^{-3}	4.25	22	
Autogenous	228×10^3	251	61×10^{-3}	16	26	
Fluid-energy*	80	3.5	421×10^{-3}	2 500	600	

* Fluid energy calculated from steam usage.

vibration mills (W_e about 20); and smallest consumers are roll crushers, ring roll and ring ball and pin mills (W_e about 10).

The Kihlstedt relationship (see p. 8), $Z^{\frac{1}{2}}S = \sim 750$, or

$$\frac{1}{Z_2^{\frac{1}{2}}}-\frac{1}{Z_1^{\frac{1}{2}}}=\frac{S_2-S_1}{750}$$

in which Z is measured in centimetres and S in square centimetres per cubic centimetre, enables the created surface to be calculated which is associated with specific power consumption. Thus, if W is the specific power consumption measured in kilowatt-hours per ton or, within experimental error, kilowatt-hours per tonne, then this is associated with the creation of $750[(1/Z_2^{\frac{1}{2}})-(1/Z_1^{\frac{1}{2}})$ cm^2/cm^3, or, if ρ is the density of the material measured in g/cm^3,

$$\frac{750}{\rho}\left(\frac{1}{Z_2^{\frac{1}{2}}}-\frac{1}{Z_1^{\frac{1}{2}}}\right)\times 10^6\ \text{cm}^2/\text{t}$$

Then

$$\frac{W}{\dfrac{10^6 \times 750}{\rho}\left(\dfrac{1}{Z_2^{\frac{1}{2}}} - \dfrac{1}{Z_1^{\frac{1}{2}}}\right)} = \text{kilowatts per square centimetre of created surface}$$

and, since 1 kilowatt $= 3.6 \times 10^{13}$ erg,

$$\frac{W \times 3.6 \times 10^{13} \times \rho}{10^6 \times 750(Z_2^{-\frac{1}{2}} - Z_1^{-\frac{1}{2}})} = \text{ergs per square centimetre of created surface}$$

This reduces to

$$\frac{W \times 10^4 \times 4.8 \times \rho}{(Z_2^{-\frac{1}{2}} - Z_1^{-\frac{1}{2}})}$$

which could be called the Bond–Kihlstedt specific surface energy or BK index. But

$$\frac{W}{10(P^{-\frac{1}{2}} - F^{-\frac{1}{2}})} = W_i(\text{ Bond's index})$$

and since P and F are measured in microns while Z_2 and Z_1 are measured in centimetres,

$$\frac{10W}{(Z_2^{-\frac{1}{2}} - Z_1^{-\frac{1}{2}})} = W_i$$

and BK index $= W_i \times 10^3 \times 4.8 \times \rho$ erg/cm^2 (1 erg/cm^2 = 10^{-3} μJ/mm^2). The corresponding index for a particular piece of equipment is $4.8 \times 10^3\ \rho W_e$.

Now the surface energy of a material such as limestone is of the order of 100 erg/cm^2. It has been shown above (p. 52) that the strain energy at fracture is more than 50 times the surface energy. Thus the ratio between energy consumed by the equipment and that which might be considered the theoretical energy consumption is

$$\frac{4.8 \times 10^3\ \rho W_e}{5 \times 10^3} = \rho W_e \text{ approximately}$$

For limestone ($\rho = 2.8$) and using the figures obtained above:

Jaw, gyratory crushers, pan and tumbling mills	$30 \times 2.8 = 84$ times theoretical
Rotary impactors and vibration mills	$20 \times 2.8 = 56$ times theoretical
Roll crushers, ring roll, ring ball and pin mills	$10 \times 2.8 = 28$ times theoretical

These differences between types of equipment are relatively small compared with the huge factors of multiplication of the theoretical, but the smallest consumers of energy are those types of equipment which apply a steady continuous compressive stress on the material, a result in keeping with theoretical considerations (see p. 35).

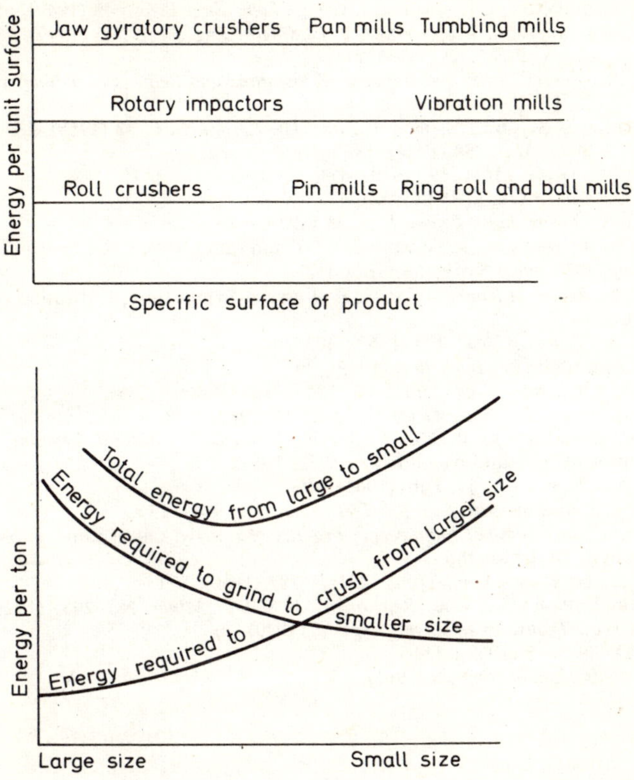

Figure 4.4. Energy optimisation by changeover from one mill to another

Curves can be drawn[22, 25] of the type shown in *Figure 4.4* for individual cases and assessments made as to the energy usage economics of routing and changeover points in comminuting from one size to another. In big operations such an assessment is clearly worth while.

In general, jaw, gyratory and roll crushers are normally operated under free crushing conditions and pan, ring roll, ring ball, vibration and tumbling mills perforce operate under packed and choked conditions. Energy usage by the equipment for the duty performed is more for choked than for free crushing conditions and, having regard to the additional advantage of better particle-size control which can be exercised under free crushing conditions, the choice of changeover point can require careful judgement. As much as twice the energy per unit of new surface created can be absorbed in choke feed compared with free crushing[22].

REFERENCES

1. VON RITTINGER, P. R., *Lehrbuch der Aufbereitungs Kunde*, 1867. Berlin; Ernst and Korn
2. MARTIN, G., BOWES, E. A. and TURNER, F. B., *Trans. Ceram. Soc.*, **25** (1926), 63
3. KWONG, J. N. S., ADAMS, J. T., JOHNSON, J. F. and PIRET, E. L., *Chem. Eng. Prog.*, **45** (1949), 508
4. ADAMS, J. T., JOHNSON, J. F. and PIRET, E. L., *Chem. Eng. Prog.*, **45** (1949), 655
5. JOHNSON, J. F., AXELSON, J. and PIRET, E. L., *Chem. Eng. Prog.*, **45** (1949), 708
6. FAIRS, G. L., *Trans. IMM*, **63** (1953), 211
7. KICK, F., *Das Gesetz der Proportionalem Widerstand und Seine Anwendung*, 1885. Leipzig; Felix
8. ANDREASEN, A. H. M. and JENSON, I. H., *Ber. Dt. Keram. Ges.*, **32** (1955), 232
9. TAPLIN, T. J., *Min. Mag.*, **50** (1934), 18
10. STADTLER, H., *Trans. IMM*, **19** (1910), 471
11. BOND, F. C., *Trans. AIME*, **193** (1952), 484
12. HOLMES, J. A., *Trans. Instn Chem. Eng.*, **35** (1957), 125
13. WALKER, W. H., LEWIS, W. K., MCADAMS, W. H. and GILLILAND, E. K., *Principles of Chemical Engineering*, 1937. New York; McGraw-Hill
14. DOBIE, W. B., *Recent Advances in Mineral Dressing* (Symposium, Institution of Mining and Metallurgy), 1952
15. HUKKI, R. T., *Trans. AIME*, **223** (1962), 403
16. BORN, M. and STERN, O., *Berl. Ber.* (1919), 901
17. VERWEY, E. J. W., *Rec. Trav. Chim.*, **65** (1946), 521
18. KUSNEZOW, W. D., *Z. Phys.*, **42** (1927), 302; **44** (1927), 228
19. CAREY, W. F. and STAIRMAND, C. J., *Recent Advances in Mineral Dressing* (Symposium, Institution of Mining and Metallurgy), 1952, 10
20. BOND, F. C., *Chem. Eng.*, **69**, Feb. (1962), 103
21. BENNETT, J. G. and BROWN, R. L., *J. Inst. Fuel*, **14** (1941), 135
22. HEYWOOD, H., in *Chemical Engineering Practice* (ed. H. W. CREMER and T. DAVIES), Vol. 3, 1957. London; Butterworths
23. BOND, F. C. and WANG, J. T., *Trans. AIME*, **187** (1950), 871
24. GROSS, J. and ZIMMERLEY, S. R., *Rep. Invest. U.S. Bur. Mines*, No. 2952 (1929)
25. FARRANT, J. C., *Trans. Instn Chem. Eng.*, **18** (1940), 56
26. A.S.T.M., D.409 51, **5** (1952), 840
27. BOND, F. C., *Br. Chem. Eng.*, **6** (1961), 378

5

The Size and Shape of Comminuted Particles

After a lump of solid has been shattered or crushed, it usually breaks into pieces apparently irregular in number, shape and size. Clearly, if care is taken in the process, pieces of predetermined size can be produced, as is witnessed by countless pieces of masonry and sculptures, but in the equipment so far devised for breaking large quantities of material the question may be asked, 'Do the product particles have any predictable features?' It is proposed to explore the question in this chapter.

SIZE DISTRIBUTION

The feed to a piece of comminuting equipment will very often be itself the result of some comminuting process, but first let the size range be artificially imposed by a classifying process and let it be narrow. On entry of the feed to the machine, the largest pieces might be thought most susceptible to attack and the smallest the least susceptible, because they can most easily evade the comminuting agents whatever they might be. The intermediate sizes might be thought to be affected according to their size, but not necessarily directly proportionally.

The general effect of comminution would therefore be expected to be a narrowing of the size range, with remnants of the largest particles gradually disappearing but still forming the size through which 100% passes though with decreasing claims to significance. A series of changes such as are shown in *Figure 5.1a* might be thought possible as the product is charged and re-charged to the machine. The feed graph is represented as a straight line for simplicity. The first run through (I) contains a significant part of the original larger fractions, the second (II) less and so on.

Some equipment is deliberately or unavoidably size-selecting. This is clearly so for the upper size of the feed in all equipment; but placing a grid over the discharge could regulate further the top size, whereas air separation could regulate the bottom size of the product. Graphs such as those shown in *Figure 5.1b–d* could be obtained. As more and more of the original particles are broken, so the shape of the original distribution curve gradually disappears and the character of the equipment increasingly imposes its pattern. Some investigators[1] have called the original particles the residue and the newly broken particles the complement. This process is illustrated in *Figure*

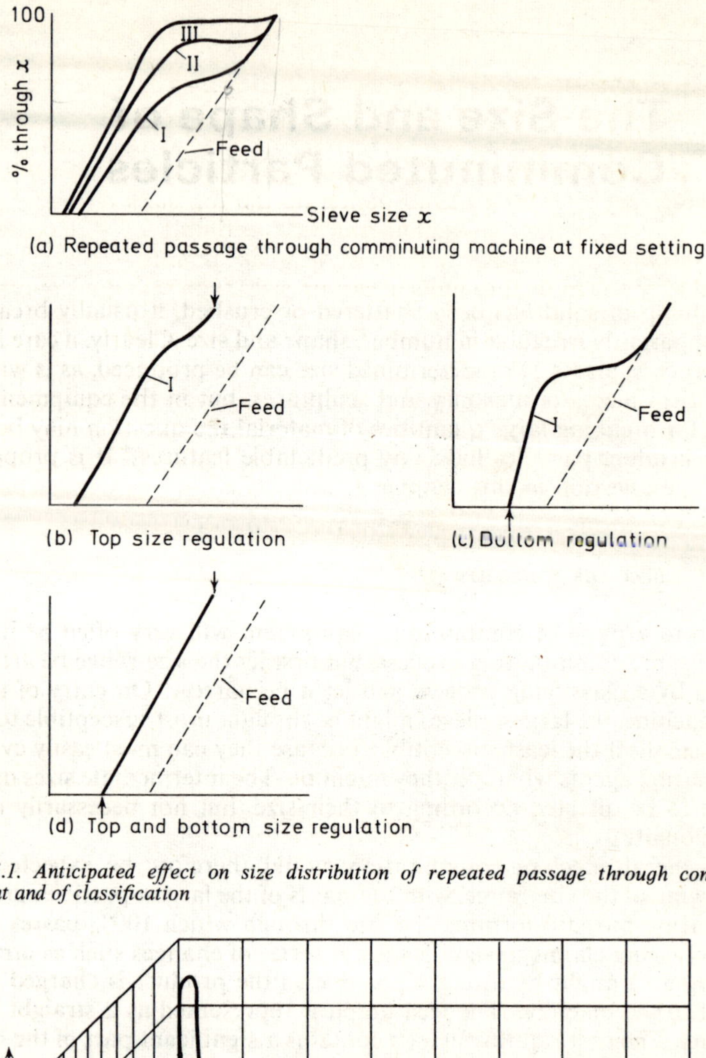

Figure 5.1. Anticipated effect on size distribution of repeated passage through comminution equipment and of classification

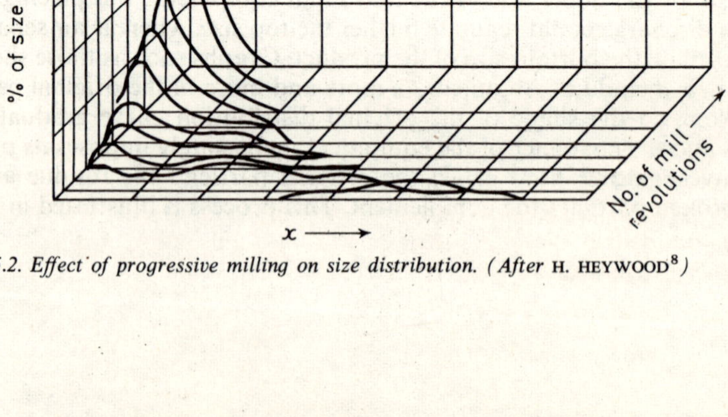

Figure 5.2. Effect of progressive milling on size distribution. (After H. HEYWOOD[8]*)*

5.2, in which the distribution of the feed is seen to haunt the distribution of the product, at first strongly and then diminishingly so as hardly to be perceptible according to the degree of size reduction that has occurred. Unless there is regulation of the top size of the product, the upper sizes linger on most and the figure corresponding to 100% passing in the product is often close to the same measurement in the feed; size reductions greater than about 10 are necessary in most cases to obliterate the top sizes of the feed.

Feeds which consist of agglomerations, assemblages of crystals, and porous, foliated or fibrous materials of weak bondings which are predisposed to break more into units of certain sizes than into units of others are special cases of the general solid condition. They will be discussed in more detail later.

For the more general cases the expectation that the breakdown is not entirely unpredictable is encouraged by the common experience that when a quantity of material large in proportion to the quantity in a single particle is comminuted, the cumulative quantities passing a series of sieve sizes usually lie on a smooth graph when plotted against sieve size on simple arithmetic coordinates; examples of such curves are shown in *Figure 5.3*.

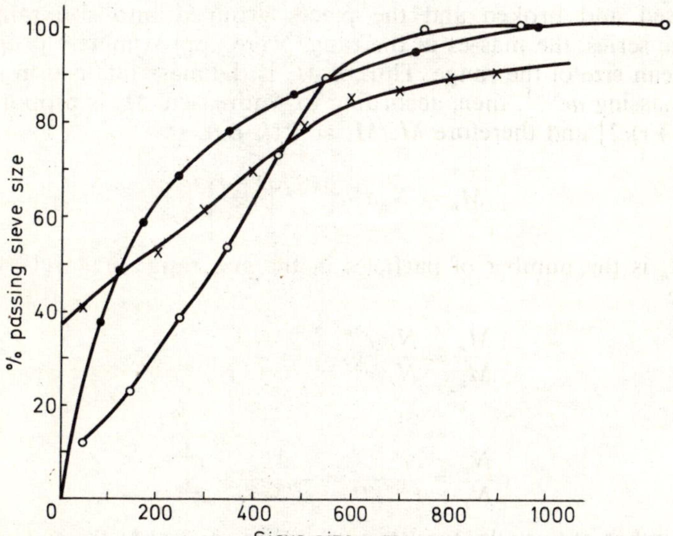

Figure 5.3. Examples of cumulative size distributions. They usually produce smooth curves

The Quantification of Size Distribution

Early work on quantifying size distribution was by STADTLER[2] and MARTIN, BOWES and TURNER[3], who suggested that, if a series of screen apertures were arranged so that their sizes were in a geometric series, the masses or volumes retained in the size intervals were related geometrically and that the number of crushed particles in each size range was constant. Thus, if the apertures were a, ar, ar^2, ar^3, ..., ar^n, then, according to Martin, the number of particles passing one and being retained on the next is constant for a comminuted product. Now let this number be N; then the volume (or mass) series would be

$$N\left(\frac{a+ar}{2}\right)^3, N\left(\frac{ar+ar^2}{2}\right)^3, N\left(\frac{ar^2+ar^3}{2}\right)^3 \ldots$$

and the cumulative volume up to and including the screen range $(ar^{n-1}+ar^n)/2$ would be summing backwards to infinity

$$N\left(\frac{ar^{n-1}+ar^n}{2}\right)^3 \frac{1}{1-(1/r)^3}$$

If this cumulative volume is represented by M and the sieve size passing M by x, then $M = Kx^3$ or $\log M = 3 \log x + c$, which when plotted as $\log M$ versus $\log x$ would give a straight line. Such in fact is found approximately to be the case.

ANDREASEN[4] took cubes of earthenware and showed that, when these were compressed and broken and the pieces grouped into size ranges in a geometric series, the masses in the range were approximately proportional to the mean size of the range. Thus, if M_n is the mass retained in sieve size ar^n and passing ar^{n-1}, then, according to Andreasen, M_n is proportional to $ar^{n-1}[(1+r)/2]$ and therefore $M_n/M_r = r^{n-r}$. But

$$M_n = N_n a^3 a^{3(n-1)}\left(\frac{1+r}{2}\right)^3$$

where N_n is the number of particles in the size range $ar^{n-1}[(1+r)/2]$ and therefore

$$\frac{M_n}{M_r} = \frac{N_n}{N_r} \cdot \frac{r^{3n-3}}{r^{3r-3}} = \frac{N_n r^{3(n-r)}}{N_r}$$

or

$$\frac{N_n}{N_r} = \frac{r^{n-r}}{r^{3(n-r)}} = \frac{1}{r^{2(n-r)}} = \frac{r^{2r}}{r^{2n}}$$

or the number of particles was inversely proportional to the square of their size (see *Table 5.1*). This relationship was found approximately to be maintained over a range of grinding time.

GAUDIN[5] proposed a more general form of the geometric law, $M = (x/D)^m$, in which M is % passing sieve size x, D is the sieve size of the largest particle and m is a constant.

SCHUHMAN[6] proposed the relationship $100 - M = R = \exp[-(x/\bar{x})^n]$, in which R is % retained by sieve size x, $\bar{x}$ is arithmetic mean size of distribution and n is a constant. This equation can be transformed by taking double logarithms: $\log \log (1/R) = n \log (x/\bar{x})$. It will give a straight line when plotted in this form.

ROSIN and RAMMLER[7] suggested the form $R = 100\exp(-bx^n)$, in which b and n are constants. From this equation,

$$\frac{dR}{dx} = 100\, nbx^{n-1} \cdot \exp(-bx^n)$$

which HEYWOOD[8] put in the gamma form

$$100 \cdot b^{1/n} \cdot \Gamma\left(\frac{n-1}{n}\right)$$

The gamma form can only have a finite value if n is greater than 1. Heywood suggested the form $dN/dx = a\exp(-bx^n)$, in which dN is the number of particles in the size range x and from which $dW/dx = ax^3 \cdot \exp(-bx^n)$. Here dW is the mass in the size range x. This equation has the advantage that it fulfils the requirement that $W = 0$ when $x = 0$ and d^2W/dx^2 tends to zero, i.e. when W is maximum.

Table 5.1. WEIGHTS OF MATERIAL IN SIZE RANGES OF PARTICLES IN A BROKEN CUBE

	Weight of material in size ranges (mean size)			
Time	0–2 (1)	2–6 (4)	6–26 (16)	26–102 (64)
1	1	3	20	55
2	4	6	32	54
3	6	11	43	40
4	7	15	58	20

Other equations[9–11] of similar form have been proposed. The rate equation (see p. 118) is of the form $P = F(1-e^{-Kt})$ and again the similarity is obvious. This equation has been used to determine the kinetics of grinding (see p. 91).

Size Distribution and Probability

The whole matter seems to be a probability problem. The Gaussian relationship has the form:

$$y = \frac{1}{h\sqrt{\pi}} \exp(-x^2/h^2)$$

in which y is the number of items having the dimension x in the quality being considered; h is a constant and π has its usual constant value. Such a relationship, when plotted as a graph, produces a bell-shaped curve, as shown in *Figure 5.4*. The integrated form of this equation, i.e. the summation of the numbers of items between two limits of the dimension x, e.g. x_1 and x_2, is given by

$$Y = \int_{x_1}^{x_2} \frac{1}{h\sqrt{\pi}} \exp(-x^2/h^2\, dx) = \frac{1}{h\sqrt{\pi}}\left[\left(\frac{\tan^{-1} x/h}{2}\right)^{\frac{1}{2}}\right]_{x_1}^{x_2}$$

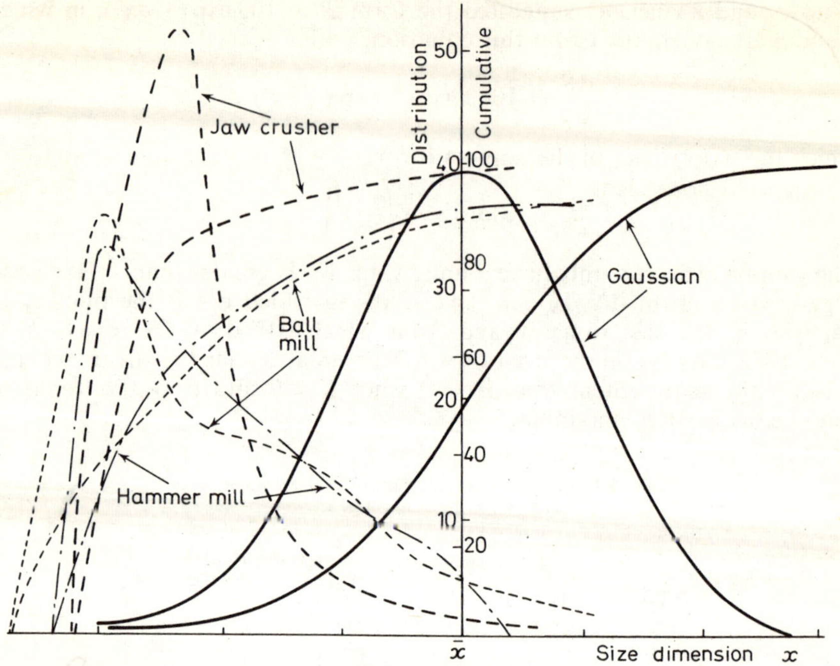

Figure 5.4. Particle size distribution and cumulative curves for Gaussian and actual cases

This equation when plotted as a function of x has the form also indicated in *Figure 5.4*.

If in the size distribution of particles in a product of comminution, y is taken to be the weight or volume of material in the mean of a narrow sieve range defined by x, x may be plotted against y and a distribution curve obtained. If the integrated form is taken to be the sum of the weights or volumes of the particles less than a certain size, say x_2, i.e. if the probability equation is the equivalent of the integral between the limits of zero size and x_2, and this integrated form is plotted against x, a cumulative curve is obtained (see *Figure 5.4*). A comparison can then be made between these curves obtained in practice and the theoretical distribution and cumulative Gaussian curves.

The limitation to the expression of practical size distributions is that sieving cannot be carried out satisfactorily below sizes of about 50 μm, which means that for size ranges which do not extend more than 400 μm (4 standard deviations) a distorted distribution is likely to be measured even for a distribution which is truly normal (symmetrical).

In fact (as is shown in *Figure 5.4*) for products from three quite different types of machines, viz. jaw crusher, hammer mill and ball mill, skew distributions are obtained. This is not surprising when the processes of selection in all comminuting equipment are remembered: the size of feed the equipment can receive; and the size of particle which can pass from one part of the equipment to the next and which can be released from the equipment.

The expansion of $\exp(-x^2/h^2)$ in the probability relation equation above is

$$1-\frac{x^2}{h^2}+\frac{x^4}{2h^4!}-\frac{x^6}{3h^6!}+\frac{x^8}{4h^8!}\cdots$$

or, in its integrated form,

$$x-\frac{x^3}{3h^2}+\frac{x^5}{5(2h^4!)}-\frac{x^7}{7(3h^6!)}+\frac{x^9}{9(4h^8!)}\cdots$$

Gaudin's equation can have the expanded form (using the Binomial Theorem)

$$M = \frac{(x)^m}{D} = 1-R$$

or

$$\frac{x}{D} = (1-R)^{1/m} = (1-R)^z = 1-zR+\frac{z(z-1)R^2}{2!}-\frac{z(z-1)(z-2)R^3}{3!}$$

Schuhman:

$$\exp\left[-(x/x)^n\right] = 100-M$$

or

$$\frac{M}{100} = \left(\frac{x}{\bar{x}}\right)^n-\frac{x^{2n}}{2\bar{x}^{2n}}+\frac{x^{3n}}{3\bar{x}^{3n}}-\frac{x^{4n}}{4\bar{x}^{4n}}\cdots$$

Rosin and Rammler:

$$R = 100\exp(-bx^n)$$

or

$$M = \frac{1}{100}\left[bx^n-\frac{b^2x^{2n}}{2!}+\frac{b^3x^{3n}}{3!}-\frac{b^4x^{4n}}{4!}\cdots\right.$$

Heywood:

$$y = ax^3\exp(-bx^n) = a\left[x^3-bx^{n+3}+\frac{b^2x^{2n+3}}{2!}-\frac{b^3x^{3n+3}}{3!}\cdots\right]$$

the integrated form of which is

$$M = a\left[\frac{x^4}{4}-\frac{bx^{n+4}}{n+4}+\frac{b^2x^{2n+4}}{(2n+4)2!}-\frac{b^3x^{3n+4}}{(3n+4)3!}\cdots\right]$$

The similarity of form of these expressions is obvious.

BENNETT, BROWN and CRONE[12] developed the probability equation

$$M = 1-\exp(-x/\bar{x})$$

and more recently GILVARRY[9] has arrived at a similar expression by an entirely different route. Gilvarry discriminates between three different kinds of cracks—those in the volume, those near the surface and those on the edges. If these three kinds of cracks occur according to a Poisson distribution, which suggests that if z is the average number of occurrences the expectation that 0, 1, 2, 3, 4, ... events will occur is proportional to e^{-z}, ze^{-z}, $z^2e^{-z}/2!$, $z^3e^{-z}/3!$, $z^4e^{-z}/4!$, ..., and if the average number of volume, surface and edge cracks is z_1, z_2 and z_3, then the probability of formation of a particle volume v, surface area s and length l is

$$\exp(-z_1 v)\exp(-z_2 s)\exp(-z_3 l)[z_1\,dv + z_2\,ds + z_3\,dl]$$

which may be put into the form $-e^{-Q}dQ$ and integrates to $M = 1 - \exp(-Q)$, and, for geometrically similar particles of size passing sieve size x, $M = 1 - \exp(-x/k)$, where k is a constant.

PRESCOTT and WEBB[13] investigated the size distribution produced in a hammer mill on the basis of Gilvarry's work. Gilvarry's derivation was based upon the material in each particle in the feed suffering only one fracture event and Prescott and Webb considered that particles of less than 1.5 cm fed to a hammer mill would be so treated. The same investigators eliminated from the product particles below 63 μm (the lower limit of their sieves) because they considered Gilvarry's equation did not cover the products of abrasion. They showed that the equation applied to dry processing but that a quadratic term entered the exponent when wet processing was used, which Prescott and Webb considered a manifestation of viscous forces from the water assisting comminution at high speeds.

The Gaudin and the Rosin–Rammler equations are the most different and they diverge most either when the residue (R) is large and particle size (x) is small, i.e. with very fine-grained material, or when the residue is small and the particle size large, i.e. with coarse-grained material. Such is in fact found to be the case in some substances. Coal, for example, appears to follow the Gaudin equation in its coarsest sizes; as crushing proceeds, the Rosin–Rammler equation more closely corresponds. TAYLOR[1] found that as the residue of the original feed became fractured so the size distribution changed from a Gaussian to a Gaudin relationship and eventually to a Rosin–Rammler one.

Table 5.2. TYPICAL GAUDIN INDICES FOR VARIOUS TYPES OF EQUIPMENT AND BROAD CLASSIFICATION OF CORRESPONDING SIZE RANGES

Type of equipment	*Approximate Gaudin index (m)*	*Size range*
Roll crusher	1.3	Narrow
Jaw crusher Gyratory crusher Rod mill	1.0	Intermediate
Hammer mill Vibration mill Ball mill Pan mill	0.5	Wide

The Gaudin equation is the simplest and gives a straight line when put into logarithmic form:

$$\log M = m \log \frac{x}{D} = m(\log x - \log D)$$

$$= m \log x + \text{constant}$$

A straight line is in fact obtained in many cases when such a relationship is plotted, except that above the 80% passing coordinate there is a curving towards a larger particle size. The slope of this line gives the value of *m*. Thus,

$$\log M_1 - \log M_2 = m(\log x_1 - \log x_2)$$

and

$$m = \frac{\log M_1 - \log M_2}{\log x_1 - \log x_2}$$

where M_1, x_1 and M_2, x_2 are coordinates on the straight part of the graph.

The value of *m* changes only slightly with successive comminution stages and when the feed residue is eliminated, the value becomes very nearly constant so that successive graphs are nearly parallel (see *Figures 8.29* and *9.6*). The significance of the index *m*, and indeed of the Rosin–Rammler index, usually denoted by *n*, is that they represent the size spread of the distribution. A large index indicates a narrow spread and a small index indicates a wide size difference between the largest and smallest particles. For maximum packing density[14] the widest range achievable in unclassified comminution equipment is usually required: the Gaudin index should be between 0.3 and 0.5, while the Rosin–Rammler index should be between 0.6 and 0.8. In general, the finer the average degree of grinding, the closer the distribution approaches the state of maximum packing.

Size distribution after comminution depends on both the material and the equipment but the indices are characteristic more of the equipment than of the material. A general indication of the indices likely to be obtained from particular pieces of equipment is provided by *Table 5.2*. Abrasion tends to give a narrow size distribution, compression a broad one and impact an intermediate one (see *Figure 5.5*). Materials do influence the situation.

BREAKAGE AND SELECTION FUNCTIONS

An approach to the subject of size distribution has been to distinguish between selection of material by the equipment and the breakdown of the material so selected. The idea was probably originated by EPSTEIN[15] and developed by BROADBENT and CALLCOTT[16], and is now receiving world-wide attention.

A selection factor or function $S(x)$ is postulated as the proportion of those particles delivered to the mill which are of sieve size x and which are broken in what can be called unit comminuting action. Unit comminuting action

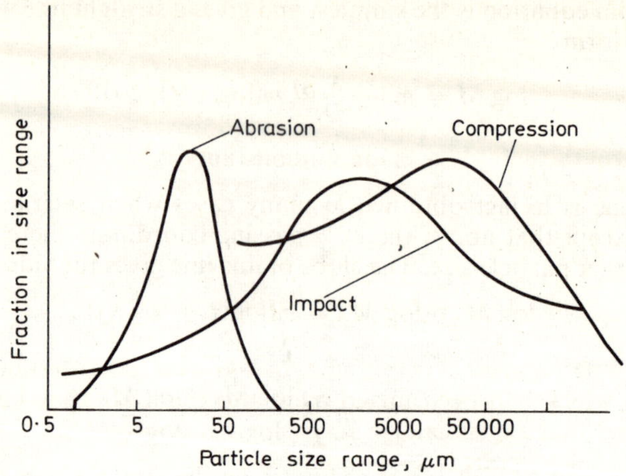

Figure 5.5. Characteristic size distributions from comminuting action

has to be defined for each piece of equipment: broadly, the idea is of a single identifiable and measurable cycle, e.g. one revolution of a hammer-mill impeller, one revolution of a ball mill, one opening and shutting action of a jaw crusher, one passage through the jet of a fluid-energy mill.

The breakage factor or function $B(yx)$ is that fraction of those particles delivered to the mill, which are of sieve size x, which are broken and which break to particles smaller than sieve size y. Thus, of the particles delivered to the mill of sieve size x, $S(x) \cdot B(yx)$ are broken to particles smaller than sieve size y.

Questions arising from such concepts are:

(1) Does the quantity of a given size present in a mill affect selection?
(2) Does the presence of other sizes affect selection or breakage?
(3) Does the selection function vary with size?
(4) Does the breakage function vary with size?

Attempts[17,18] to answer such questions and to derive the actual functions for particular materials have been made along lines about to be indicated.

If the total mass of the feed is H and $\delta H(u)$ is that part of the feed which lies between the sizes u and δu, and if the selection function for this particular size is represented by $S(u)$, then the mass broken is $S(u)\delta H(u)$ and $[1-S(u)]\delta H(u)$ is unbroken and remains between the sizes u and $u+\delta u$.

If the proportion of the material broken from between u and $u+\delta u$ into sizes below x is $B(x, u)$ then the weight of such material from the original mass $\delta H(u)$ is

$$B(x, u)S(u) \cdot \delta H(u)$$

and the total mass from the original total feed H which is broken and is less than size x is

$$\sum_{u=x}^{u=\infty} B(x, u)\cdot S(u)\cdot \delta H(u)$$

The material at the end of unit comminuting action below size x is composed of: (1) that broken from sizes greater than or equal to x to sizes less than x; and (2) that originally in the feed H which was less than size x or is composed of

$$\sum_{u=x}^{u=\infty} B(x, u)\cdot S(u)\cdot \delta H(u) + H(x)$$

GARDNER and AUSTIN[17] traced the breakdown of a certain irradiated size fraction which had been mixed back into the feed. They found for coal and limestone that the breakage function was constant for the same size-reduction ratio independently of what absolute sizes are involved. If this were universally true, it would mean that a particle of the same material in the same machine would break into the same proportional parts independently of its absolute size—e.g. a particle x will break into $B\%$ less than $x_1/2$ and a particle x_2 will break into $B\%$ less than $x_2/2$. Thus, if the screen sizes were arranged in geometric progression, successive millings screened by them should produce parallel size distributions or masses passing:

$$\frac{\text{screen size 2}}{\text{screen size 1}} = \frac{\text{screen size 3}}{\text{screen size 4}}, \text{ etc.}$$

If the breakage functions are $B(2, 1)$, $B(3, 2)$, etc., which are all equal, milling must result in proportions being shifted by the same amount at each milling cycle, so that if, for example, x_1 shifts to position 2, x_2 will shift to position 3 and x_n to position $n+1$. Such parallel size distributions are in fact frequently experienced, particularly after several stages of comminution, i.e. when the residual feed sizes have largely disappeared; for examples see *Figures 8.12*, *8.25* and *9.6*.

Table 5.3. BREAKAGE FUNCTION: EFFECT OF REDUCTION RATIO

Reduction ratio	0.1	0.2	0.3	0.4	0.6	0.7	0.8
Coal[17]	0.02	0.06		0.16	0.22		0.5
Limestone[17]	0.07	0.10		0.16		0.22	
Fertiliser granules[18]							
Feed 3 mm		0.15		0.35	0.53		0.72
6 mm		0.32		0.60	0.78		0.90
8.5 mm		0.38		0.68	0.88		0.97
2.70 mm, 1 000 rev/min		0.01		0.02	0.03		0.08
2 000 rev/min		0.04		0.12	0.21		0.35
3 100 rev/min		0.15		0.30	0.51		0.62
5.2 mm, 1 000 rev/min		0.07		0.15	0.28		0.51
2 000 rev/min		0.11		0.26	0.46		0.70
3 100 rev/min		0.24		0.46	0.63		0.83

However, GARSIDE and WILDSMITH[18], working on fertiliser granules, have found that the breakage function for a fixed size reduction is not independent of size. Sets of results from the two investigations are compared in *Table 5.3*. The two sets of results arise from different circumstances. The Garside and Wildsmith figures were produced on an industrial hammer mill and those of Gardner and Austin in a laboratory Hardgrove machine. The former investigators considered unit action to be once through the machine, whereas the latter considered equal numbers of rotations of the mill.

The values obtained as selection functions by the same two sets of investigations are shown in *Table 5.4*. Generally, as might be anticipated, the larger the particle the greater its chance of selection. HENDRY[19] also found

Table 5.4. SELECTION FUNCTION AND PARTICLE SIZE

Particle size, μm	100	200	400	1 000	3 000	5 000	7 000	8 000	9 000
HARDGROVE MACHINE									
(per rev)									
Coal, S. Nicholas	0.005	0.006	0.009	0.07					
Coal, Trevorton	0.008	0.01	0.01	0.10					
Coal, Pittsburgh	0.008	0.01	0.01	0.10					
Limestone	0.008	0.01	0.012	0.035					
SWING HAMMER									
(per pass)									
Granules, 3 100 rev/min				0.64	0.80	0.90	0.97	0.98	0.99
Particle size, μm					[2 700]	[5 200]			
1 000 rev/min					0.35	0.50			
2 000 rev/min					0.57	0.85			
3 000 rev/min					0.77	0.92			

this to be true for a fluid-energy mill. Hendry showed that there was an optimum feed rate which was the same for all sizes at which the selection function was a maximum.

Garside and Wildsmith made their equation easier to apply by using a feed with narrow size distribution; thus $H(x)$ was eliminated and the remainder of the equation

$$\sum_{u=x}^{u=\infty} B(x, u)\cdot S(u)\cdot \delta H(u) = G(x)$$

was then plotted against x. At the value of x equal to the feed size, the value of $G(x)$ equals the selection function, because the breakage function is equal to unity at the feed size. By covering a range of feed sizes the values of $S(x)$ could be obtained.

The same investigators found that the breakage function for individual feed sizes could be represented by the Gaudin equation, $M = (x/D)^m$, or, as they symbolised the equation, $B(y, x) = 1-(1-y/x)^m$. They found that m varied with the feed size:

Feed size (mm)	3	6	8.5
m	0.82	1.55	2.15

Both teams of investigators were able to reproduce, from the selection and breakage functions they had determined, actual size distributions from typical feed-size distributions. They used computer techniques to do this.

DEGREE OF FINENESS

FAIRS[20] used the concept of degree of fineness as the proportion of the product from a fixed feed size that was below a certain size. He found for limestone, barytes, anhydrite—and also talc (private communication)—that the degree of fineness for the same orders of throughput was relatively independent of the nature of the material and depended only on the equipment. The results are shown in *Table 5.5*.

Table 5.5. COMPARISON OF VARIOUS TYPES OF EQUIPMENT WITH A TEST UNIT IMPACT CRUSHER FOR DEGREE OF FINENESS ACHIEVED AND ENERGY CONSUMED PER UNIT OF NEW SURFACE PRODUCED

Type of mill	*Fineness ratio*	*Ratio of energy per unit of new surface,* erg/cm^2
Unit test impact crusher	1	1
Ball mill	1.2	0.99
Swing hammer mill, coarse clearances	1.6	0.96
Swing hammer mill, fine clearances	3.6	1.67
Attrition mill (rotor, grid discharge)	5.9	2.78

BLANC[21] refers to a method of calculating 'fineness modulus' by taking the separate results of sieving a material on a series of sieves 4, 8, 16, 30, 50 and 100 mesh, determining the residues on these sieves always from the complete sample, adding these residues as percentages together and dividing by 100 to obtain the modulus.

NATURAL GRAIN SIZE

The nature of the material affects the size of the product of comminution to some extent. Amorphous materials tend to chip and break locally, whereas crystalline materials distribute the stress more evenly and break more uniformly. The structure of the material affects the size. Materials consisting of grains, having pores, composed of fibres or leaves, or consisting of particles which are agglomerates or particles of one material imbedded in a matrix of another, may well be disposed to break into units of certain sizes rather than others. Materials having this disposition are said to have natural grain size. The occurrence of natural grain size can be detected in the $\log M$ versus $\log x$ graph, from a bulging out of the line (see *Figure 5.6*).

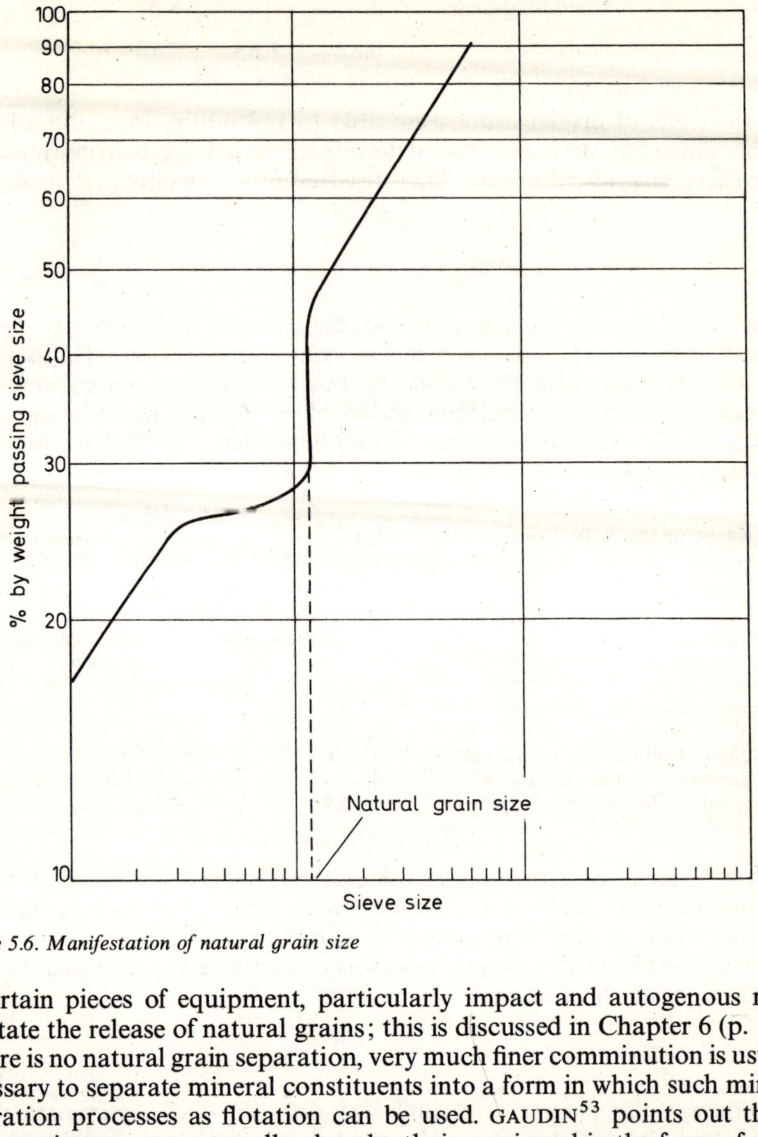

Figure 5.6. Manifestation of natural grain size

Certain pieces of equipment, particularly impact and autogenous mills, facilitate the release of natural grains; this is discussed in Chapter 6 (p. 112). If there is no natural grain separation, very much finer comminution is usually necessary to separate mineral constituents into a form in which such mineral separation processes as flotation can be used. GAUDIN[53] points out that if two constituents occur equally abundantly in a mineral in the form of equal cubical grains, and the mineral is haphazardly broken into a size corresponding to half grain size, only $\frac{1}{8}$ of the particles would contain pure constituent, breakage to $\frac{1}{32}$ of the grain size would be necessary to release 90% into particles composed of single constituents, and reduction to $\frac{1}{512}$ of the original grain size would be necessary to ensure that nearly 100% was so liberated. If the minerals are unequally abundant, the same figures apply for the less abundant mineral, but less reduction is necessary to liberate the more abundant one, e.g. if the weight ratio is 10/1, breakage to grain size would liberate 50% of the more abundant mineral and reduction to quarter original grain size will liberate 90%. The economic advantages of selecting equipment which effects natural grain separation can therefore be very great.

GEOMETRY OF BROKEN PARTICLES

It has been fairly well established that the average surface area and the average volume of a large number of particles of the same material, all within a close sieve range, can be represented by expressions such as

$$\text{surface area} = K_1 b^2; \qquad \text{volume} = K_2 b^3$$

where b is the arithmetic mean of the close sieve range. MARTIN and BOWES[22], for example, found that ground sand varying in size from 2188 to 116 520 particles per gram gave a figure for K_2 very close to 0.277 over the whole range. This suggests that particles of the same material comminuted in the same equipment are geometrically similar.

Certain equipment is disposed to break material into its grains, particularly rotary impact and autogenous mills, while vibration mills release material with crystalline form which becomes more evident as the particle size is decreased. Cubical product shape can be imposed[23] by impact mills with certain design features, as is discussed in Chapter 9 (p. 195), by metal contact rather than cushioned particle contact, by direct blows rather than glancing blows, by keeping moisture content low and by increasing the reduction stages and decreasing the reduction ratio, as in closed-circuit operation.

The product from pieces which break down into material grains might be expected to show geometric similarity. But the property of crystals of fracturing along planes of cleavage might also be expected to produce a tendency towards definable shapes. Indeed, it might be expected that only in amorphous substances or those in arrested states of solidification, such as glasses, is the inbuilt anisotropy and orderliness of solids blurred and obscured by randomness. Cleavage could be expected to take place in most solids over the planes which require least surface energy, even in the microscopic stages of Griffith cracks. The laminated way in which mica breaks is well known and oolitic material often easily breaks into the small egg-like particles from which it derives its name. Kaolin breaks along weak hydroxyl bonds and graphite along van der Waals linkages to form flaky materials. In each case particles tend, in general, to retain their geometry during comminution: slabs stay as slabs; needles produce needles; and cubes, cubes.

BORN and STERN[24] calculated the surface energies on various faces of a rock salt crystal and showed that surface energy should depend upon the kind of face. *Table 5.6* shows some of their results. KUSNEZOW[25] measured the work

Table 5.6. SURFACE ENERGIES ON CRYSTALLINE FACES OF A ROCK SALT CRYSTAL

Face	111	110	100
Surface energy ratio	5.81	2.71	1

required to grind away unit volumes of the faces of a crystal of the same material and found that in fact the work required was approximately in the same ratios as Born and Stern had calculated.

ZWICKY[26] suggested that crystals had certain structural weaknesses in patches in and near the surface and that these patches had linear dimensions of the order of 100 Å (0.01 μm) and were in blocks of about 10 000 atoms, i.e. cubes of side about 40 Å. SMEKAL[27] suggested that during crystallisation groups of 1000 atoms (cubes of side about 18 Å) formed themselves directly into single crystals and that groups did not coalesce until they reached this size. RANDALL, ROOKSBY and COOPER[28] suggested that glasses up to 1000 Å show an ordered crystalline structure. These sizes are all much smaller than those normally able to be measured directly in particle-size determinations, but many size-reduction operations produce particles less than 1 μm and some materials are reduced to sizes which their surface areas imply are of the order which Zwicky describes (see p. 89).

Hardness vectors can be constructed for single crystals and a figure formed from the vectors (*Figure 5.7*). The figure is imagined to be constructed

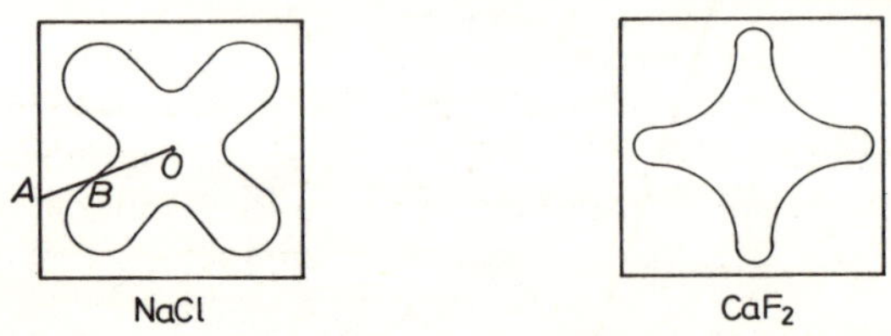

Figure 5.7. Hardness vectors for sodium chloride and calcium fluoride

on the section through the points on the crystal surface to the hardnesses of which the figure refers. Then the hardness of any point on the periphery of that section on the crystal surface is found by joining that point to the centre of the section, and the hardness of that point is the length of the joining line from the centre of the section cut by the figure. Thus the hardness at a position *A* on the crystal is represented by the distance *OB*. It can be seen that the variation in hardness is quite considerable. Cleavage is found more likely to occur in positions of least hardness than elsewhere[29].

If the crack patterns of a solid piece just prior to breaking can be likened to a reticulation of foam or an assemblage of grains[34], certain topological rules can be employed to define the particle shapes which might result from a process of comminution. In a foam, as a requirement of minimum surface energy, the cell walls must join in groups of three at 120° to one another; the curvature of the walls must be proportional to the pressure across the walls; and the three centres of curvature of the three films must lie on a straight line. Euler's rule connecting the numbers of points (*A*), lines (*B*), faces (*C*) and (solid) polyhedra (*D*) in such a simply connected array as a foam (i.e. an array without holes) is

$$A - B + C - D = 1 \ldots \quad (5.1)$$

If the surface of a cracked piece of material is so considered just prior to falling into individual polyhedra, the superficial faces, if they were like those of a reticulation of foam, would be formed from lines meeting three at a point and therefore each face (or polygon) of n sides would contribute $n/3$ corners or points. Thus $3A = \sum nP_n$ where P_n is the number of polygons with n sides. Each line or side of a polygon is shared by two polygons and therefore $2B = \sum nP_n$. Substituting these relationships in equation 5.1 for one polygon (i.e. the material piece),

$$\tfrac{1}{3}\Sigma nP_n - \tfrac{1}{2}\Sigma nP_n + \Sigma P_n - 1 = 1$$

or

$$\Sigma(6-n)P_n = 12$$

and

$$\Sigma(6-n)P_n - 12 = 0$$

If $\Sigma P_n = P$,

$$\frac{\Sigma(6-n)P_n}{P} - \frac{12}{P} = 0$$

and if P is large,

$$\frac{\Sigma(6-n)P_n}{P} \approx 0$$

Since no polygon can have fewer than 3 sides and polygons greater than 10 are rarely met in nature, it is seen that the factor $(6-n)$ produces a series (by substituting 3, 4, 5, . . . 10 for n) the members of which are symmetrical about 6, positive up to 5 sides ($n = 5$) and balancing negatively up to 9 sides. Furthermore, it is only regular six-sided figures which with straight sides include an angle of 120° at all vertices. Thus it can be conjectured that regular six-sided polygons exhibit greatest stability. The Giant's Causeway in Northern Ireland shows cooled basalt formed into a reticulation of hexagons on a gigantic scale.

The same principles show that in the solid form the stable shape corresponds with a polyhedron with an average of 5.1044 sides which is approximated by a tetrakaidecahedron or fourteen-faced figure made up of eight hexagons and six squares (see *Figure 5.8a*). Such polyhedra pack together to form voidless structures; but in nature, in human fat tissue, in mineral polycrystals, in polycrystalline alloys such as brass and aluminium–tin alloy[34] polyhedra with many five-sided faces are found to be very prevalent, the average number of sides in the faces being intermediate between the theoretical 5.1044 and the 5.1429 of the tetrakaidecahedron. Pieces of material from a comminution process also often have shapes which include a number of five-sided faces. *Figure 5.8b* is an outline sketch of a random selection of limestone particles in five size ranges produced from a cone crusher. The particles were placed with a face on the paper and an outline of the face sketched round it. The average number of sides of these figures is 5.6.

HEYWOOD and PRYOR[31] found that, when cubic lumps were first broken, they produced a proportion of flaky shapes but that these tended to break down into a shape common to the rest as comminution proceeded. They examined siliceous rocks and limestones and found that these tended to a mean shape. Laminated materials such as mica retained their flaky shape.

(a)

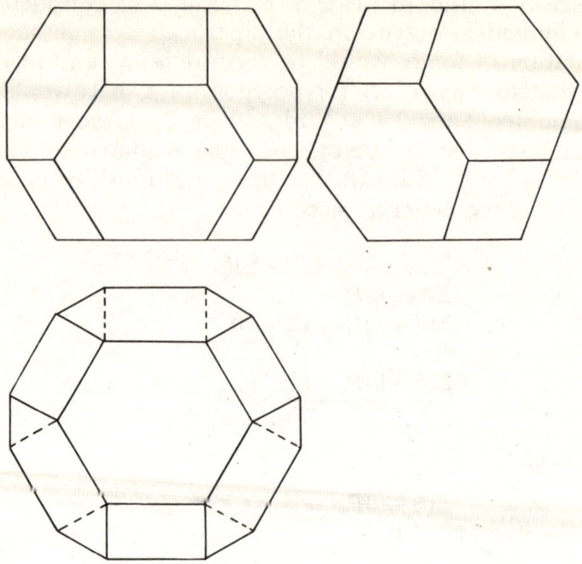

(b)

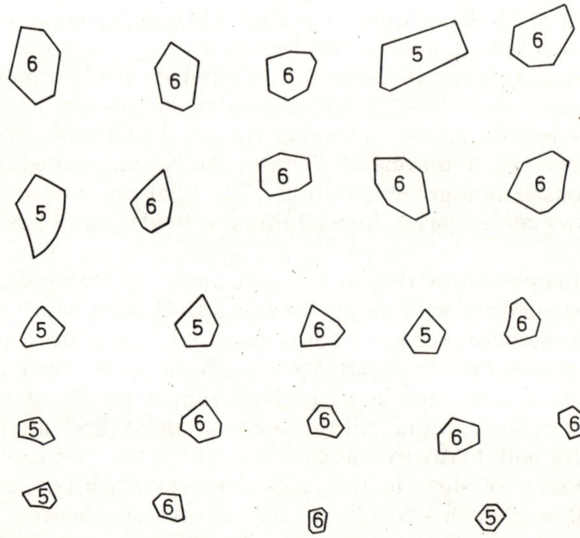

Figure 5.8. (a) Tetrakaidecahedron; (b) sketch of silhouettes of sample of limestone pieces (number of sides indicated inside figure)

The proportions of this mean shape to which the former kind of material approached correspond to a truncated tetrahedron. The approximate ratios of the sides are shown in *Table 5.7* together with the corresponding values for a tetrakaidecahedron. The surface-to-volume ratios are also shown. Assuming that these particles went through sieves 50% in the larger way and 50%

in the smaller and making the crude calculation that the mean sieve size is half-way between the largest and smallest sides, and with the volume assumed to be the cube of this mean sieve size, the surface–volume relationship is calculated (*Table 5.7*). *Figure 5.9* shows the relationship between surface area and mean sieve aperture size, as defined in this paragraph, for a variety of shapes.

Table 5.7. RATIOS OF DIMENSIONS OF MEAN PARTICLES. (x = SIEVE SIZE THROUGH WHICH PARTICLE WOULD PASS)

	Length	*Breadth*	*Thickness*	$\frac{Surface}{Volume}$, cm^2/cm^3 *True*	$\frac{Surface}{Volume}$, cm^2/cm^3 *According to sieve test*
Heywood and Pryor[31]	2.5	1.7	1	$8.2/x$	$2.8/x$
Tetrakaidecahedron	1.25	1.18	1	$7.4/x$	$1.7/x$
Steinherz[32]	3	2.5	1		

STEINHERZ[32] found a set of ratios for Portland cement, quartz and glass (*Table 5.7*). *Figure 5.10* represents the Heywood–Pryor particle. BOND's[23] average particle is indicated on the same diagram. The coincidence of the broad dimensions of these particles is noteworthy. GREGG and HICKMAN[33] showed that the length-to-thickness ratio for graphite decreased considerably with progressive comminution, so that the ratio altered with specific surface area, as in *Table 5.8*.

DEFINITION OF SHAPE

Various attempts have been made to define the shapes of particles. British Standard Specification 812 refers to a flakiness index: a flake is there defined as a particle with its least dimension less than $\frac{3}{5}$ of its mean dimension or, if a, b and c are the three dimensions of a particle chosen so that c is the least dimension of the particle, then the particle is a flake if c is less than $\frac{3}{5}$ of $(a+b+c)/3$, from which c is less than $(a+b)/4$. The flakiness index is the percentage of flake particles.

If the three dimensions of a particle are considered to include its largest dimension (a) and the smallest dimension (c), which is at right angles to the largest dimension, so that b is the remaining third dimension at right angles to both a and c, then the ratio between a and b is sometimes called the elongation of the particle. When the elongation is greater than 2.0, the particle is said to be elongated. Another criterion for elongation is that maximum dimension exceed 5 times minimum dimension.

A method of determining these ratios has been suggested by BOND[23]. The sample is screened with a set of ordinary squared-meshed sieves into close size ranges. A particular size range is said to measure the intermediate

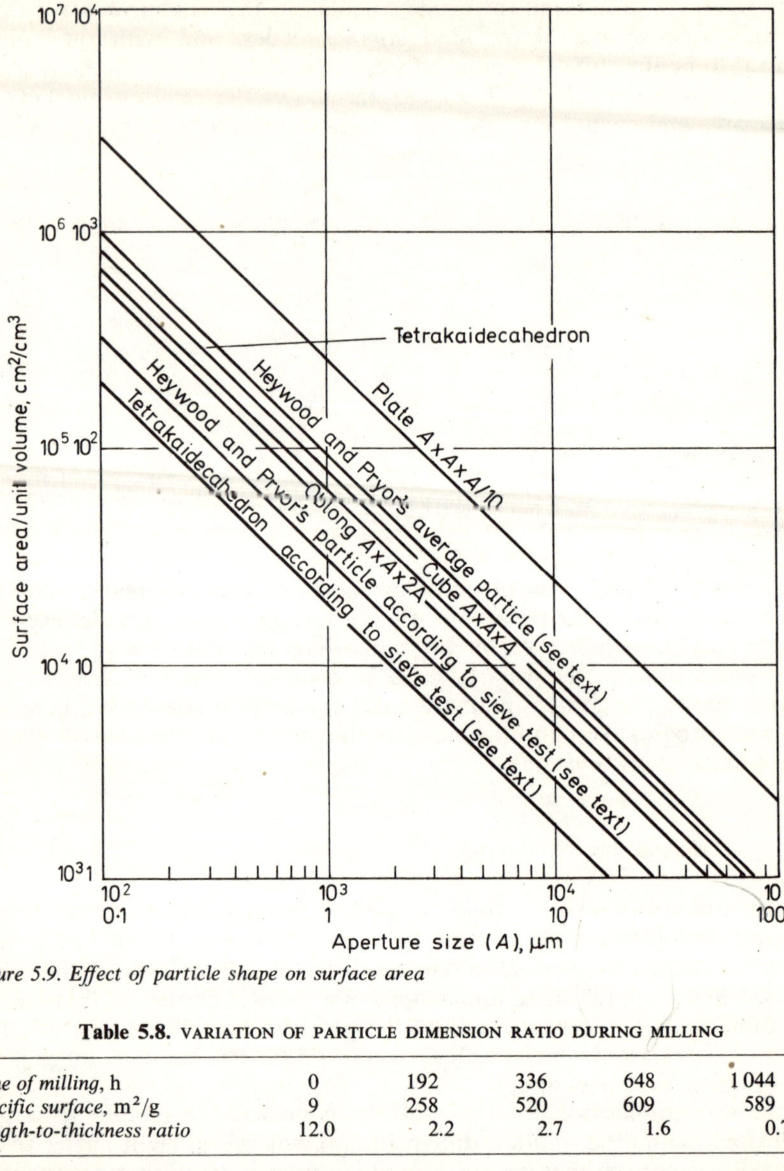

Figure 5.9. Effect of particle shape on surface area

Table 5.8. VARIATION OF PARTICLE DIMENSION RATIO DURING MILLING

Time of milling, h	0	192	336	648	1 044
Specific surface, m^2/g	9	258	520	609	589
Length-to-thickness ratio	12.0	2.2	2.7	1.6	0.7

dimension *b*. The fractions are then plotted on a graph of log (% passing) versus log (screen size, μm). The size fractions from this first screening are then recombined and passed through a set of screens with slotted openings, one dimension of which corresponds with the square-mesh set and the other being, say, one-third of that dimension. The fractions are then plotted on the same graph using the screen size corresponding with the smaller dimension. Then at any % passing ordinate the ratio of *b* to the shortest dimension *c*

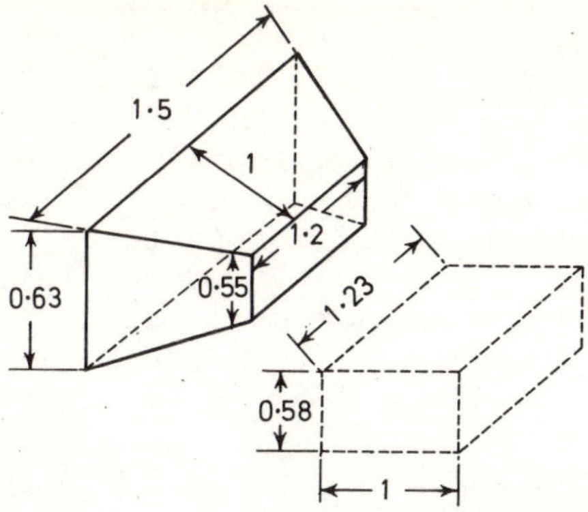

Figure 5.10. Heywood and Pryor's average particle dimension ratios: Bond's average particle indicated dotted

is the ratio between the antilogarithms of the abscissae on the two graphs for that ordinate. The determination of the longest dimension (*a*) is less exact and uses particles screened between two close square-mesh screen sizes. The particles are laid in an area defined by bars forming a square with sides equal to 10 times the size of this close sieve range; e.g. if the size of the sieve range were 5 mm, the square would be 5 cm × 5 cm. The particles are laid as flat as possible with thinnest dimension vertical, as close as possible to one another, and the number which can be packed in one layer in this area is counted. If this number is *N*, then *N*/100 is the ratio of *a* to *b*.

Another method of determining these dimensions is microscopically, by measurement of individual particles. STEINHERZ[32] used this method.

EFFECT OF CONTINUED COMMINUTION

The effect of continued recycling of material through comminution equipment depends upon whether particles are induced to escape when they have attained a certain degree of reduction or not. In this section consideration will be given to the effects of continued comminution without separation; the aspects associated with separation will be discussed below (p. 93).

The question arises, 'To what degree of subdivision may a solid be taken by comminuting process?' Clearly, in the extreme the solid could be evaporated into single molecules, but, as has been discussed above (p. 82), the smallest assembly of particles which still behaves in the manner associated with solids is in the region of 1000–10 000 molecules, which for elements would mean cubes not smaller than about 20–50 Å or 0.002–0.005 μm. Anything reduced smaller than this order of size would be disposed to join up and grow.

In the comminution process coarser particles might be expected to break more easily because they are more prominent targets for any breaking surface. In fact, coarser particles do disappear more quickly than finer ones, as is illustrated by an analysis of the breakdown of size grades during a process of grinding (*Figure 5.11*). Many investigators have explained this as being due to the fact that the Griffith cracks responsible for the weakness of the particles are eliminated during the grinding process as a result of breaking taking place along them and their then forming the surfaces of the smaller particles. SMEKAL[27], who supported this explanation, went so far as to suggest that, at the stage at which particles were being produced without flaws or cracks, further comminution should be induced by imperfections being introduced into the particles with X-rays or radioactivity. So far as can be discovered, there is no evidence to suggest that difficulty in comminuting small particles is anything other than difficulty in isolating them and making them targets for attack. Whenever the equipment is changed to improve the exposure of the smallest particles to attack and freeing them from a bed of mixed sizes, as in the change from a ball mill to a fluid-energy mill, further size reduction thereby occurs.

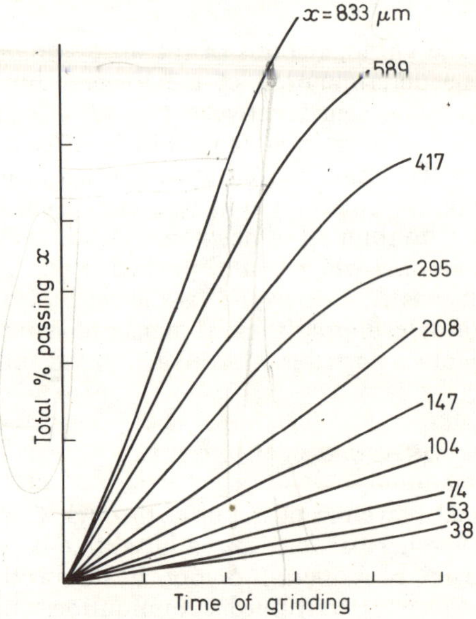

Figure 5.11. Effect of continued grinding on the proportion of the size grades. Percentage passing plotted against size. (After N. ARBITER *and* U. N. BHRANY[50]*)*

Finer particles do tend to aggregate under pressure. TRZEBIATOWSKI[35] showed that small particles of metal under a very large pressure, of the order of 15 000 atm, formed material which was hardly distinguishable in physical properties from cast masses. BOWDEN and TABOR[36] have shown that impacted or sliding solids become welded together. Agglomeration has been variously ascribed to removal of protective air films by impact, to electrostatic attraction, to being induced by friction and to crystal growth. Agglomeration such as that occurring in a ball mill has been expressed[37] by $A = 1 - 1/e^b \cdot D \cdot d^m$, where A is the proportion of surface lost by agglomeration of the over-all surface expected from Rittinger's law, D is the diameter of the mill drum, d is the diameter of the balls and b and m are constants.

Continued comminution[33] of graphite, kaolin, molybdenum sulphide and magnesium hydroxide in a ball mill shows first an increase in surface area roughly proportional to time or number of revolutions of the mill and then a gradual falling off in rate of production of new surface, leading eventually to an actual decrease of surface area (see *Figure 5.12*). The shape

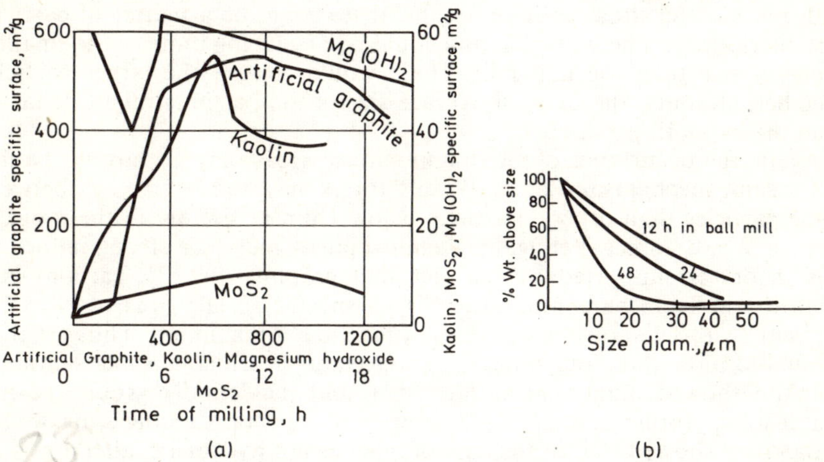

Figure 5.12. Effect of continuous milling on (a) surface area of different materials (note maxima), (b) size distribution of particles of porcelain enamel (note insignificant effect on small sizes)

of the particles of these materials is flaky or foliated. Calcite, the particles of which are not flaky or foliated, did not in these experiments behave in the same way, at least down to the particle size shown in *Table 5.9*. The particles of graphite at the maximum degree of size reduction had approached that suggested by ZWICKY[26] and by SMEKAL[27] as being the limit for the separate existence of solid particles. Particles of molybdenum sulphide appeared to be much larger, as indeed did particles of calcite, which did not show any signs of a maximum.

Table 5.9. AGGLOMERATION: PROPERTIES OF MATERIALS AT MAXIMUM SURFACE COMPARED WITH CALCITE NOT AT MAXIMUM

Material	*Molecular weight*	*Density,* g/cm³	*Properties at maximum surface area* *Surface area,* cm²/cm³	*Average particle size,* μm	*No. of molecules in particle*
Graphite	12	2.25	25×10^5	0.004	7×10^3
Kaolin ($Al_2(OH)_4 \cdot Si_2O_5$)	258	2.6	2×10^5	0.05	750×10^3
Marble	100	2.7	0.14×10^5	0.77	$7\,300 \times 10^6$
Molybdenum sulphide	128	4.8	0.2×10^5	0.5	$2\,760 \times 10^6$
Calcite (maximum not reached)	100	2.7	0.3×10^5	0.3	430×10^6

However, a comparison with the behaviour of marble[38] in a vibration mill shows that marble, which is a close chemical relative of calcite, reached a maximum surface area corresponding to a much larger average particle size (× 18) than that at which calcite showed no signs of reaching a limit in a ball mill. The work with marble was carried out with a series of different vibration mill balls: steel, tungsten carbide and ceramic balls were all tried and the same maximum surface area was reached with all three kinds of ball. A possible explanation of the difference between the results achieved with marble and those achieved with calcite could be a matter of particle-size distribution. The vibration mill could be producing more of the smallest range of size than the ball mill. These would cement the larger particles together, although the over-all average size of the particles might be larger than the ball mill product.

Again, the occurrence of maximum surface areas may not always be due to the same mechanism. Van der Waals' forces may well be stronger between some particles than others. Particles below 1 μm in size are in the range of van der Waals' forces. Yet again, chemical purity may be a strong influence; this theory is supported by the fact that graphite (99.97% carbon) in a vibration mill can be comminuted[39] to a specific surface area of 114×10^5 cm^2/cm^3 without showing any signs of reaching a maximum. Thus graphite ground to more than four times the surface area at which another sample of graphite showed maximum surface area, and incidentally ground so fine that it was pyrophoric in air, itself showed no signs of reaching a maximum.

GREGG[33] showed that in the case of magnesium hydroxide, after a certain degree of size reduction had been reached, the crystal broke along one particular face (001); the platelets were then broken with some dehydration; reabsorption of water took place; and the smaller hydrated platelets re-assembled along the same crystal (001) face. A very complicated reaction for what is ostensibly particle-size reduction.

HEAVER, MCGRATH and SMALLEY[54] have demonstrated that even at sizes of the order of 250 μm freshly ground quartz shows cohesive forces at least sufficient to arrest flow.

Thus aggregation may be the result of any combination of macroscopic forces, van der Waals' or chemical forces.

Chemical reactions sometimes occur as grinding proceeds. Potassium chloride[40] and mercurous bromide, on being ground, react to form a mixture which contains mercurous chloride and potassium bromide; and GREGG[33] showed that kaolin reacted more and more with hydrochloric acid as grinding proceeded and in greater proportion than would correspond with increased surface area. Gregg quotes instances of mica[41] increasing in cationic replacement capacity as fineness increased and carbonates[42] having lower decomposition temperatures after grinding. Quartz[43] becomes more reactive to sodium hydroxide but only to a depth below the surface of about 100 atoms; this surface gives an amorphous X-ray diagram. Beilby (ref. 25, Ch. 3) suggested that this superficial change often occurred when substances were rubbed or impacted. The increased surface activity of freshly broken quartz has been associated with the cause of silicosis as fresh finely produced silica is much more reactive on lung membranes than aged fine silica.

α-alumina[44] will hydrate if it is finely ball milled in the presence of water, whereas no reaction can be detected if coarse particles of alumina are stood indefinitely in water. Other evidence of change of crystal structure on grinding is calcite changing to aragonite[45], α- to β-quartz and crystalline to vitreous silica[46]. Some dyestuffs lose their tinctorial value after grinding.

Much of the chemical reaction described can be explained as being the result of high temperatures achieved superficially and locally in the rubbing action, which have been shown by BOWDEN and TABOR[36] to be very high, in some instances reaching over 1000°C depending upon the melting point of the material. The superficial high temperature so achieved could be expected to have an increasing effect on the body of the particle as the diameter of the particle decreased, i.e. as a greater proportion of molecules occupied superficial positions and were directly affected.

KINETICS OF GRINDING

Values for the rate of grinding over the range in which no aggregation occurs can be obtained from the relationship derived from likening comminution to a reaction of the first order or one which depends only on the proportion of one reactant:

$$\frac{\mathrm{d}P}{\mathrm{d}t} = K(F-P) \qquad \text{or} \qquad P = F(1-\mathrm{e}^{-kt})$$

in which P is the product, F is the feed and t is the time.

This relationship has been developed for batch grinding and BATRA and BISWAS[47] have modified it to the form

$$\frac{\mathrm{d}W(xt)}{\mathrm{d}t} = -K_x W^{\beta}(xt)$$

where $W(xt)$ is the quantity of material coarser than x at time t, K_x is assumed to depend only on size and β is a constant. Integration between the times 0 and t gives

$$\frac{1}{1-\beta}[W^{1-\beta}(x,t) - W^{1-\beta}(x,0)] = -K_x t$$

BATRA and BISWAS then follow KAPUR[55] and propose $K_x \doteq -K_0 x^{\alpha}$, and find for the behaviour of quartz in a rod mill that $\beta = 0.6$ and $\alpha = 0.988$ closely fits the facts.

The same authors show that their equation leads to equations for size distribution and for energy consumption. Thus, for a particular time t,

$$W^{1-\beta}(x,t) - W^{1-\beta}(x,0) = -(1-\beta)K_0 t x^{\alpha}$$

which can be written in the form $(1-R^{1-\beta}) = (x/x_0)^{\alpha}$ (where x_0 is some reference size). This is very similar to some of the relationships postulated by Gaudin and Schuhman. Again, if the energy consumption E is made proportional to the time t, then $E = C/x^{\alpha}$ and $\mathrm{d}E/\mathrm{d}x = C/x^{\alpha-1}$, which is similar to the Walker, Lewis and Gilliland relationship (see p. 54).

Several relationships between mill output and size of product have been proposed. Rosin and Rammler suggest $T = CR^x$ or $R = (T/C)^{1/x}$, where T is the rate of output, C is a constant and R is the residue on screen size x. ALJAWDIN[52] found the relationship $R = \exp(-Kt^n)$ at a fixed screen size x, in which both K and n are constants. ARBITER and BHRANY[50] found $1 - R = M = Kt(x/x_0)^n$, in which x_0 is a reference size. This is seen to be Gaudin's size distribution law with direct proportionality to time grinding. Arbiter and Bhrany obtained the results shown in *Figure 5.11* using a ball mill charged with quartz. BOWDISH[51] gives for a ball mill

$$-\left(\frac{\partial R}{\partial t}\right)_x = K \cdot R^a A^b$$

in which a and b are constants and A is the ball area. ARBITER and HARRIS[52] from dimensional analysis give

$$R = C\left(\frac{x}{x_0}\right)^{\alpha}\left(\frac{Et}{E_0}\right)^{\beta}\left(\frac{Et^2}{x^2}\right)^{\gamma}$$

in which E and E_0 are, respectively, energy consumed and an energy reference value, e.g. energy required to produce unit weight to reference size x_0.

SIZE CONTROL DURING MILLING

Ideally material would cease to be comminuted as soon as it reached the size desired.

In open-circuit milling the feed may be screened prior to entering the mill to remove material already sufficiently small (or scalped, as it is called). But whether it has been scalped or not, the feed is put through the mill and such controls as feed rate, retention time and internal size classification are relied on for the mill to produce the size range required (see *Figure 5.13*). Manifestly, in such a process some overgrinding usually takes place because some material stays in the mill while the more evasive particles are being broken; thus there is the double problem of expending power unnecessarily and of producing material unnecessarily small. If, for example, a selection function of 0.7 is chosen, then to achieve 98% below a certain size in the whole quantity without separation at least four presentations to the comminuting action are necessary, which means that the vast majority of the particles have had to be broken three times unnecessarily and that the equipment has been occupied four times too long, i.e. that the capacity has been reduced to a quarter of what it could theoretically be. In some equipment such prodigal use of capacity is practically unavoidable in open circuit because, once the material has entered the equipment, it suffers a number of comminuting events regardless of its size—e.g. in a jaw crusher the successive compressions inevitably occur and to avoid such unnecessary size reduction in open circuit would need a large number of machines each fed with a close size range or restricted to a single feed with very close size range.

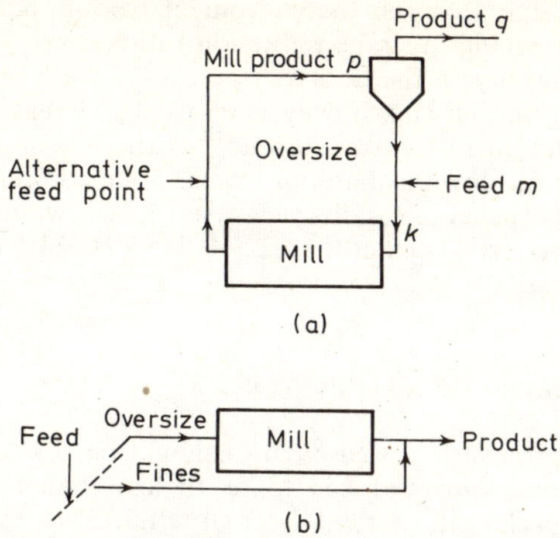

Figure 5.13. Open (a) and closed (b) grinding circuits showing, in each case, method used to classify feed prior to entry into mill

As is discussed in the chapters on comminuting, equipment is structured, or can be readily adapted, so that the ideal is very nearly realised; roll crushers, rod mills and impact mills with grids are examples of one kind and ring roll and ring ball mills with air separation are examples of the other kind. Much equipment, however, is not so structured or so adaptable, and to produce material all of which is below a certain size means that a considerable part is usually much below this size, i.e. that a considerable amount of overmilling has to be done, which is prodigal of both capacity and power consumption. Such equipment often has another defect: the fine material forms a matrix or cushion for the coarser fraction so that it is protected from comminuting attack.

CLOSED-CIRCUIT MILLING

Closed-circuit milling is, strictly, the deliberate removal of material from the influence of comminuting action before all the material has reached the desired size, separating that which is of a suitable size, and returning that which is not, to the mill together with the fresh feed. The description is, however, often applied to any milling process which includes a loop, even if it only returns air.

Either an open- or a closed-circuit system can treat the feed and remove from it that fraction which is already of a suitable size. The two kinds of circuit are compared in *Figure 5.13*.

The economics of the two kinds of circuit have to be assessed for each particular case; they depend not only on the extra cost of equipment compared with the greater utilisation of capacity and the saving in power, but also upon the precision with which material can be removed from the comminuting action before it has been oversize-reduced. The effectiveness of closed-circuit depends upon the equipment responding to a larger load when milling to a smaller average reduction ratio (see p. 10) while reducing a considerable quantity to the desired size.

Closed-circuit grinding can, however, be used to break down a material over a large reduction ratio. An example[48] is the use of a 60 cm diameter roll crusher to break down a material from 2.5 cm to 0.14 cm (a reduction ratio of 18). A roll crusher of this diameter (see p. 164) would not be expected to reduce 2.5 cm material smaller than $2.5–0{\cdot}015 \times 60 = 1.6$ cm. A circulation ratio of 5/1 was used.

CLOSED-CIRCUIT CALCULATIONS

A very simple approach to closed-circuit calculation is to consider a certain fraction of the feed being reduced to or to less than the maximum size required in one cycle with a remainder, fraction r, of oversize. Thus (see p. 75)

$$1 - r = \text{selection function} \times \text{breakage function}$$

Then, as the oversize is recycled, it becomes successively $r, r^2, r^3, \ldots, r^{\infty}$. The summation of this convergent geometric series to infinity is $r/(1-r)$. It represents the circulating load compared with the feed rate required to reduce all feed to product size, and is seen to be numerically equal to

$$\frac{1 - \text{selection function} \times \text{distribution function}}{\text{selection function} \times \text{distribution function}}$$

In *Table 5.10* the recirculating load ratio is given for various values of r. If the separating equipment has an efficiency η, then the summation of the recirculating load ratio becomes $r/(\eta - r)$ and the total load on the crusher compared with the feed rate is $1 + r/(\eta - r) = \eta/(\eta - r)$.

Table 5.10. RECIRCULATING LOAD REQUIRED FOR VARIOUS OVERSIZE LOADS

Fraction of feed which would appear as oversize in open circuit (r)	0.9	0.8	0.5	0.25	0.10
Recirculating ratio to feed rate necessary in closed circuit to obtain product in range $[r/(1-r)]$	9	4	1	0.33	0.12

It may be useful to elaborate on what is supposed to be happening in this simple closed circuit compared with an open circuit. Suppose that, of a feed material, this same proportion r of greater size than is required passes through the open-circuit equipment. A simple way of dealing with this would be to pass the product through separating equipment, which might be a screen or air separator, depending upon the sizes involved, to accumulate the oversize and then at a convenient time to pass it through the equipment again, when again a proportion r of the oversize would fail to break below the desired size. The capacity of the equipment would be decreased in its effect, to a fraction $(1-r)/1$ of its nominal capacity. In closed circuit the same thing could happen in effect, and no capacity advantage is gained unless overcrushing is avoided by removing the material from the equipment as quickly as possible after it is ground.

HENDRY[19] has shown that in a fluid-energy mill the selection function for material passing once through the jets is about 0.1 for a 10 µm particle. Thus, for such a group of particles to be broken down to any size, they would have to circulate (counting their initial feedthrough) ten times; this ignores the further factors of breakage function and classifying efficiency.

For real feeds, consideration has to be given to the behaviour of each size grade. It has already been shown that both the selection and the breakage function depend upon the size range, and many types of classifiers tend to behave differently towards different size ranges.

Such considerations have been allowed for in the more refined methods of BROADBENT and CALLCOTT[16] and of BROWN[49]. The procedure suggested by the latter will be described in this section. The assumptions made are:

(1) The size distribution of the product of the comminuting equipment from a narrow size range of feed is unaffected by the presence of the products from other size ranges of feed, provided the total load on the mill remains constant, i.e. provided the selection and breakage functions remain constant.
(2) The efficiency of the separating process remains constant for a given size range and is independent of the rate at which material in that size range is fed to the equipment and of the presence of particles in other size ranges.

There is some evidence that both of these assumptions are true.

Referring to *Figure 5.13*, if k is the total load on the mill, p is the product from the mill, q is the separated material leaving the system and m is the feed, then, in the steady state, $k=m+p-q$. Now consider a narrow size range of K_1 and let it in the comminution process be broken into fractions a_{11}, $a_{21}, a_{31}, \ldots, a_{j1}$ so that a_{j1} is the fraction in size range j from the feed range 1. Then p will be made up of

$$K_1(a_{11}+a_{21}+a_{31}+\ldots a_{j1})+K_2(a_{22}+a_{32}+a_{42}+\ldots a_{j2})+$$
$$+K_3(a_{33}+a_{43}+a_{53}+\ldots a_{j3})\ldots+K_j(a_{jj})$$

or, in matrix form,

$$\mathbf{P} = \mathbf{K}\begin{pmatrix} a_{11} & & & & & \\ a_{21} & a_{22} & & & & \\ a_{31} & a_{32} & a_{33} & & & \\ a_{41} & a_{42} & a_{43} & a_{44} & & \\ a_{51} & a_{52} & a_{53} & a_{54} & a_{55} & \\ a_{61} & a_{62} & a_{63} & a_{64} & a_{65} & a_{66} \end{pmatrix}$$

Thus $\mathbf{P} = \mathbf{A}\cdot\mathbf{K}$.

All the products from the different size ranges of K will approach the classifier and be treated according to their size (j) by a separation efficiency c or, in general, $q = cp$, and $p-q$ or $p(1-c)$ will pass on to join the feed m. Thus $K = m+p(1-c)$. The quantities of K in the different size ranges 1, 2, 3, ... and $K_1, K_2, K_3, \ldots$ will be made up of the parts in these size ranges from m and those not removed from p.

The equation $k = m+p(1-c)$ can be written in matrix form:

$$\mathbf{K} = \mathbf{M}+\mathbf{P}-\mathbf{CP}$$

and multiplying through by $\mathbf{A}$ to bring all matrices to the same order,

$$\mathbf{AK} = \mathbf{AM}+\mathbf{AP}-\mathbf{ACP}$$

Now from equation

$$\mathbf{P} = \mathbf{AM}+\mathbf{AP} = \mathbf{ACP}$$

it follows that

$$\mathbf{AM} = (\mathbf{I} - \mathbf{A}+\mathbf{AC})\mathbf{P}$$

and if $\mathbf{I} - \mathbf{A}+\mathbf{AC} = \mathbf{G}$

$$\mathbf{P} = \mathbf{G}^{-1}\mathbf{A}\cdot\mathbf{M}$$

or, combining this with the matrix form of the equation $q = cp$ deduced above,

$$\mathbf{Q} = \mathbf{C}\cdot\mathbf{G}^{-1}\cdot\mathbf{A}\cdot\mathbf{M}$$

BROWN[49] gives the example set out in *Table 5.11*. From these figures

$$\mathbf{AC} = \begin{pmatrix} 0.40 & & & & & \\ 0.02 & 0.59 & & & & \\ 0.02 & 0.04 & 0.65 & & & \\ 0.09 & 0.04 & 0.08 & 0.91 & & \\ 0.14 & 0.08 & 0.07 & 0.02 & 1.00 & \\ 0.33 & 0.25 & 0.20 & 0.07 & 0.00 & 1.00 \end{pmatrix} \begin{pmatrix} 0.00 & 0 & 0 & 0 & 0 & 0 \\ 0 & 0.05 & 0 & 0 & 0 & 0 \\ 0 & 0 & 0.14 & 0 & 0 & 0 \\ 0 & 0 & 0 & 0.39 & 0 & 0 \\ 0 & 0 & 0 & 0 & 0.80 & 0 \\ 0 & 0 & 0 & 0 & 0 & 0.90 \end{pmatrix}$$

In order for this multiplication to be possible the number of size ranges in the product analysis has to be the same as that in the classifier analysis. The multiplication comes to

$$\begin{pmatrix} 0 & & & & & \\ 0 & 0.03 & & & & \\ 0 & 0.00 & 0.09 & & & \\ 0 & 0.00 & 0.01 & 0.35 & & \\ 0 & 0.00 & 0.01 & 0.01 & 0.80 & \\ 0 & 0.01 & 0.03 & 0.03 & 0 & 0.90 \end{pmatrix}$$

The addition $\mathbf{G} = \mathbf{I} - \mathbf{A} + \mathbf{AC}$ may then be performed and $\mathbf{G}^{-1}$ determined; $\mathbf{G}^{-1}\mathbf{AC}$ is then obtained from

$$\begin{pmatrix} 0 & & & & & \\ 0 & 0.07 & & & & \\ 0.01 & 0.02 & 0.20 & & & \\ 0.13 & 0.08 & 0.14 & 0.79 & & \\ 0.26 & 0.19 & 0.17 & 0.05 & 1.00 & \\ 0.61 & 0.62 & 0.48 & 0.17 & 0 & 1.00 \end{pmatrix}$$

and from a stated feed **M**, **Q** can be determined. Brown assumes a feed of composition (55, 21, 13, 11, 0, 0 lb/h), which may be set out in the form

$$\mathbf{M} = \begin{pmatrix} 55 & & & & & \\ 0 & 21 & & & & \\ 0 & 0 & 13 & & & \\ 0 & 0 & 0 & 11 & & \\ 0 & 0 & 0 & 0 & 0 & \\ 0 & 0 & 0 & 0 & 0 & 0 \end{pmatrix}$$

On multiplication by $\mathbf{G}^{-1} \cdot \mathbf{A} \cdot \mathbf{C}$, this gives

$$\mathbf{Q} = \begin{pmatrix} 0 & & & & & \\ 0 & 1.47 & & & & \\ 0.55 & 0.42 & 2.60 & & & \\ 7.15 & 1.68 & 1.82 & 8.70 & & \\ 14.30 & 4.00 & 2.20 & 0.55 & 0 & \\ 33.50 & 13.10 & 6.25 & 1.87 & 0 & 0 \end{pmatrix}$$

Summing across gives the composition of the material leaving the system

$$\begin{array}{r} 0 \\ 1.47 \\ 3.57 \\ 19.35 \\ 21.05 \\ 54.72 \\ \hline 100.16 \end{array}$$

The total mass leaving the system checks, within the accuracy of the arithmetic, with the mass entering the system or the feed rate of 100 lb/h.

Thus, from a knowledge of the behaviour of a set number of size ranges of a particular material at a particular total load, in a particular piece of comminuting equipment, and a knowledge of the behaviour of the separating

Table 5.11. DATA FOR CLOSED-CIRCUIT CALCULATION (AFTER BROWN[49])

Mean size grade of feed to mill, μm		+1264	1264	638	284	114	−114
Analysis of product	+1264	0.40					
	1264	0.02	0.59				
	638	0.02	0.04	0.65			
	284	0.09	0.04	0.08	0.91		
	114	0.14	0.08	0.07	0.02	1.00	
	−114	0.33	0.25	0.20	0.07	0.00	1.00
Efficiency of classifier		0.00	0.05	0.14	0.39	0.80	0.90

equipment towards such material, the product leaving the system can be calculated from a given feed composition and settings of separating equipment or the feed rate of the equipment can be determined to give the size distribution of product which is required.

Closed-circuit milling is often applied to jaw and gyratory crushers and to ball mills.

REFERENCES

1. TAYLOR, J., *J. Inst. Fuel*, **27** (1954), 249
2. STADTLER, H., *Trans. IMM*, **19** (1910), 471
3. MARTIN, G., BOWES, E. A. and TURNER, F. B., *Trans. Ceram. Soc.*, **25** (1926), 63
4. ANDREASEN, A. H. M., *Trans. Ceram. Soc.*, **29** (1930), 239
5. GAUDIN, A. M., *Trans. Am. Inst. Min. Metall. Eng.*, **73** (1926), 253
6. SCHUHMAN, R., *AIME Tech. Publ.*, No. 1189 (1940)
7. ROSIN, P. and RAMMLER, E., *J. Inst. Fuel*, **7** (1933), 29
8. HEYWOOD, H., in *Chemical Engineering Practice* (ed. H. W. CREMER and T. DAVIES), 1957. London; Butterworths
9. GILVARRY, J. J., *J. Appl. Phys.*, **32** (1961), 391
10. EVANS, I., *Proc. Conf. Science in Use Fuel*, Institute of Fuel, 1958
11. MANNING, A. B., *J. Inst. Fuel*, **25** (1952), 31
12. BENNETT, J. G., BROWN, R. L. and CRONE, H. G., *J. Inst. Fuel*, **14** (1941), 111
13. PRESCOTT, T. W. L. and WEBB, F. C., *Trans. Instn Chem. Eng.*, **50** (1972), 21
14. ANDREASEN, A. H. M. and ANDERSON, J., *KolloidZeitschrift*, **50** (1930), 217
15. EPSTEIN, B., *Ind. Eng. Chem. Ind. Edn*, **40** (1948), 2289
16. BROADBENT, S. R. and CALLCOTT, T. G., *J. Inst. Fuel*, **29** (1956), 191
17. GARDNER, R. G. and AUSTIN, L. G., *Dechema Zerkleinern Symposion, 1962*, 218
18. GARSIDE, J. and WILDSMITH, J. A., *Trans. Instn Chem. Eng.*, **47** (1969), 285
19. HENDRY, R., *Dechema-Monogr.*, **57** (1967), 695
20. FAIRS, G. L., *Trans. IMM*, **63** (1953), 211
21. BLANC, E. C., *Aufbereitungstechnik*, **3** (1962), 101
22. MARTIN, G. and BOWES, E. A., *Trans. Ceram. Soc.*, **27** (1927/8), 247
23. BOND, F. C., *Chem. Eng.*, August (1954), 1
24. BORN, M. and STERN, O., *Sber. K. Preuss. Akad. Wissenschaften, Phys.-Mat. Kl., Berlin*, (1919), 901
25. KUSNEZOW, W. D., *Z. Phys.*, **42** (1927), 302; **44** (1927), 228
26. ZWICKY, F., *Phys. Z.*, **24** (1923), 131
27. SMEKAL, A., *Phys. Z.*, **26** (1925), 707
28. RANDALL, J. T., ROOKSBY, H. P. and COOPER, B. S., *Z. Krist.*, **75** (1930), 196
29. EXNER, F., *Untersuchungen über die Härte an Krystall Fläschen: Vienna* (1873)
30. THOMSON, W. (LORD KELVIN), *Phil. Mag.*, **24** (1887), 53
31. HEYWOOD, H. and PRYOR, E. J., *Bull. IMM* (1946), 477
32. STEINHERZ, A. R., *J. Soc. Chem. Ind., Lond.*, **65** (1946), 314

33. GREGG, S. J., *Chemy Ind.* (1968), 611
34. SMITH, R. W., *Min. Eng.*, **13**(4), (1961), 390
35. TRZEBIATOWSKI, W., *Z. Phys. Chem.*, **24** (1934), 75
36. BOWDEN, F. P. and TABOR, D., *Friction and Lubrication of Solids*, 1950. Oxford University Press
37. BEKE, B., *Principles of Comminution*, 1964. London; Maclaren
38. ROSE, H. E., *Dechema Zerkleinern Symposion, 1962*, 439
39. SMITH, E. A., *Chemy Ind.* (1967), 1436
40. VEGARD, L. and HAUGE, T., *Z. Phys.*, **42** (1927), 1
41. MACKENZIE, R. and MILNE, A. A., *Min. Mag.*, **30** (1953), 178
42. NAESER, G. and SCHOLZ, W., *Ber. Dt. Keram. Ges.*, **39** (1962), 106
43. TALBOT, J. H. and KEMPIS, E. R., *Nature, Lond.*, **197** (1966), 66
44. FRISCH, B., *Ber. Dt. Keram. Ges.*, **42** (1965), 189
45. BURNS, J. H. and BREDIG, J., *Chem. Phys.*, **25** (1956), 1281; JAMIESON, J. C. and GOLDSMITH, J. R., *Am. Miner.*, **45** (1960), 818
46. CLELLAND, D. W. and RICHIE, P. D., *J. Appl. Chem.*, **2** (1952), 42
47. BATRA, N. K. and BISWAS, A. K., *Chem. Eng. Sci.*, **26** (1971), 997
48. COGHILL, W. H. and DEVANEY, F. D., *Trans. Am. Inst. Chem. Eng.*, **34** (1938), 113
49. BROWN, R. L., *Br. Chem. Eng.*, **4** (1959), 463
50. ARBITER, N. and BHRANY, U. N., *Trans. AIME*, **217** (1960), 245
51. BOWDISH, F. W., *Trans. AIME*, **217** (1960), 194
52. ARBITER, N. and HARRIS, C. C., *Br. Chem. Eng.*, **10** (1965), 240
53. GAUDIN, A. M., *Principles of Mineral Dressing*, 1939. New York; McGraw-Hill
54. HEAVER, A. A., MCGRATH, L. and SMALLEY, I. J., *Chemy Ind.* (1967), 621
55. KAPUR, P. T. and AGRAWAL, P. K., *Chem. Eng. Sci.*, **25** (1970), 1111

6

The Behaviour of Mixtures During Comminution

MIXTURES OF TWO SOLID COMPONENTS

Investigations of mixtures of two components ground in a ball mill[1] showed that the rate of grinding of each of the components (in the particular case, limestone and quartz) invariably decreased from what it would have been if the ball mill had been loaded to the same total weight with the single component. This general conclusion was supported by work[2] on potassium dichromate ground with, in turn, silica sand, dolomite and limestone. A consideration of the power advantage from grinding the components separately would have to take account of the power consumed in mixing the components.

HOLMES and PATCHING[1] showed that for the materials they examined the rate of production of new surface was inhibited in a given mixture in about the same ratio for each component, but that the absolute individual rates depended upon the proportions in the mixture. A set of figures taken from the graphs in their paper, and therefore only approximately the values they actually obtained, is given in *Table 6.1*. K_L and K'_L represent rates in the mixture and alone for limestone and K_Q and K'_Q have similar respective meanings for quartz; Z is the sieve size through which 80% of the material passes. Results obtained by TANAKA[2] do not confirm that the degree of inhibition is the same for each component, but his work was carried out on much finer material. Tanaka's results are also shown in *Table 6.1*, the figures again being taken from the published graphs.

Product size (a) affects the rate of production of new surface. The ratio K_L/K_Q was found to be about $8a^{-0.3}$ while the ratio K'_L/K'_Q was found to be about $11a^{-0.3}$. HOLMES and PATCHING[1] found that the greatest inhibition of rates was obtained with the lowest sizes of product (see *Table 6.1*). TANAKA[2], grinding to much smaller sizes, found that the relationship was cyclical and in a size interval of about 5–30 μm there were two to four cycles of values. Tanaka relates the number of cycles to the hardness ratio of the materials and suggests that the greater this ratio, the greater the number of cycles.

The hardness ratio of the components, according to Tanaka, affects the relative rate of grinding. When the hardness ratio is low, the tendency is for the rate of grinding of the harder material to be more inhibited in the mixture compared with its rate when alone, than is the softer material, but when the hardness ratio is high, the effect tends to reverse, particularly when the harder material is present in the lesser proportion (see *Figure 3.15*).

Table 6.1

(a) SHOWING RATIO OF RATE OF GRINDING ALONE (K') AND IN MIXTURE (K) FOR VARIOUS MIXTURES OF LIMESTONE (L) AND QUARTZ (Q) AND AT VARIOUS PRODUCT SIZES

Product size, μm (*Z*)	*Proportion of limestone in mixture,* % 22		32		56		71		76	
	K_L/K'_L	K_Q/K'_Q	K_L/K'_L	K_Q/K'_Q	K_L/K'_L	K_Q/K'_Q	K_L/K'_L	K_Q/K'_Q	K_L/K'_L	K_Q/K'_Q
100	0.63	0.75	0.78	0.85	0.72	0.85	0.75	0.82	0.75	0.85
150	0.73	0.73	0.83	0.83	0.83	0.80	0.80	0.78	0.83	0.80
200	0.80	0.75	0.85	0.85	0.83	0.83	0.90	0.85	0.85	0.82
250	0.82	0.80	0.88	0.90	0.83	0.83	0.90	0.87	0.88	0.87
300	0.80	0.85	0.97	0.93	0.82	0.82	0.85	0.83	0.90	0.90
350	0.80	0.87					0.85	0.83		
400	0.85	0.88					0.83	0.83		

(b) SHOWING RATIO OF GRINDING RATES ALONE (K') AND IN MIXTURE (K) FOR 50% MIXTURES WITH POTASSIUM DICHROMATE (P), SILICA (S) AND DOLOMITE (D)

Average particle size, μm	*Proportion of constituents — 50%* *Dolomite and potassium dichromate* K_D/K'_D	K_P/K'_P	*Silica and potassium dichromate* K_S/K'_S	K_P/K'_P
30	0.9	0.7	1.0	0.6
15	0.7	0.6	0.9	0.5
10		0.7		0.4
8		0.9		0.6

The components retain their size distribution characteristics in the mixtures. Thus, separately, limestone and quartz in a ball mill followed approximately the Schuhman formulae (see p. 70):

$$\text{for limestone} \quad M = 80\,(x/a)^{0.59}$$
$$\text{for quartz} \quad M = 80\,(x/a)^{1.07}$$

In mixtures the formulae for the components were:

$$\text{for limestone} \quad M = 80\,(x/a)^{0.66}$$
$$\text{for quartz} \quad M = 80\,(x/a)^{1.1}$$

Grindability, on the other hand, as determined by the Bond ball mill method (see p. 58), gave quite unpredictable results for the components. Thus, for mixtures of between 100 and 40% limestone, the grindabilities were intermediate between those for limestone and quartz alone, while in mixtures containing less than 40% limestone the grindabilities were all lower than those of either limestone or quartz.

Materials are also fragmented selectively in an impact mill. FERRARA and DEL FA[3] have shown that by adjustment of any of the factors (1) peripheral speed, (2) position of impact plate relative to rotor and (3) size-range selection for the feed, differentiation between species at certain size ranges of product could be clearly distinguished and that by successive open-circuit milling this differentiation could be significantly increased. They found that the best differentiation was obtained with impact plates set from the rotor at slightly more than the feed size and argued that this allowed the more tenacious species to pass unchanged. They also found that relatively small peripheral speeds gave the best selection. Closed circuits were inimical to selection because by definition they did not allow particles to escape unless they had been reduced below a certain size.

When mixtures containing constituents of different hardnesses are being ground in a closed circuit, it is important to realise that the initial product will often contain a dominating proportion of softer material and that equilibrium conditions corresponding with the feed composition will take some time to establish. The condition inside the mill circuit will be such that the circulating load will contain a greater proportion of hard material than does the feed and there is the possibility of excessive wear occurring. Such a situation might make more economic a different type of equipment from that thought suitable when only the feed composition was considered.

WET AND DRY COMMINUTION

If wetting of the material is acceptable before, during and after grinding, then there are certain advantages in wet grinding. There are some materials which are changed in an undesirable way by wetting, or by wetting and drying, and it is necessary to ascertain the effects of wetting on the particular material before wet grinding is adopted.

The wetting agent is usually water (see p. 106) but other substances are sometimes used, particularly when a physical or chemical process is included in the whole operation which can advantageously take place simultaneously with a comminution process, e.g. a process in which advantage may be taken of the mixing qualities possessed by many types of comminution equipment and material may be ground and dispersed in a liquid simultaneously or in which a chemical reaction with a liquid can be accelerated if the solid component is more finely ground and dispersed in the liquid. In such circumstances, an investigation of the use of wet comminution is indicated.

Wet grinding is the usual method with stamp mills and is frequently the method used with rod and ball loaded tumbling mills. It is sometimes used with vibration mills.

By wet grinding is meant the grinding of material containing about 50% of water by volume in the uncombined state. If water is added slowly to the dry milling operation, e.g. in a ball mill, the discharge will first be seen to decrease and even stop completely between about 4 and 15% moisture. Then, as more water is added, the discharge begins again and issues as a pulp or slurry. Pastiness begins according to the degree of subdivision of the solid particles and may be considered to be the state of solid particles adhering to one another by virtue of the uncombined water which is present. A simplified approach to the problem is to regard adherence as being due to the line of maximum total surface tension being long enough to support the weight of one particle. Thus, in *Figure 6.1*, if two spherical particles each of

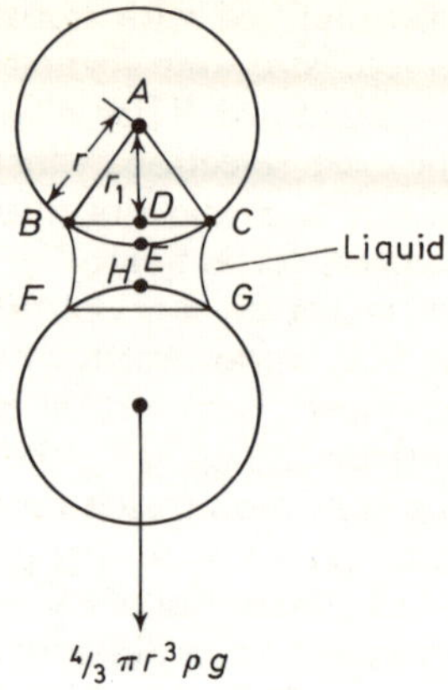

Figure 6.1. Coherence of particles owing to moisture

radius r are considered, the fundamental condition is that the interfacial surface tension between the liquid and the particle along the length of the longest arc BEC is at least equal to, i.e. will support, the weight of one particle. If σ is this interfacial surface tension per unit length and the length of the arc BEC is l and g is the acceleration due to gravity, then

$$l\sigma \geqslant \tfrac{4}{3}\cdot \pi r^3 \rho g \tag{6.1}$$

If the maximum possible length of arc is taken, i.e. πr, σ is assumed to be the same as that for water at 20°C, i.e. 73 mN/m, and a typical value for the density of a mineral particle is taken to be 2500 kg/m^3, and these values are inserted in equation 6.1, the maximum value of r is found to be 1.5 mm. In other words, the maximum size of particle of such a material which can be made to adhere through surface tension to a similar particle is one with a diameter of about 3.0 mm.

The minimum volume of liquid which will make two particles adhere is the volume of the cylinder $BCGF$ (assuming that the sides BF, CG are perpendicular to the base PG: in fact they will be slightly curved owing to surface tension) from which has been subtracted the volume of the solid segments BEC and FHG and for which the circumstance has been taken of the particles practically touching. If the distance AD is equal to r_1, then this volume will be

$$2\pi\left[\frac{r^3}{3}+\frac{2r_1^3}{3}-r_1^2\cdot r\right]$$

The ratio of half this volume to the volume of one particle is

$$\frac{r^3+2r_1^3-3r_1^2 r}{4r^3}$$

and represents the minimum volume relationship between moisture and the solid in order to obtain adherence. This expression may also be written

$$\frac{1}{4}\left[1+2\left(\frac{r_1}{r}\right)^3-3\left(\frac{r_1}{r}\right)^2\right] \qquad (6.2)$$

This can be transformed into a mass relationship by multiplying it by the ratio between the densities of the liquid and the solid. Taking the liquid to have the density of water (1000 kg/m^3) and the density of the solid to be 2500 kg/m^3, expression 6.2 becomes

$$\frac{1}{10}\left[1+2\left(\frac{r_1}{r}\right)^3-3\left(\frac{r_1}{r}\right)^2\right] \qquad (6.3)$$

If the angle subtended by the arc of length l is θ, then $l = r\theta$ or $l/2r = \theta/2$, and from equation 6.1

$$\frac{l}{2r} = \frac{2\pi r^2 \rho g}{3\sigma} = \frac{\theta}{2} \qquad (6.4)$$

By substituting the value of surface tension for water at 20°C (σ = 73 mN/m) and of a typical density for a mineral (ρ = 2500 kg/m^3), equation 6.4 becomes

$$\frac{\theta}{2} = 70 \times 10^4\, r^2$$

in which r is measured in metres. But $r_1/r = \cos\theta/2$ and therefore for various values of r, expression 6.3 can be determined.

Examples are shown in *Table 6.2*. As r becomes smaller, so does θ

Table 6.2. MOISTURE REQUIRED TO MAKE PARTICLES (ρ = 2.5 × 10^3 kg/m^3) ADHERE IF WETTED WITH WATER

Diameter of particle	cm	0.3	0.2	0.15	0.10				
	mm	3.0	2.0	1.5	1.0	0.2			
	µm						127	63	46
Moisture (% wt. of solid)	*calculated*	10	1.4	0.14	0.007	0.000 1			
	practical	2					1	0.5	0.2

become smaller and therefore so does r_1/r approach unity and expression 6.2 approach zero. The calculated results are compared with some results obtained practically by NORTH[5]. The calculated moisture is the minimum required to cause two particles to cohere. Pastiness is caused by the presence of more moisture than that, but the calculation indicates that less and less moisture is necessary to cause coherence as the particle size decreases and that it is therefore important, particularly in ball milling and in some instances in rod milling, to decide whether a dry or wet process is to be used. If a dry process is chosen and the materal is not dried sufficiently, pastiness can result during the size reduction process and the equipment can be clogged up.

If the choice is wet grinding, then the slurry should not be thinned excessively and the capacity of the machine thereby unnecessarily reduced. A thin slurry causes more wear to the lining and grinding media. A suitable proportion is volume of water equivalent to that of the loosely compacted bulk volume of the solid, e.g. a material of density 2700 kg/m^3 required 0.6 m^3 of water per 1000 kg of material: 1000 kg of bulk material (40% voids) would occupy about 0.6 m^3. It may well be necessary for a subsequent part of the process to add more water, but this should be done after the grinding stage of the process is complete. A test for optimum liquid content commonly used on the plant is to dip a ball into the slurry; the liquor should adhere to the ball but it should not be sticky.

The influence of the various properties of the liquid, as RUMPF[6] has pointed out, will depend upon the volume of liquid present relative to the material being ground and upon the total void space between the grinding media. If the void space is not filled, then the sticking of particles to the media is important: if there is insufficient liquid to permit all the coarsest particles to be moved into favourable comminuting positions by adhesion of particles to the media, the coarsest particles will tend to be isolated, the very finest particles will tend to agglomerate and form cushions, while the particles receiving the most useful attention will be those in an intermediate range of size because they will be small enough to adhere to the media and be too large to agglomerate. In these circumstances surface tension will be the important property of the liquid, although viscosity, which would inhibit sliding of the particles on the media, could play a part in improving the comminution of larger particles, because friction could then be of greater influence than the relatively ineffective forces of surface tension when the particles are too large, as has been shown in the above calculation.

When the quantity of liquid is sufficient to fill the voids, then surface tension is unimportant and viscosity plays an important, if somewhat tergiversating role as an aid to comminution. High viscosity can keep the movement of the material particles slow while the balls are moving and thus provide a greater relative chance of attack, but (1) reduces the chances of turbulence and of flushing out of the finer particles (compare air sweeping in dry mills), (2) provides more resistance to the movement of the media and to their impact, (3) provides more cushioning for the beds of material and (4) provides less grinding for more power.

The quantity of liquid, besides having the effects outlined, can also have a considerable effect on the residence time of the material in the equipment. A thinner slurry will usually flow more quickly through the equipment and the product is likely to be coarser.

Water has a very high surface tension and very low viscosity, and there does not seem to be any strong evidence that it can be improved on as the liquid phase in wet grinding unless the chemical properties of water are inimical to the material being ground.

Advantages claimed for wet grinding compared with dry grinding are:

(1) Power consumption per tonne is lower. The Bond index is about 75% of the dry index.

(2) Capacity of the mill is increased by as much as 15%, especially if the mill is lined with a soft material such as rubber.
(3) Dust is not produced as a smoke or suspension in air, a feature which is particularly valuable when the material being ground is harmful to health.
(4) Higher speeds of rotations of tumbling mills are possible and 75–80% of the critical speed (see p. 231) can be used without the balls or rods being held against the shell. This percentage compares with 65–70% for dry grinding.
(5) Surface-active agents can be more easily introduced and dispersed to assist rate of grinding (see p. 108).
(6) As the material can be introduced as a slurry, the material washes through the mill, feed control is easily automated and pumps and pipelines can be used for distribution.
(7) Wet screening, which is more efficient than dry screening, can be used and thus stricter control of particle size can be exercised.
(8) Materials which have to be wetted in a subsequent part of the process may prove resistant in a finely divided state and may not only be much more easily ground wet but also be produced in a form amenable to further wetting later.

Disadvantages compared with dry grinding are:

(1) The wear on the media and the liners is usually considerably more (see *Table 6.3*).
(2) Less material in the range of the very smallest particle sizes is produced.
(3) There is sometimes more flocculation of particles.
(4) Drying a ground material usually necessitates more expensive dust-collecting equipment than drying raw unground material.
(5) A wet process cannot be instantaneously shut down as can a dry one and is therefore less convenient for day or single-shift operation. To shut down a wet process means cleaning out the system and it takes some time to re-establish equilibrium conditions after restart. Broadly speaking, a wet process should not be operated for less than a continuous full week.

Table 6.3. WEAR IN TUMBLING MILLS IN WET AND DRY CONDITIONS (WEIGHT OF METAL PER UNIT WEIGHT GROUND)

Media and liners	*Wear ratio* *wet*	*dry*
Balls: ball-loaded mill	1	0.14
Liners: ball-loaded mill	0.1	0.014
Rods: rod-loaded mill	1	0.1
Liners: rod-loaded mill	0.075	0.01
Liners: autogenous	0.2	0.03

The most important advantages of wet grinding are increased capacity for a given size of equipment and less power consumption per tonne of product. Typically, capacity is increased by 66% and power consumption per tonne is 75% of that in dry grinding. Against this has to be set the additional wear on media and liners, and modern costs are such as to leave the field open to a simple choice based on convenience: Is the product in more desirable form wet or dry? Is the feed already wet or dry? Is it susceptible to pastiness and therefore better taken over to the wet state? These are the kinds of questions, the answers to which will decide the form of process used.

ADDITIVES TO AID GRINDING

Reference has already been made to the effect of substances, particularly of liquids, on the strength of solids (p. 31). Sometimes this effect is to increase the strength, sometimes to decrease it. So far as the process of comminution is concerned, the effect sought is the weakening of the solids and a decrease in the power necessary to break them to the sizes required. Since the process of comminution includes the creation of new surface and therefore the creation of surface energy, the possibility suggests itself that if the amount of this surface energy were decreased, less energy would be required to create the new surface. Many substances are known which decrease the surface energy of liquids and they have considerable effect even when added in very small concentrations, e.g. of the order of 0.01%. It has been shown that many of these substances can have a striking effect on the energy required to grind some solids. One explanation for the effect of these so-called additives is that they are able to enter the micro-cracks of the solid, possibly by way of the macro-cracks; and, because of the highly reactive surfaces produced in breaking, these substances move with velocities akin to the velocities of the crack propagations and provide a barrier to the dissipation of energy. REHBINDER, SCHREINER and ZHIGACH suggested that the effect was that of the additive filling the cracks during stress relief and preventing them sealing up during the comminution process. Gaseous substances are able to function as additives, water vapour and carbon dioxide being examples, which lends support to the theory that chemisorption is involved.

The effect of favourable additives is either to decrease the grinding time to reach a certain degree of subdivision and thus decrease the power usage or to increase the degree of subdivision reached in a certain time.

Another effect these additives may have is to prevent agglomeration of the finest particles (see p. 88) and thus remove the need for the repetition of breaking effort.

Yet another suggestion is that favourable additives decrease coating of media and linings, which tends to decrease rate of grinding. Particularly in the cement industry[8], materials are added to the extent of about 1% for this very purpose—e.g. fish oil stearate, tallow, Vinsol resin, colloidal carbon and TDA (a proprietary reagent which appears to be naphthalene sulphonic acids condensed with triethanolamine and converted into the sodium salt and mixed with calcium ligninsulphonates). Such treatment is said to decrease grinding time by 20–30%, but the effect of coating has been disputed[9].

Attempts have been made to classify additives according to what is judged to be the active constituent, i.e. according to whether it is cationic, anionic or nonpolar. GILBERT and HUGHES[10] could not find any improvement in grinding with a number of additives which previous investigators had found very effective. The only differences which they could discover in their experiments was that in the successful experiments stamp mills and closely graded feeds had been used, whereas they used tumbling mills and run of jaw crusher or cone crusher as feed.

Some most informative work on cement clinker by GHIGI and RABOTTINO[9] has shown that the effect of additives can change from being favourable to being distinctly unfavourable, depending upon the concentration in which they are present. *Figure 6.2* is taken from their paper and illustrates this discovery. They show, further, that while a given concentration of an additive (0.1% sodium stearate) can considerably improve the rate of grinding in a rod mill of cement clinker with 25 mm diameter rods, for 6 mm rods the rate

Table 6.4. ADDITIVES USED IN COMMINUTION

Additive	*% added*	*Material ground*	*Wet (W) or dry (D)*	*Grinding rate factor*	*Reference*
Water	0.06	ceramic enamels	D		
	0.06	marble	D	1.6	
	0.4	cement clinker	D	1.3	
Alcohols and phenols					
methanol		quartz			11
isopentanol		quartz		1.29	12
isopentanol		iron powder		20	
s-octanol		quartz		1.4	12
series normal alcohols		soda lime glass	D	20	13
glycerol		iron powder		5	
phenols and polyphenols	0.01–0.25	cement			
phenols and polyphenols		gypsum			
Ketones					
acetone	0.2	cement clinker	D	1.37	9
Amines					
triethanolamine					
Flotigan (C_{12}–C_{14} amine ex coconut oil)	0.02	quartzite		2.2	14
Flotigan (C_{12}–C_{14} amine ex coconut oil)	0.02	limestone		1.7	
Sulphonic acids					
aryl alkyl sulphonic acid (RDA)	0.06	graphite			
aryl alkyl sulphonic acid (RDA)	0.06	cement		1.3	
Fatty acids					
oleic acid	0.003	limestone		1.1	
oleic acid		zinc blende			

Table 6.4. *(Continued)*

Additive	*% added*	*Material ground*	*Wet (W) or dry (D)*	*Grinding rate factor*	*Reference*
butyric acid		quartz		1.27	12
stearic acid	1.0	pumice			
stearic acid		limestone			
sodium oleate	0.1	quartz		2.0	
sodium oleate	0.1	limestone		2.0	
vinsol resin (calcium stearate)	0.05/0.10	limestone			
sodium stearate		dolomite			
sodium stearate	0.15	cement clinker		1.2	9
aluminium stearate		cement			
caprilic acid	0.5	chrome ore–magnesite	D	1.2	
marine oil	0.5	chrome ore–magnesite	D	1.1	
beef tallow					
wool grease	5.0	limestone			
wool grease		gypsum			
Other carboxylic acids					
naphthenic acid	0.1	cement clinker	D	1.33	9
sodium naphthenate	1	quartzite	W	1.40	7
sodium sulpho-naphthenate	1	quartzite	W	1.80	7
Hydrocarbons					
n-alkanes, various		soda lime glass	D	10	13
Esters					
amyl acetate		quartz		1.23	12
Others					
carbon black	0.08	cement		1.3	
carbon black	0.32	cement			
sodium silicate	1.16	clay slip			
sodium hydroxide		magnesite			
sodium hydroxide	0.008	limestone		1.5	
sodium carbonate	0.02	limestone		2.0	
sodium chloride	0.08	quartzite		1.2	
carbon dioxide		magnesite			
carbon dioxide		dolomite			
carbon dioxide	0.03	quartzite	W	1.55	7
aluminium chloride		carbon black			
		graphite			
		talc			
ammonium carbonate		mica			
		vermiculite			
hardwood pitch		pumice			
sodium polymeta-phosphate (Calgon)	0.01/0.04	lead zinc ore		1.3	15
kaolin	10	sulphur	D		
thallium chloride	0.02	quartzite	W	1.65	7

can be decreased. It is possible that, as in certain instances discussed in Chapter 3 (p. 31), additives can assist the healing of cracks; but why they should change from being maintainers to healers of cracks with change in concentration of additive is puzzling. *Table 6.4* lists some of the additives

20

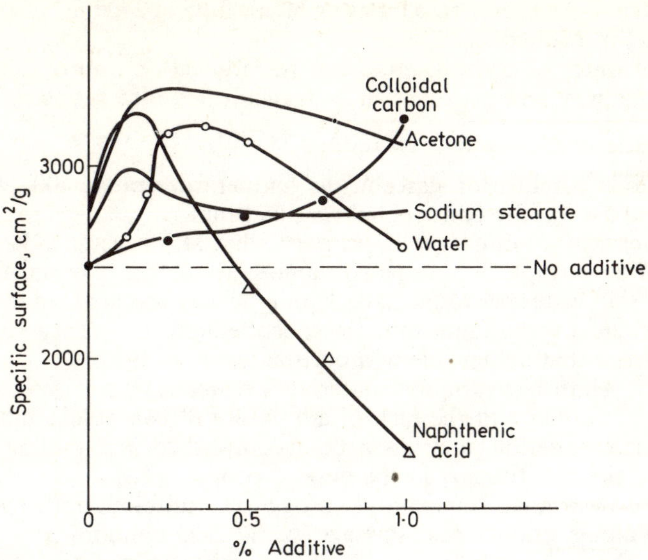

Figure 6.2. Effect on specific surface of ground cement clinker of certain proportions of additives. (After G. GHIGI *and* L. RABOTTINO[9]*)*

which have been used, the material upon which they had an effect and an estimate where available of the extent of their effect. The grinding rate factor is measured by:

$$\frac{\text{new surface produced with additive}}{\text{new surface produced without additive in the same time}}$$

In the absence of quantitative information, the appropriate column has been left blank.

CONCOMITANT EFFECTS OF COMMINUTION

A number of effects can accompany comminution. Some of these are purposely sought; strenuous efforts are usually made to avoid some others.

Drying. Perhaps the most usual concomitant effect, and one which is often deliberately sought, is drying. The process is linked with equipment which deliberately or incidentally causes large flows of air (e.g. ring roll and ring ball, rotary impactor and fluid-energy mills) or which can easily be coupled to air flow (e.g. tumbling mills). The flow of large volumes of air will of itself cause drying, but the effect is improved by heating the air; temperatures in excess of 350°C are used and some mills incorporate parts in heat-resisting metals so that they can withstand the high temperatures. As an example, clay for brickmaking was extracted with a moisture content of 11% and was comminuted to 80% passing 780 μm in a rotary impactor and simultaneously dried down to 0.9% moisture by using an air flow of 2400 m^3 of air per tonne of product introduced at 80°C.

Loss of water of crystallisation can be induced on appropriate materials in the same way and gypsum can be transformed into plaster of Paris,

$$CaSO_4 \cdot 2H_2O \rightarrow CaSO_4 \cdot \tfrac{1}{2}H_2O$$

with exhaust combustion gases in the temperature range 400–700°C introduced into a specially constructed ring ball mill.

Washing. Rotary impact mills, in particular, allow water to be added as a scourer. For example, limestone containing up to 15% clay was treated with about 6 m^3 of water per tonne of rock and the clay was scoured from the rock in the turbulent water in the mill, the effect being that capacity was decreased by 25% from that achievable without the water addition.

Mixing. Almost invariably, materials are made more homogeneous by being comminuted and the task of admixture of two solids, both of which require size reduction, may often be accomplished in a milling process by feeding in the constituents in the desired proportions.

Mineral separation. Impact and autogenous tumbling mills tend to break material along grain lines. Passage of a rock containing a mixture of minerals through such mills can cause separation of the grains of the different minerals with a degree of precision which varies with the material, the mill type and operation and the reduction ratio, but which can be such that a subsequent process such as magnetic, gravity or flotation separation can effect neat division into a rich fraction. For example, some Blakedene rock containing 56% fluorspar, when passed through an impact crusher, produced gradings with the fluorspar content shown in *Table 6.5*. When the plus 40 fraction was rejected, the minus 1 fraction put on one side and the remainder put in float–sink equipment with liquid of density between 3.15 and 3.35, the material which sank had the size range and composition also shown in *Table 6.5*. When the minus 1 fraction was added to the material which had sunk, the fluorspar recovered was 91.5% of that in the original Blakedene rock, and its concentration was 81.5%. Thus the separation of the grains must have been very clean to allow the float–sink method to effect this very considerable increase in concentration.

Table 6.5. SEPARATION OF FLUORSPAR FROM BLAKEDENE ROCK BY IMPACT CRUSHING AND A FLOAT-SINK PROCESS

Sieve size, mm	+40	25–40	19–25	12–19	3–12	1–3	−1	*Total*
(a) % original before f–s	2.3	4.8	6.7	8.1	36.1	13.0	29.0	100
(b) CaF_2 analysis % before f–s	8.2	15.0	27.0	32.0	56.0	67.0	75.0	
(a × b) units of CaF_2 before f–s	0.19	0.72	1.8	2.6	20.2	8.7	21.8	56.0
(c) % original after f–s	↓	1.0	1.9	2.1	21.5	8.0	↓	
(d) CaF_2 analysis % after f–s	↓	60.2	78.0	82.3	88.9	86.3	↓	
(c × d) units of CaF_2 after f–s	reject	0.60	1.48	1.73	19.15	6.90	21.8	51.7

Surface covering. Mills which incorporate a rubbing action in their mechanism, of which pan and tumbling mills are examples, can be used to effect colouring of particles by incorporating a dye in the grinding process or, in reverse, to affect the shade of a dye by adding white ingredients which are covered by the dye. These mills will break down the particles of the dye and of the other constituent and cause the surface covering at the same time.

Chemical. Certain materials change chemically during comminution. Examples of these effects are discussed in Chapter 5 (p. 90).

Among the less desirable concomitant effects of comminution are the following.

Noise. Noise invariably accompanies comminution, its amount varying with the equipment and the duty. Many companies now offer special services to combat noise in their equipment. Noise is used to control the feeding of some machines (see p. 126).

Burning and explosion. One of the principal objects of comminution is to increase the reactivity of materials, and many materials become pyrophoric and explosive when they are very finely ground. It may be assumed that anything that will burn in air or anaerobically with whatever difficulty will burn more easily when it is finely divided; examples are sulphur, starch, sugar, flour, pyrites, metals and (of substances which will burn anaerobically) any nitrate with organic matter. The problem of substances which will burn anaerobically cannot be discussed in detail, but it can be stated that it may well not be possible to size-reduce such substances in a mixture and that another route of manufacture may be necessary. The lower inflammability limit for most combustible dusts requiring oxygen to burn is about 20 g per cubic metre of air, which means that any dust concentrations above this are inflammable up to the upper inflammability limit. As the upper inflammability limit has not been determined for many substances and as settling of dust invariably occurs, it should be assumed that any dust with a lower inflammability limit of 20 g/m^3 is a hazard at any concentration greater than this. Safety precautions which may be adopted are the use of inert gas—gases which are usually inert (but not always) are carbon dioxide or nitrogen—in concentration greater than 10%, together with the provision of explosion panels in appropriate parts of the equipment and, in certain instances, explosion vents, blast walls, earth-mounded bateaux and a lightly constructed roof. The Factory Inspector in the U.K. or his equivalent elsewhere should always be consulted before equipment is erected for the size-reduction of material which has the slightest disposition to combust in air or anaerobically.

REFERENCES

1. HOLMES, J. A. and PATCHING, S. W. F., *Trans. Instn Chem. Eng.*, **35** (1957), 111
2. TATSUO TANAKA, *Dechema Zerkleinern Symposion, 1962*, 360
3. FERRARA, G. and DEL FA, C., *Dechema-Monogr.*, **57** (1967), 791
4. ANON, *Tonindustr.-Ztg*, **60** (1936), 6
5. NORTH, R., *Quarry Mgrs' J.*, Oct. (1969)
6. RUMPF, H., *Dechema-Monogr.*, **57** (1967), 423
7. REHBINDER, P. A., SCHREINER, L. A. and ZHIGACH, K. F., *Hardness Reducers in Drilling*, 1944. Moscow; Academy of Sciences
8. TAGGART, A. F., *Handbook of Mineral Dressing*, 1953. New York; Wiley

9. GHIGI, G. and RABOTTINO, L., *Dechema-Monogr.*, **57** (1967), 427
10. GILBERT, L. A. and HUGHES, T. H., *Dechema Zerkleinern Symposion, 1962*, 170
11. MALLIKARJUNAN, R., *et al.*, *Trans. Ind. Inst. Metals*, **18** (1965), 79
12. VON ENGELHARDT, W., *Öl Kohle*, **39** (1943), 707
13. SELIM, A. A., *et al.*, *J. Soc. Petrol. Eng. (AIME)*, **9**(4) (1969), 425
14. VON SZANTHO, E., *Z. Erzbergb. Metall Hüttwes*, **2** (1949), 353
15. SIEBEL, J. and ZEISEL, H. G., *Deutsches Patent*, 1 071 455 (1959)

7

The Classification, Selection and Layout of Comminution Equipment

EARLIEST CLASSIFICATION

A distinction was made in very earliest times between crushing and grinding. Thus Abraham (Genesis 18:6) enjoins Sarah, his wife, to make fine meal from which to make cakes for the angels, thus making a distinction between coarse meal made by pounding, as in a pestle and mortar, and finer meal made by grinding in a quern. And in Numbers (11:8) Israelites 'went about and gathered (manna) and ground it in mills and beat it in a mortar'.

A more recent distinction[1] between a crusher and a grinding mill is in the fact that a grinding mill is a machine in which the comminuting elements are free to make direct contact with each other, but are kept from doing so by the material to be crushed. It is sometimes claimed that fine grinding cannot be accomplished in machines in which the component parts move in definite restricted paths with respect to one another.

CLASSIFICATION ACCORDING TO PURPOSE

The usual further broad classification of equipment is in accordance with the maximum size of product from it. Examples of such classifications are given in *Table 7.1.*

Table 7.1. CLASSIFICATION OF EQUIPMENT BY SIZE OF PRODUCT

			Taggart	*Hukki*
Explosive shattering				∞–1 m
Primary crusher		–10 cm	15–10 cm	1m–10 cm
Secondary crusher (intermediate)		10 m–1 cm	1.25–1 cm	10 cm–1 cm
Grinding mills	coarse	–1 cm	3.5 mm–900 μm	1 cm–1 mm
	fine		600–200 μm	1 mm–100 μm
	very fine		–150 μm	100–10 μm
	superfine			10–1 μm

Table 7.2. SHOWING VARIOUS CLASSIFICATIONS WHICH HAVE BEEN PROPOSED FOR COMMINUTING EQUIPMENT

Kanowitz[5]	*Kelleher*[2]				*Andreasen*[3]	*Farrant*[4]	*Coghill and Devaney*[1]
Continuous pressure	Compression	Pressure	Continuous	Slow compression	Stressing by external mechanical means	Slow speed (Tumbling)	Parts move in definite or restricted paths
	Shredding	Shear	Reciprocating	Cutting or tearing	Stressing by gravitational forces	Medium speed (roller and ring roll or ring ball mills)	
Blow or impact	Impact	Impact	Hammering	Impact	Stressing by inertial forces produced by mechanical means	High speed (Impact)	
			Tumbling		Stressing by inertial forces produced by gravity		Parts are free to make direct contact with each other, but are kept from doing so by material to be comminuted
Grinding or Abrasion	Attrition	Attrition		Attrition		Fluid-energy mills	

CLASSIFICATION ACCORDING TO COMMINUTION PROCESS

Types of equipment may be designed to give products in two or more size categories. If this is so, a useful classification, for the purpose of discussing types of equipment, is by the comminuting action of the equipment. The various kinds of severing action which can take place are illustrated in *Figure 3.1*.

Some of the classifications of equipment which have been proposed are given in *Table 7.2*. The one selected for use in this book is that in which the principal comminuting action of the equipment is by:

Compression or nipping
Blow or impact
Tumbling or projection
Cutting or shredding
Attrition
Other methods

The various types of equipment available can then be classified as shown in *Table 7.3*.

Table 7.3. CLASSIFICATION OF EQUIPMENT

Compression or nipping	*Blow or impact*	*Tumbling*	*Cutting or shredding*	*Attrition*
Jaw	Hammer	Rod-loaded	Shears	Buhrstone
Gyratory	Pin	Ball-loaded	Dicers	Colloid
Roll	Stamp	Autogenous		Fluid-energy
Disc	Vibration			Sand
Pan				
Ring roll				
Ring ball				

MECHANICS OF COMMINUTING EQUIPMENT

Various attempts have been made to analyse the mechanics of the kinds of actions of the principal classes of equipment. Thus JOISEL[6] has proposed

Jaw crushers:	Pressure
Hammer mills:	Kinetic energy
Ball mills and rod mills:	Projection

Others have proposed that the whole action can be likened to a first-order (i.e. dependent only upon the concentration of one substance) chemical reaction. At first there are presented to the equipment a number of large pieces which are gradually broken down. Thus fewer large pieces are available to be broken down as the action proceeds and the rate of change follows such a law as

$$\mathrm{d}P/\mathrm{d}t = K(F-P) \qquad \text{or} \qquad P = F(1-\mathrm{e}^{-Kt})$$

which shows the proportion of the feed (F) which is converted to product P in time t. Curves of rate of production of new surface in a ball mill follow just such a law (see *Figure 10.9*).

ANDREASEN[3] has suggested dimensional analysis as the basis for predicting performance. He classifies equipment into that which functions by using:

(1) Stressing by external mechanical means, e.g. jaw, gyratory and roller crushers.
(2) Direct stressing of material by gravitational forces, e.g. pan mills.
(3) Inertial forces produced by mechanical means, e.g. impact mills.
(4) Inertial forces produced by gravity, e.g. tumbling mills.

Indicating *dimensional* ratios with bold face type, Andreasen's method of analysis may be indicated as follows.

Two objects are said to be geometrically similar if the ratios between all corresponding linear dimensions on each are constant. Two geometrically similar objects are said to be dynamically similar if all the forces on one bear a constant ratio to all the forces on the other and if the areas over which the forces act in one body are geometrically similar to the areas over which the forces act in the other body. Such relationships are shown conventionally by such symbols as $L_1/L_2 = \boldsymbol{L}$, in which $\boldsymbol{L}$ represents the ratio between the geometrically similar lengths in two objects. Dynamic similarity can be indicated by $F_1/F_2 = \boldsymbol{F}$, in which $\boldsymbol{F}$ represents the ratio between dynamically similar forces acting on two objects.

Andreasen's class (1) mills, of which jaw and gyratory crushers are examples, are similar because they exert the same kinds of compressive stresses on the material being comminuted, the dimensional ratio for which is represented by $\boldsymbol{F}/\boldsymbol{L}^2$, in which the ratio of dynamically similar forces is shown to act on the ratio of geometrically similar areas. Assuming that the breaking strain is the same in each material, this stress ratio is equal to the ratio of their Young's moduli. If the Young's moduli of the materials are the same, then this stress ratio will equal unity, or $\boldsymbol{F}/\boldsymbol{L}^2 = 1$.

To scale up mills in this class corresponding forces should be changed in proportion to the square of the corresponding linear dimensions.

Two particles will behave similarly in compressive machines if the force exerted on them is proportional to the square of their corresponding lengths. Andreasen took two cubes related in length by, say, $L_1 = KL_2$ and found that the compressive forces necessary to break them were related by $F_1 = K^2 \cdot F_2$.

An example of class (2) mills is a pan mill. Andreasen postulated that the essential quantity was force due to weight, Mg, or, since g may be considered constant, due to mass M. The ratio of the rates of throughput of the machines is proportional to $\boldsymbol{M}/\boldsymbol{T}$ or $\boldsymbol{\rho L}^3/\boldsymbol{T}$ if $\boldsymbol{\rho}$ is ratio of densities. For the same peripheral speeds the ratio L/T is 1 and the throughput ratio is then proportional to the square of an appropriate linear dimension, e.g. diameter of pan. The ratio of amounts of work done by the rollers on the material in the pan is M, the total number of revolutions n and the area of the pan, because the radius of revolution of the rollers is usually proportional to and a very large fraction of the radius of the pan, or ML^2n. This work is exerted on the contents of the pan which is the charge loading multiplied by the area of the pan, or $L^2 \cdot M/L^2$. Thus ML^2n is proportional to L^2M/L^2, or Mn is proportional to M/L^2. *Table 8.9* shows this to be approximately the case for figures taken from *Table 8.8* (p. 181).

For class (3), an example of which is a hammer mill, the essential quantity is kinetic energy, which has the dimensions of any form of energy of course, ML^2/T^2. Thus two hammer mills will act comminutionally in proportion to the kinetic energy they impart to a material. The power ratio of such machines will be proportional to ML^2/T^3. Since mass is proportional to ρL^3, the power ratio is proportional to $(\rho L^3 \cdot L^2)/T^3$. For constant velocity ratio L/T power ratio is proportional to ρL^2, i.e. the square of the size ratio of the mills. $\rho L^3/T$ is proportional to the capacity ratio and, again for constant velocity ratio L/T, the capacity is proportional to the square of the size ratio.

For class (4), examples of which are tumbling mills, the essential quantity is potential energy, which has the dimensions ML^2/T^2. For the scaling factor of the media to remain the same as the scaling factor for the shell, ML^2/T^2 or $(ML \cdot L)/T^2$ must remain constant, where L/T^2 are the dimensions of gravity. Thus $\rho L^2 \cdot Lg$ remains constant. But the diameter ratio of the media must be proportional to the scaling factor L and therefore for geometrical similarity ρ must vary inversely with the linear scaling ratio L.

PRELIMINARY CONSIDERATIONS FOR SELECTION OF EQUIPMENT

Before deciding upon the comminution equipment to be used, it is advisable to clarify the nature of the task to be performed. The following paragraphs are intended to provide a check list of the items which might be influential in such a decision and the answers to which could form a background against which subsequent sections can be read and used in the process of selection. Size reduction is invariably an expensive operation and careful preliminary consideration is nearly always economically worth-while.

General considerations. Is size reduction really necessary? What are the various forms in which raw material can be received? Can the material be used directly as received and how do the costs of any inefficiencies of such an operation compare with the costs of obtaining and operating and maintaining size reduction equipment? Does it pay to buy the material in a given size range? Is it cheaper to grade the material as received and dispose of the fractions outside the size range required? Can size grades be allocated to parts of the process, so that coarser grades can be used in some parts and finer grades in other parts? Can graded fractions of the material be used in other processes or profitably sold to other manufacturers?

Degree of size reduction. Why is size reduction required? Is speed of chemical reaction involved and, if so, has the influence of various size ranges on total costs been assessed? Are there other variables which could be altered more economically to achieve the required reaction rate? Is the product required within close size limits and what are these? Is it wasteful to have material too fine? Are there advantages in having a wide size range, e.g. to obtain a higher bulk density? Are stages of size reduction implicit in the process so that off-takes are required at various size ranges? Is there an optimum size at which material is released from the comminution system?

Rate of comminution. What is the rate of usage at each stage of size reduction? Is the usage steady or fluctuating in quantity? Can the material be stored before processing and afterwards? Should the process be batch or continuous? Is the tonnage to be produced large enough to support a large installation of equipment and building? Which is the most expensive in the locality concerned—labour, power or equipment? Are there limitations on the degree of complexity of equipment imposed by the standard of labour available? Which is likely to be the more economic—long hours of operation or short bursts? Can the comminuting process mesh into existing operations from the point of view of management, labour, power, equipment, buildings? Can underutilisation of existing labour, etc., be absorbed, having regard to such agreements as have been made, including those for power consumption? Will the operation support stage-by-stage reduction or is a multipurpose machine likely to be more economic? Has the whole process, from start to finish, been considered, including ancillary stages to supply and support the comminuting process?

Size of feed. What is the size distribution of the feed? What is its particle shape—is it foliated, plates, cubes, etc.? What is its homogeneity? Is tramp material likely to be present and, if so, what is its nature? What is the crystal structure and fabric of the material? Is the material a weak agglomeration or highly porous, and does it break down easily to certain sizes, larger or smaller than those being considered for the process? Has the material a natural grain size? Is size reduction to one limit likely to be much cheaper than another, and can the process be altered to accommodate such economies?

Properties of material. What are its mechanical properties, such as hardness, abrasion, toughness? Should the material be cut or torn rather than crushed or ground? Is the material fibrous? Should it be shredded? Is it friable? Is it plastic? Should the material be chilled before comminuting? What are its density, compressive strength, coefficient of friction? Can the material be gripped in a continuous compressive or nipping action? Does it tend to agglomerate under pressure? Do important properties change during comminution, e.g. colour?

Moistness. Is the feed moist? Does this impede the flow of the material? Does moisture content matter in the product? Is the material adversely affected by moisture? Is the material hydroscopic? Are the economies of wet grinding sustained for the finished product? Is the material soluble, and would there be subsequent losses in the processing? Are there water of crystallisation problems? Does the material part with its water of crystallisation during comminution? Has the material an oil or fat content? Is the material sticky or likely to become sticky during the process? Has the material thixotropic properties?

Temperature. A temperature rise must be expected in the comminution process, because the destiny of nearly all energy used in the comminution process is heat. Thus an approximate equation holds of the type:

$$K \times 3.6 \times 10^6 = \Delta T[M \times S_m + A \times S_a - R]$$

in which K is energy consumed by electric motors driving comminuting equipment (kWh); ΔT is rise in temperature of system (°C); M is weight of material treated (kg); S_m is specific heat of material (J/kg °C); A is weight of air entering system (kg); S_a is specific heat of air (J/kg °C); and R is losses due to radiation from the equipment per °C of temperature of the equipment from ambient temperature (J/°C). From this the temperature rise may be determined by inserting values in consistent units for the appropriate quantities.

What is the effect of this rise on the material? Are there likely to be stickiness, decomposition of product, evolution of offensive, corrosive or poisonous fumes, explosion or fire, change in colour, change in important electrical or magnetic properties?

Chemical reactivity. Is the material likely to corrode any of the equipment? Will it change during comminution? Is the material combustible? Will it explode or inflame? What is the effect of large volumes of air accompanying the process? What is the behaviour of the material in anaerobic conditions? What is the behaviour of the material as a dust or smoke? Is it toxic? Will it agglomerate or flocculate during the process? Will it pick up any important contaminants from the equipment? What is the limit of the contamination the product will tolerate? Are there relevant national or industrial standards for the chemical purity of the products (British Pharmacopoeia, Admiralty specifications, etc.) which are likely to be affected by the process?

Location of equipment. Where is the equipment going to be placed? Are there materials' receipt and despatch considerations? Are labour, power, water likely to be factors? Are there climatic problems? Would mobile equipment justify its extra cost? Are there environmental problems which dust and noise could affect?

After consideration of such matters, *Table 7.4* may be found useful.

After selection of some apparently suitable forms of comminuting equipment, the behaviour of the material in that kind of equipment should be discovered by actual trials: the equipment may appear to fulfil all the requirements revealed by the preliminary investigations and yet idiosyncratic effects may have been overlooked, e.g. the particular colour required may be lost in one kind of equipment and retained in another.

Table 7.4

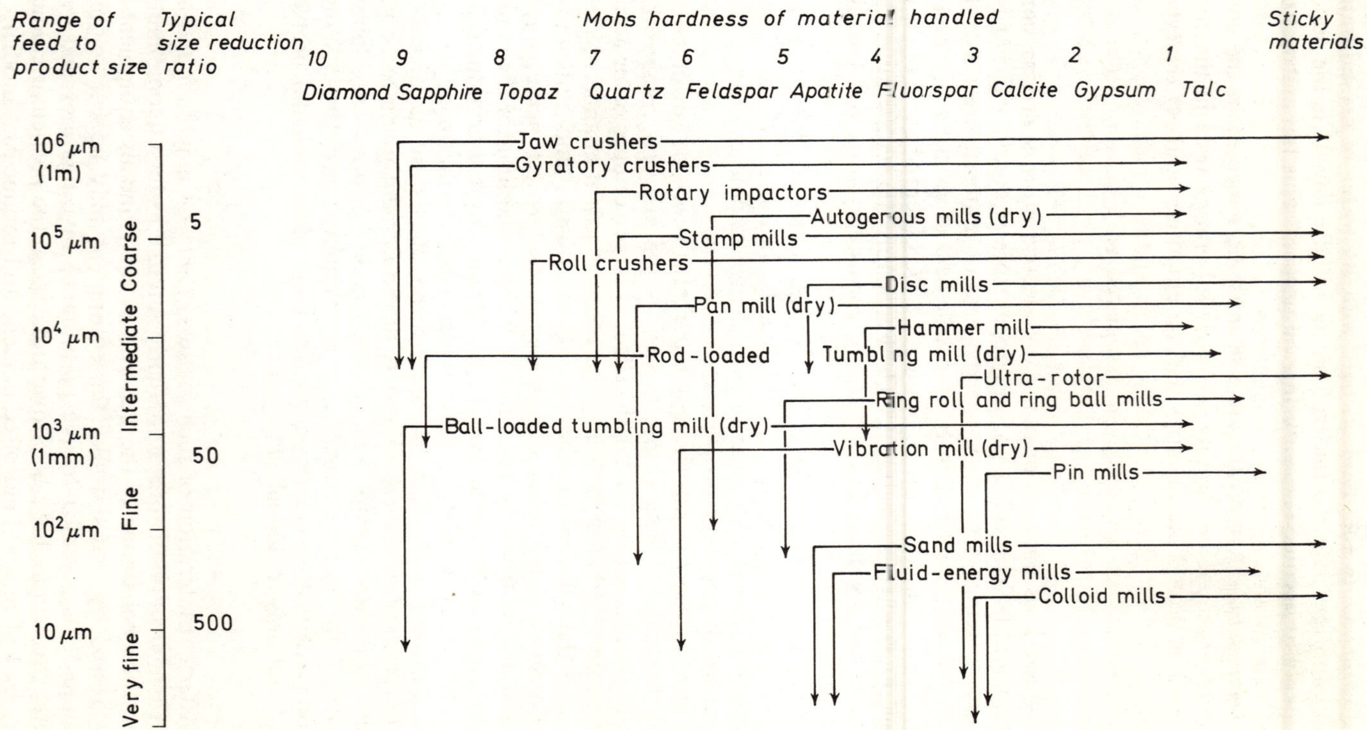
Range of feed to product size
Typical size reduction ratio
Mohs hardness of material handled
10 Diamond
9 Sapphire
8 Topaz
7 Quartz
6 Feldspar
5 Apatite
4 Fluorspar
3 Calcite
2 Gypsum
1 Talc
Sticky materials
10⁶ μm (1m)
10⁵ μm
10⁴ μm
10³ μm (1mm)
10² μm
10 μm
Coarse
Intermediate
Fine
Very fine
5
50
500
Jaw crushers
Gyratory crushers
Rotary impactors
Autogerous mills (dry)
Stamp mills
Roll crushers
Disc mills
Pan mill (dry)
Hammer mill
Rod-loaded
Tumbl ng mill (dry)
Ultra-rotor
Ring roll and ring ball mills
Ball-loaded tumbling mill (dry)
Vibration mill (dry)
Pin mills
Sand mills
Fluid-energy mills
Colloid mills

PLANT LAYOUT AND DESIGN

In considering the installation of plant for comminution, the flow diagrams should be prepared for the entire process of which comminution forms a part. Locations which are considered fixtures should be identified; these might include the place of receipt of raw materials and the place from which the finished product is to be despatched. Buildings which common-sense dictates should be retained should be identified, and any restriction which such buildings impose should be ascertained—for example, possibilities of extension; effects of nuisances associated with comminution such as dust and noise; floor loading limitations; availability of services such as power substation supply; and the maximum demand situation. The location and availability of supervision and of operators with low work loads should be investigated.

All of this should integrate to a background against which the flow diagram may be placed. Typical arrangements for the larger kinds of installation are shown in *Figures 7.1* and *7.2*. The care which such arrangements

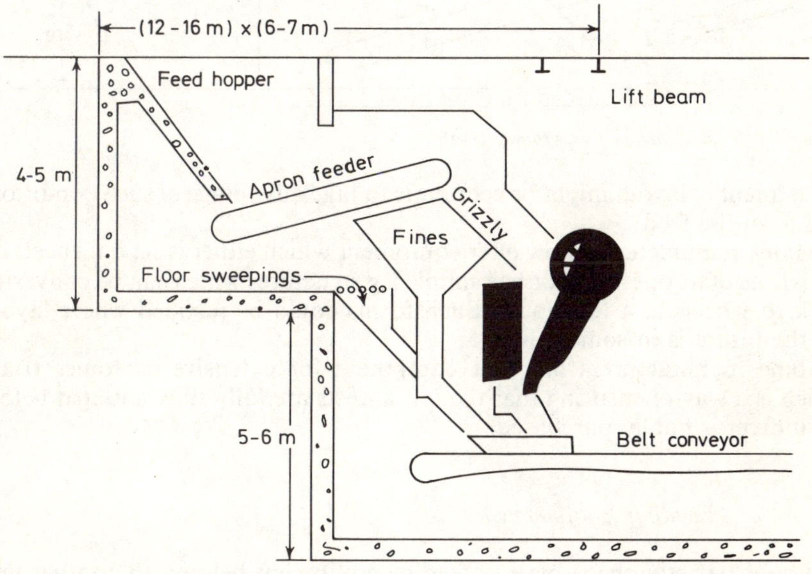

Figure 7.1. Typical layout for jaw crusher

require must also be given to smaller installations, feeding arrangements, access for plant control and maintenance, provision for the removal and substitution of the biggest plant items, and facilities for restricting spillage but providing for its disposal should it occur. For the primary size of equipment it is seen that considerable depth from feed to delivery point is proposed; although this is achievable by elevation equipment, if the site lends itself to such a construction by slope of ground or the existence of excavation, such as might have arisen from a quarry operation, then some

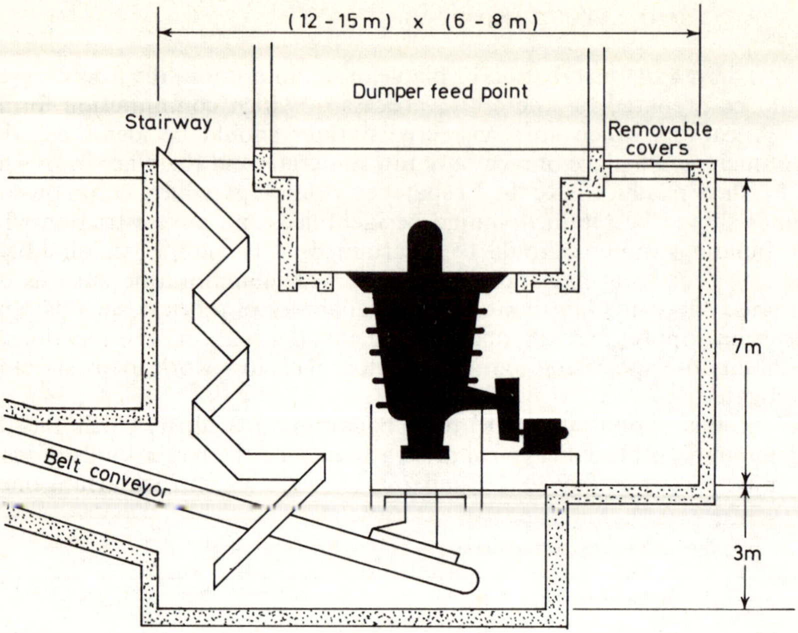

Figure 7.2. Typical layout for gyratory crusher

adjustment of layout might be economic to take advantage of such conditions and to utilise feed.

Many manufacturers now offer equipment which either is set on wheels for the whole of its operation or has simple leg structures which can be converted back to wheels in a few hours. Such forms could be justified where layout for the future is in some doubt.

Some manufacturers hire out equipment for extensive customer trials, which is a way of ensuring that the initial tests are fully substantiated before equipment is finally purchased.

Feeding Equipment

Although, strictly, the subject of feeding equipment belongs to another unit operation, the efficiency of comminuting equipment is so dependent upon smooth uniform feeding that a few notes on feeding equipment will be useful. Feeding is here considered to be the material-conveying operation immediately preceding entry to the comminuting equipment. Feeding equipment therefore abstracts material from a place of accumulation, e.g. a storage hopper.

Feeding of the larger-size material is by apron feeder, which consists of a chain-based steel-slatted conveyor belt; the slats or flights overlap and in fact form the chain links (see *Figure 7.3*) for the largest sizes. Versions for smaller size of material have the flights carried on chains. Apron feeders are

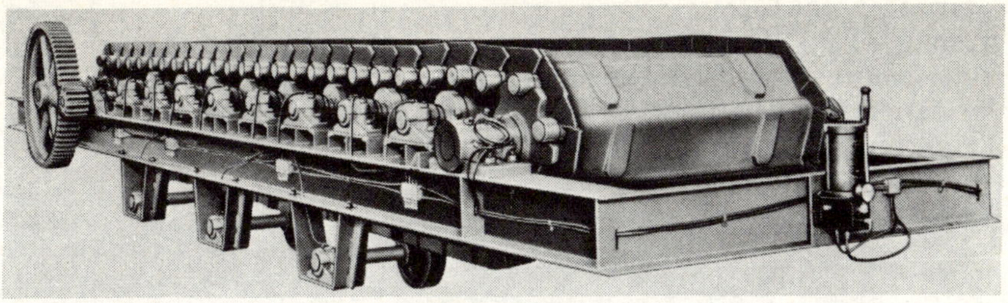

Figure 7.3. Heavy duty apron feeder with overlapping steel flights joined by hinge pin and thus distributing pull across width of apron. (Courtesy Pegson Ltd)

used for jaw, gyratory and roller crushers. When they are fitted with skirt plates, the feeders are known as pan feeders.

A device used on large crushers is the Ross chain feeder (see *Figure 7.4*). It is available in a range of sizes to suit crushers from 300 mm (12 inches) × 200 mm (8 inches) to 2.1 m (84 inches) × 1.5 m (60 inches). It consists of loops of anchor type chain usually about 5 suspended from a head drum. Its action is to restrain the material from flowing down the feed chute and then by rotation of the head drum to permit the material to flow at the speed of the chains. It is suitable for material which is flowing on a slope of about 40°, but not for sticky material.

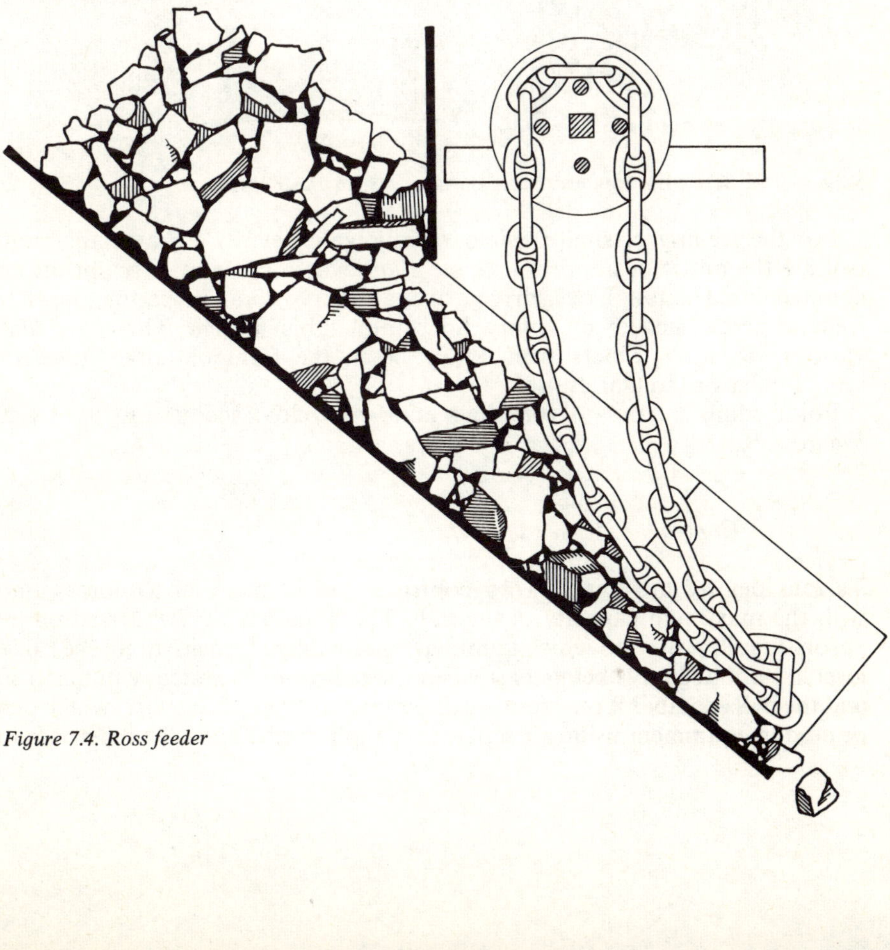

Figure 7.4. Ross feeder

Grizzlies which couple feeding with separation of fines are used in large crushing plants. They consist of a set of parallel shafts on each of which is set a number of bevelled wheels. The shafts are driven and, as they revolve, the bevelled wheels push forward the feed; at the same time the bevels permit fine material to fall between them (see *Figure 7.5*).

Figure 7.5. Grizzly screening apparatus. (Courtesy Newell Dunford Ltd)

For the feeding of smaller sizes various gate devices are available controlling the outlet of hoppers (e.g. see *Figure 7.6*). These may be sprung or counterpoised gates, slides, driven or stationary; rotating paddles or star feeders; screw feeders or rotary horizontal table feeders. There are also various vibrating feeders which may be in the form of tables, pans or semicircular or circular chutes.

For feeding slurries to tumbling mills epicycloid scoops are used (see *Figure 7.7*).

Control

Various devices are available to control input of material to correspond with the maximum capacity of the mill. These include devices based on (a) absorbed mill power, to which hammer-mill loading is sensitive; (b) the noise level in tumbling mills being least when the ball voids are filled with material and the balls do not hit one another directly; and (c) fluid pressure, which can be used in equipment using air separation and is based upon the difference in

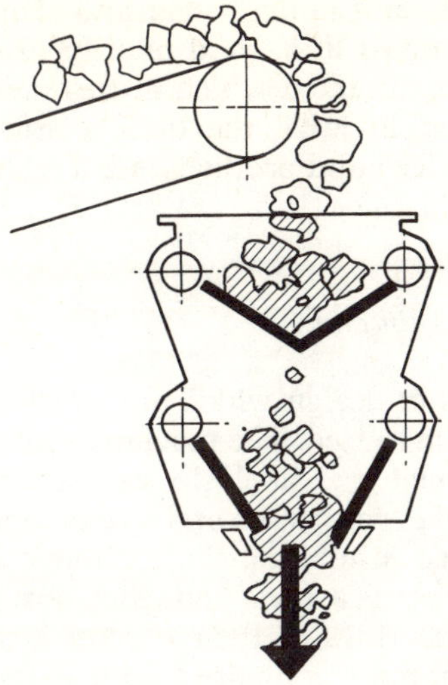

Figure 7.6. Double pendulum gate valve. (Courtesy W. C. Holmes Ltd)

Figure 7.7. Typical overflow rod mill with scoop feeder. (Courtesy Allis-Chalmers Ltd)

pressure between the air entraining much and that entraining little solid, so that when the air contains little solid in suspension, i.e. little grinding is taking place, bellows are actuated which in turn add more feed.

Sampling is important in mill control. Provision should be made for obtaining samples of the feed and product and, if classification is used, for the inlet and outlet of the classifier.

Design and Maintenance

There are many features of design which are worth considering when comminuting equipment is purchased. The machine should be robust and able to withstand all the treatment it is likely to receive, and it should be easy to adjust and maintain on the factory site and workshops. Cost of storing spares should not be overlooked in making the economic assessment.

Among design features, bearings and gear-boxes are worthy of some consideration. Rolling bearings in their correct location can undoubtedly effect considerable economies compared with plain bearings in saving of power and downtime for repairs; but they may well not be suitable in conditions of vibration, or when they have to stand idle for lengthy periods, or when their location makes removal more difficult than for plain bearings (rolling bearing manufacturers use some ingenious devices to remove their bearings and these should be investigated).

Gear-trains, besides being adequately designed for the job, should be located so that they are free from vibration, a sometimes difficult task with comminution equipment.

Wear is the great bugbear of the comminution engineer and the choice of materials is nearly always a compromise. There is undoubtedly the need for a really thorough metallurgical study of lining and moving crusher parts. Manganese steel develops good wear and impact resistance but it has to be work-hardened; during work-hardening it 'grows' and its expansion can cause considerable stresses in bolts, etc. Chrome molybdenum steels have good impact resistance but less good wear resistance. Ni-hard has excellent wear resistance but poor impact resistance, and is said to be best when poured from an electric furnace and chill-cast rather than from a cupola and sand-cast. The nodulised cast-irons show considerable promise for both impact and wear resistance.

REFERENCES

1. COGHILL, W. H. and DEVANEY, F. D., *Trans. Am. Inst. Chem. Eng.*, **34** (1938), 113
2. KELLEHER, J., *Br. Chem. Eng.*, **4** (1959), 467
3. ANDREASEN, A. H. M., *Trans. Instn Chem. Eng.*, **35** (1957), 104
4. FARRANT, J. C., *Crushing and Grinding*, 1958. London; HMSO
5. KANOWITZ, S. B., *Perry's Chemical Engineers' Handbook*, 1941. New York; McGraw-Hill
6. JOISEL, A., *Dechema Zerkleinern Symposion, 1962*, 49

8

Nipping Machines

Clear unambiguous classification of equipment is not possible. Equipment is developed by eclecticism and compromise as much as by new principles. In this section, however, a class of equipment will be studied which does not rely on the transfer of momentum, but mainly on the imposition of dead weight. To achieve a flow there must be movement and kinetic energy is therefore involved, though not as a major source of work. Invariably the comminuting effect is achieved by two surfaces gripping the material between them and submitting the material to a force imposed by one or both surfaces. The material may be fed between the surfaces under the influence of gravity or assisted by centrifugal force. In this class of equipment is included some capable of handling the very largest pieces of material and also equipment producing all but the very finest powder. The difference between the crushing surfaces of the equipment used for this spectrum of performances is mainly in diameter: the very largest pieces are crushed between surfaces of very large or infinite (i.e. flat) diameters and the smallest pieces between surfaces of smaller diameter.

Included in this class are:

		Crushing surface
Jaw	crushers	plane–plane
Gyratory	crushers	convex–concave
Roller	crushers	convex–convex
Disc	mills	concave–convex
Pan	mills	convex–plane
Ring Roll	mills	convex–concave
Ring Ball	mills	convex–concave

NIPPING

Each of the machines described in this section, by definition, uses nipping. Some machines described in subsequent sections use a combination of nipping with other actions to effect comminution. It is proposed to develop here general relationships which will have application, and to which references will be made, throughout the text.

A circular particle held between two circular surfaces is shown in *Figure*

8.1; the angle between the tangents at the two points of contact between the particle and the surfaces is called the angle of nip. There is an angle beyond which the surfaces fail to nip the particle, which would slip when pressure was applied to it by the surfaces and would thus not be broken. The maximum angle of nip at which a particle can be gripped between two given surfaces is therefore a determinant of the maximum particle diameter which can be comminuted by nipping between curved surfaces. *Figure 8.1* represents a

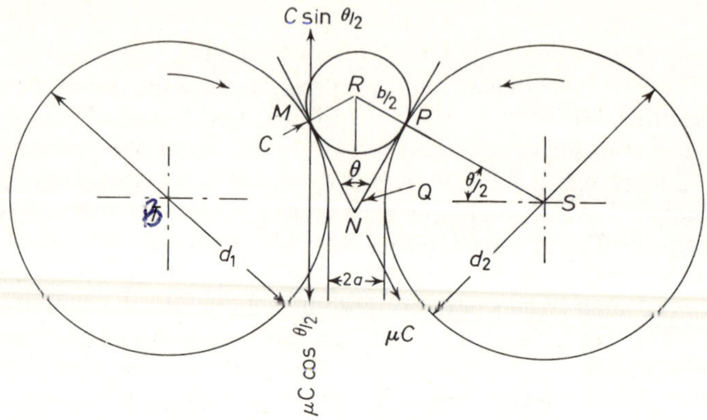

Figure 8.1. Angle of nip MNP = θ between two circular surfaces, of diameters d_1 and d_2, distance 2a apart and nipping a particle of diameter b

general condition for nipping, with the simplifying assumption that the particle is spherical or, more accurately, makes point contact with the other two surfaces and in one place on each. The circular form of the particle is idealistic: real particles will sometimes have more contact and will hardly ever be spherical, but the circular form offers the most resistance to breaking for this particular consideration.

The determination of the value of this maximum angle of nip may be approached as follows. Let the particle in *Figure 8.1* receive a compressive force C normal to the tangent at the point of contact and let the angle of nip be θ. The particle will be gripped by the surfaces if the vertical component of the compressive force C, i.e. $C \sin \theta/2$, is not greater than the vertical component of the frictional force opposing it (ignoring the contribution from the weight of the particle, which is small but would assist friction). Now the frictional force is μC, where μ is the coefficient of friction between the surfaces in the direction of the tangent to the point of contact, and the vertical component of this is $\mu C \cos \theta/2$, where $\theta/2$ is half the angle of nip. Thus, for gripping,

$$\mu C \cos \theta/2 > C \sin \theta/2 \qquad \text{or} \qquad \mu > \tan \theta/2$$

Now the static coefficient of friction has values of which the following are typical:

between steel and limestone	0.24
between steel and marble	0.17

These values would correspond with angles of nip of 27° and 20°, respectively. But this is the desired situation before the particle has moved. Once the particle begins to move, the friction is reduced. A fairly well-established relationship between kinetic friction (μ_k) and static friction (μ) is:

$$\mu_k = \left(\frac{1+0.0112v}{1+0.06v}\right)\mu$$

where v is the velocity in centimetres per second.

Thus, when a body is moving very fast, the friction can be one-fifth of the value in static conditions. The situation is also altered if the particle rolls instead of slides, but for the present purpose this possibility can be ignored.

The values of the nip angle determined from the above relationship with the static coefficient of friction must therefore be regarded as the maxima, except that feed conditions can assist the nipping. The whole subject is usually summarised by the statement that the angle of nip must be less than 30°, and in some cases this is too large an angle. Reference to circumstances peculiar to specific pieces of equipment will be made in later sections.

The method for determining the maximum angle of nip having been established, a relationship will now be developed between the curvature of the gripping surfaces and the maximum diameter of particle which can be gripped by them for a given maximum angle of nip θ.

The first circumstance to be considered is that of a particle of diameter b gripped between the curved surfaces of two dissimilar cylinders of diameter d_1 and d_2, with the smallest distance separating them equal to $2a$ (see *Figure 8.1*). Then the angle MNP is the angle of nip ($= \theta$). As MR and PR are radii of the particle, each of size $b/2$, and the angles NMR and NPR are right angles, because MN and PN are tangents, while RN is common, the triangles MNR and NPR are congruent. In particular, angles MNR and PNR are equal and therefore equal to $\theta/2$. The angle BRS is therefore $180° - \theta$ and, by the cosine rule,

$$\left(\frac{d_1}{2}+\frac{d_2}{2}+2a\right)^2 = \left(\frac{d_1}{2}+\frac{b}{2}\right)^2 + \left(\frac{d_2}{2}+\frac{b}{2}\right)^2 - 2\left(\frac{d_1}{2}+\frac{b}{2}\right)\left(\frac{d_2}{2}+\frac{b}{2}\right)\cos \angle BRS$$

But $\cos \angle BRS = -\cos\theta$. Completing the square, the right-hand side of the equation becomes:

$$\left(\frac{d_1}{2}+\frac{d_2}{2}+b\right) - 2(1-\cos\theta)\left(\frac{d_1}{2}+\frac{b}{2}\right)\left(\frac{d_2}{2}+\frac{b}{2}\right)$$

or

$$(b-2a)(d_1+d_2+2a+b) = 2(1-\cos\theta)\left(\frac{d_1}{2}+\frac{b}{2}\right)\left(\frac{d_2}{2}+\frac{b}{2}\right)$$

$$= \frac{1-\cos\theta}{2}(d_1+b)(d_2+b)$$

and

$$\frac{(d_1+b)(d_2+b)-(b-2a)(d_1+d_2+2a+b)}{(d_1+b)(d_2+b)} = \frac{1+\cos\theta}{2}$$

or

$$\frac{(d_1+2a)(d_2+2a)}{(d_1+b)(d_2+b)} = \frac{1+\cos\theta}{2} \qquad \text{(Case 1)}$$

In the special case of $d_1 = d_2 = d$, e.g. two equal rolls or balls,

$$\frac{d+2a}{d+b} = \sqrt{\left(\frac{1+\cos\theta}{2}\right)} = \cos\theta/2 \qquad \text{(Case 2)}$$

In the special case of one surface being plane or $d_1 = \infty$, $d_2 = d$, e.g. a rod or ball on a drum with a large enough diameter for its surface to be considered flat:

$$\frac{d+2a}{d+b} = \frac{1+\cos\theta}{2} \qquad \text{(Case 3)}$$

If the gripping surfaces are in contact, i.e. $2a = 0$, e.g. when two rods or balls touch,

$$\frac{d_1 \cdot d_2}{(d_1+b)(d_2+b)} = \frac{1+\cos\theta}{2} \qquad \text{(Case 4)}$$

and when the diameters of the gripping surfaces are equal, e.g. when two rods or balls of equal diameter touch,

$$\frac{d^2}{(d+b)^2} = \frac{1+\cos\theta}{2}$$

or

$$\frac{d}{d+b} = \cos(\theta/2)$$

and

$$1+\frac{b}{d} = \sec(\theta/2) \quad \text{or} \quad \frac{b}{d} = \sec(\theta/2) \qquad \text{(Case 5)}$$

JAW, GYRATORY AND ROLL CRUSHERS

The actions of jaw, gyratory and roll crushers are very similar. They rely upon two surfaces, inclined to each other and usually in the vertical plane, moving relative to each other, receiving material at the place where the two surfaces most diverge and depending upon friction for the material to be nipped until the pieces are small enough to fall through the narrowest space between the crushing surfaces. For this reason, after the three types of crushers have been individually described they will be compared.

JAW CRUSHERS

The distinctive feature of the jaw crusher is two plates which open and shut like animal jaws. The two plates are at an acute angle to each other and at least one of them is pivoted so that it swings relative to the other. Material is fed between the jaws and is alternately nipped and released to fall farther into the mouth. Depending upon the size of the bottom opening, the comminuted material eventually falls from the crusher.

Advantages

The advantages of this machine, particularly in its Blake or top-pivoted form, which is described later, is that its design permits it to be built to crush the largest lump of material, only means of construction limiting its size. If Atlas had wished to crush his burden, he might well have constructed and used a jaw crusher for the purpose. Certainly jaw crushers have been produced which can take material up to 3 m in diameter. Yet the design is so versatile that product sizes down to 5 mm and less can be produced by suitable models. Jaw crushers have the advantage over gyratory crushers that they can handle material with a high clay content.

Types

Small jaw crushers are sometimes called granulators.

Jaw crushers are classified by the method of pivoting the mobile jaw. If the jaw is pivoted at the top, i.e. in the feed area, it has a fixed receiving area and is known as a Blake crusher. If the jaw is pivoted at the bottom, i.e. in the discharge area, it has a fixed delivery area and is known as a Dodge crusher. If the jaw is pivoted in an intermediate position, it has both variable receiving and delivery areas and is known as a Universal crusher.

Essentially the mechanism required is some means of transforming circular motion into a to-and-fro one and the basis of that mechanism is an eccentric shaft to which toggles are connected. The principal methods used are shown diagrammatically in *Figure 8.2*.

The Blake form of crusher, dating from 1858 and representing a giant step forward in crushing technique, is always used for really heavy duties because of its greater leverage on the largest rock. It tends to produce pieces of less regular shape than the Dodge type. The Blake can be in the forms of the single-toggle plate (see *Figures 8.3* and *8.4*) and double-toggle types. A toggle is a plate acting as both a tie and a push bar. The power demands of jaw crushers fluctuate so much during the crushing cycle that invariably a flywheel is incorporated in the drive to smooth out the energy demands.

The single-toggle machine is driven either through a pitman (*Figures 8.2* and *8.4*) or directly by the flywheel shaft by using the shaft both as a pivot for the feed end of the mobile jaw and as an eccentric for the source of the opening and closing movement (or 'beat', as the French call it). The

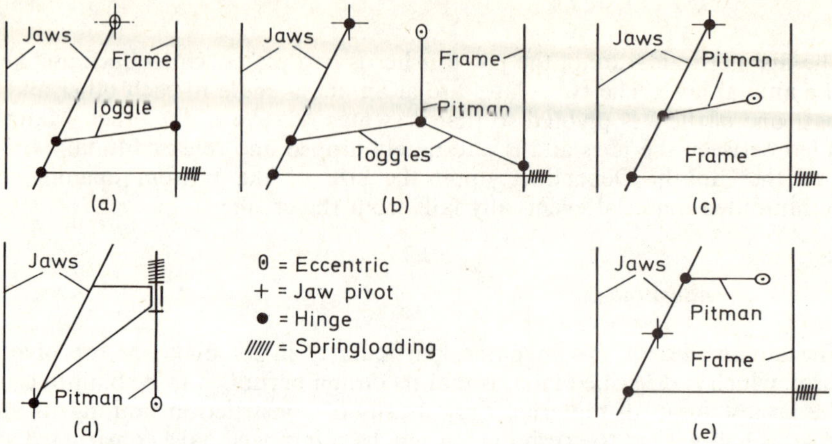

Figure 8.2. Diagrammatic representation of mechanism of jaw movement: (a)–(c) Blake, (d) Dodge, (e) universal types

modification with direct drive from the flywheel shaft is known as the overhead eccentric or Telsmith–Wheeling form.

The direct downward movement of the jaw is said to increase capacity by augmenting the flow movement of the material. The outlet end of the jaw has a grooved seating for one end of the toggle plate, the other end of which sits in a block mounted in the frame which can be adjusted so that, when it is lifted, the movement of the lower end of the jaw is increased. On some machines shims are used for this adjustment instead of the wedge. Some machines provide hydraulic equipment, so that this adjustment can be effected by the turn of a valve. Thus the whole jaw has two movements imposed upon it which manifest themselves as a circular gyratory movement. This movement is adjustable between one approximately parallel to the fixed jaw, when the toggle is set at a large angle to the horizontal, and a movement diverse from the direction of the fixed jaw when the toggle is made nearly horizontal and the crushing stroke is a minimum at the outlet point. The action is of the jaw wagging about a neutral point between the inlet and the outlet. If the size of feed is close to the size distance between the jaws at this neutral point, the jaws can be jammed. The effect of this gyratory movement is to increase the capacity of the crusher, since the material undergoes a rubbing as well as a compressive action. In this connection, the direction of the flywheel is important, since it is towards the other jaw that the energy of the flywheel augments the feeding action, whereas, if the flywheel moves in the opposite direction, the action is to disgorge the feed. Capacity can be affected by as much as 30% by the direction of the flywheel, particularly when the jaws are set at a small angle to each other.

The single-toggle machine will take material of up to 55 000 lb/in^2 (40 kg/mm^2) compressive strength. The wear and tear on the jaws of single-toggle machines is greater than in double-toggle machines, but the cost of replacement of wear parts, when materials of this compressive strength are

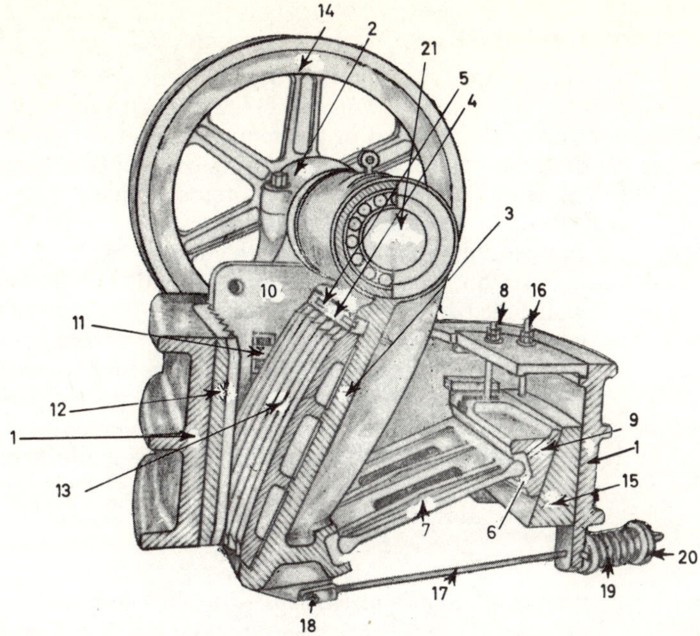

Figure 8.3. Single-toggle jaw crusher: 1, main frame; 2, main journal caps; 3, 4 and 5, swingstock, wedge and bolts; 6, 7, 8 and 9, toggle cushion, toggle, bolts and block; 10 and 11, cheek plates and bolts; 12, fixed jaw face; 13, swing jaw face; 14, flywheel; 15 and 16, adjusting wedge and bolts; 17, 18, 19 and 20, drawback rod, bolt, spring and coverplate; 21, eccentric shaft. (Courtesy Newell Dunford Ltd)

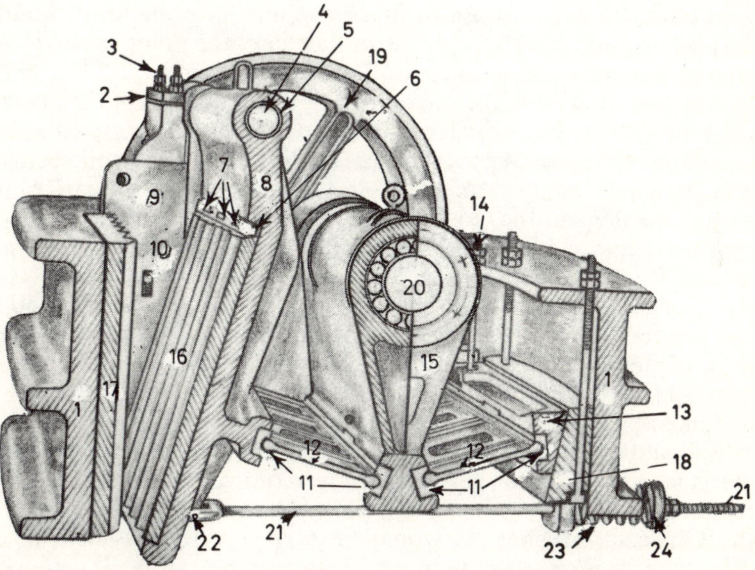

Figure 8.4. Double-toggle jaw crusher: 1, main frame; 2, 3, 4, 5, 6 and 7, cap, cap bolts, shaft, bushes, wedge and wedge bolts for swingstock (8); 9 and 10, cheek plates and bolts; 11, 12, 13 and 14, toggle cushion, toggles, block and bolts; 15, pitman; 16, swing jaw face; 17, fixed jaw face; 18, adjusting wedge; 19, flywheel; 20, eccentric shaft; 21, 22, 23 and 24, drawback rod, bolts, spring and coverplate. (Courtesy Newell Dunford Ltd)

crushed, is less than the equivalent capital cost for the double-toggle machine.

The double-toggle machine (see *Figure 8.4*) has a much simpler movement. The inlet end of the jaw is pivoted in the main frame of the crusher. A flywheel eccentric shaft has set upon it a connecting rod or plate, usually called a pitman, into the end of which are set, on opposite sides, the two toggle plates. The toggle plate on the side of the pitman away from the jaws is set in an adjustable block similar to that used in the single-toggle machine. The other toggle plate is set on the jaw near the outlet end. The movement is a simple to-and-fro one, the pitman pushing down the toggles against the frame and the pivoted jaw and lifting up to release the jaw away from the fixed one. In some machines the centre of gravity of the pitman and its movement is made to augment the closing movement so that the power given to the pitman is augmented by the weight of the pitman. The extent of the movement of the outlet end of the jaw is adjusted by the toggle block or in some cases by shims or by hydraulic devices. The double-toggle machine is suitable for tough, hard, abrasive materials.

Comparison of Single- and Double-Toggle Machines

The double-toggle machine costs about 1.5 times as much as a single-toggle machine; as the respective weights are in about the same proportion as the prices, installation costs are greater. The double-toggle machine will only justify itself for duty with tough, hard, abrasive materials.

Developments in the Blake form of machine have included modification of the pivoting and the shape of the mobile jaw: the pivot centre is approximately vertically over the centre of the bottom opening and the top part of the jaw is curved, so that the movement of the main part of the jaw is nearly parallel to the fixed jaw, which minimises plate wear. This modified machine is known as the Kue Ken. Another modification uses rollers instead of toggles, so that the movement of the wedge-shaped pitman is to push between the rollers in the closing stage and lift through them in the opening stage. Maintenance is claimed to be reduced and somewhat less power is required (20% less is claimed).

The Dodge machine rarely has a receiving area greater than 30–40 cm, but its products are of a more uniform size than are those from the Blake, because of the fixed exit size. The movement of the jaw is usually effected by means of a pitman set on a flywheel shaft, the top and bottom of the jaw being linked by arms to a spring-loaded slide on the pitman (see *Figure 8.2*). The disadvantage of the Dodge is that the leverage is weakest at the entrance to the jaws, where the largest pieces are required to be broken. It is now rarely used.

The Universal crusher, as would be expected from its design, is intermediate in performance between the Blake and the Dodge. It can operate on larger pieces than the Dodge and its products are of more uniform size than those from the Blake.

A jaw crusher which is accompanied by percussion has been reported which combines impact with compression.

Method of Rating

Jaw crushers are usually rated according to their receiving areas, i.e. in the form of width of plates × gape. The performances of the different types of jaw crushers are compared in *Table 8.1*.

Table 8.1. COMPARISON OF PERFORMANCES OF JAW CRUSHERS

Type	*Receiving area*, cm^2	*Power*, kW	t/h	*Reduction ratio*
Dodge	350	3	4	4
	500	5	7	5
	1 000	6.5	15	7
Universal	500	9	2.5	5
	1 200	35	12.5	5
	1 500	30	12.5	8
	2 000	30	8	15
	3 500	45	24	10
Blake	3 000	35	41	10
	6 500	60	70	10
	13 000	100	150	10
	20 000	140	205	10
	25 000	170		
	30 000	220	430	10
	40 000	230		

Main Frame

Main frames have to be robust and have in the past been constructed in cast-iron or cast steel and have been connected with tie-bolts. More recently they have been cut from mild steel plate and welded. Bearings are plain white metal or rolling types. Lubrication is by means of a circulating oil system.

Toggles

Toggles are made of cast-iron, cast steel or alloy steel. They are sometimes split and joined with rivets so that they form a safety release.

Jaw Plates

Jaw plates are made of chilled iron, high-carbon cast steel and manganese steel, the last being used for all but the softest materials. The plates are sometimes backed with zinc or white metal to ensure even seating. They are usually sectioned so as to be interchangeable and thus permit redistribution of plates which have suffered uneven wear. Wedges are used to hold them in place to make for ease of withdrawal and to avoid protruding bolts, which are subject to wear, leading to loosening of the plates.

Wear rates given by TAGGART[1] are:

Manganese steel	4 g per tonne of throughput
Chrome steel	6 g per tonne of throughput
Chilled iron	45 g per tonne of throughput

Plates are usually 1.3–3 times the gape in width and usually about twice the gape in length. Smooth plates provide a larger capacity than corrugated plates of the same size, but they are only suitable for soft materials; corrugated plates produce a more cubical product, of sides roughly equal to the distance between the corrugations, albeit with more fines. Convex curved plates (see *Figure 8.6*) decrease the capacity, but assist the crushing of dense materials, which tend to cake in straight-sided plates; the curved plates give a more gradual change in the distance between the jaws and therefore a more gradual crushing action, just as a smaller angle between the jaws would do, but the curved plates also permit of a wider angle for the feed.

Cheek plates are fitted on the sides of the opening to protect the main frame from wear. These usually have similar lives to the jaw plates. They are plain and bolted or wedged in place.

Size of Feed

The size of feed depends upon the opening of the mouth of the jaws, known as the gape, the width of the plates being invariably greater than the gape. The maximum size of feed is about 0.9 × gape.

Size of Product

The size of product depends upon the maximum size of the discharge and can therefore, for slabby material, be the width of the plates. Usually, however, particularly with corrugated plates, the pieces are broken down, so that the minimum distance between the plates at the discharge point is the maximum size of the product. This distance when the jaws are closest together is called the set and is determined by the size of the toggles. Blake crushers are not used for products much below 4 cm unless used in the smaller, so-called granulator form. Wear in the finer zone is accelerated as the product size is decreased. Dodge crushers, which are designed for smaller size of product, have steeper sides so that the size decrease is more gradual and the wear more evenly distributed. Examples of the size distribution of the products from a series of settings is given in *Figure 8.5*.

Method of Feeding

Jaw crushers can be fed directly from dumper trucks or by way of a feed hopper. For best capacities the chute from the feed hopper should diverge

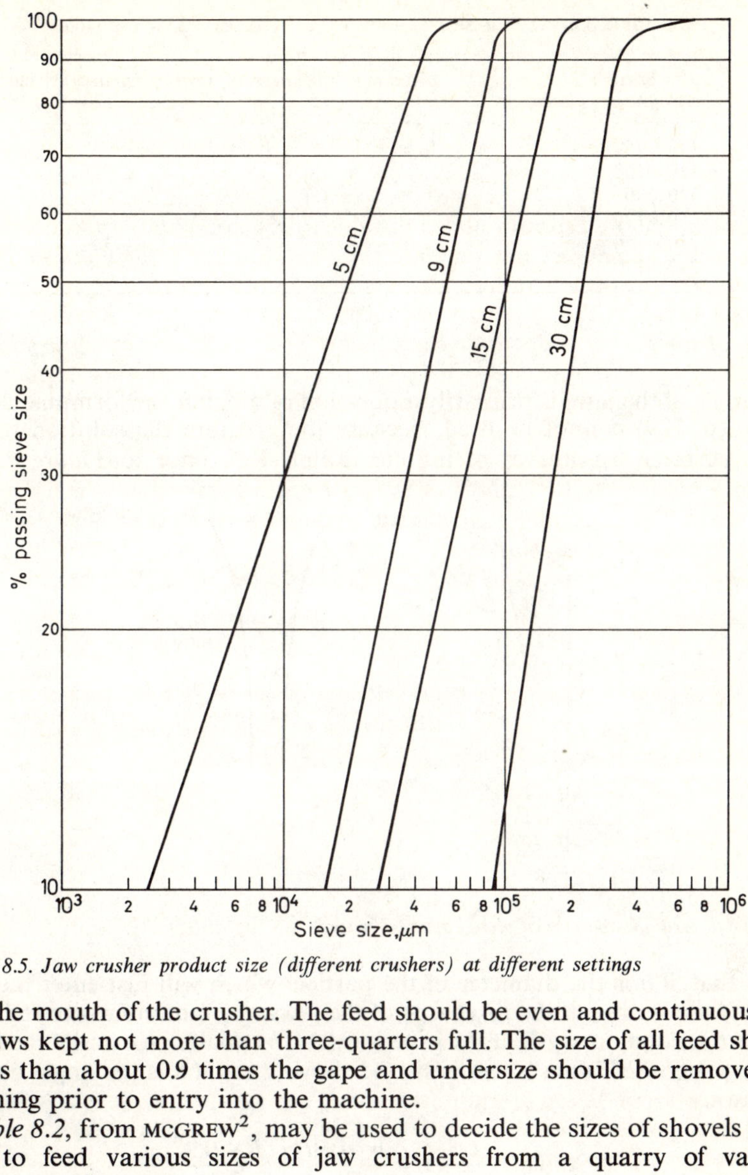

Figure 8.5. Jaw crusher product size (different crushers) at different settings

into the mouth of the crusher. The feed should be even and continuous and the jaws kept not more than three-quarters full. The size of all feed should be less than about 0.9 times the gape and undersize should be removed by screening prior to entry into the machine.

Table 8.2, from MCGREW[2], may be used to decide the sizes of shovels to be used to feed various sizes of jaw crushers from a quarry of various thicknesses of strata.

Variables in a Given Jaw Crusher

The chief variables besides the setting are the angle of the jaws, the rate of the crushing strokes and the length of the stroke of the jaw movement, or the throw.

Table 8.2. SIZES OF SHOVEL DIPPERS (m^3) TO SERVE JAW CRUSHERS

Jaw crusher straight jaw plates, cm	*Maximum thickness of stratum in quarry*, cm 60	75	90	120
110 × 75	1			
110 × 100		1		
120 × 90		1.5		
150 × 120			2	
210 × 150				3.5
210 × 170				3.5

Angle of jaws

The angle of the jaws is primarily a question of nip, but the formulae derived above (p. 129) cannot be used, because they require the solution for two plane surfaces, i.e. curves of infinite radius. Reference to *Figure 8.6* will

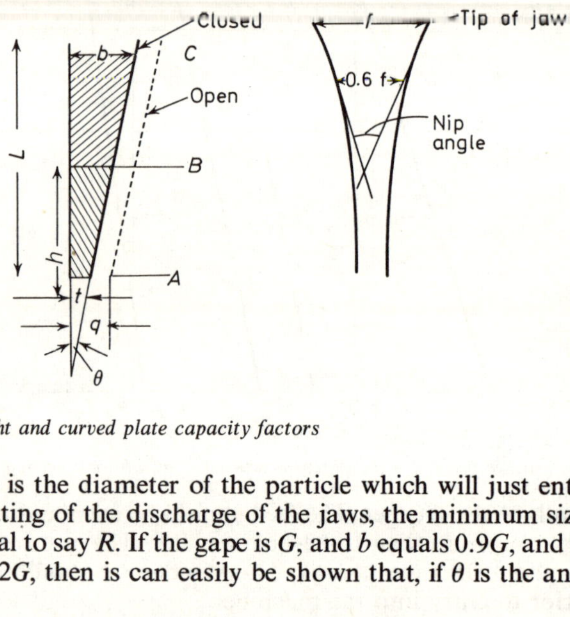

Figure 8.6. Straight and curved plate capacity factors

show that, if b is the diameter of the particle which will just enter the jaws, and t is the setting of the discharge of the jaws, the minimum size reduction equals b/t, equal to say R. If the gape is G, and b equals $0.9G$, and if the length of the plate is $2G$, then is can easily be shown that, if θ is the angle between the jaws,

$$\tan\theta = \frac{G-t}{2G} = \frac{b-0.9t}{2b} = \frac{R-0.9}{2R}$$

If values of R are assigned to this formula from, say, 2 to infinity, then the range of θ is seen to be 15.4–26.6°, values within the range of the angles corresponding to typical coefficients of friction (see p. 130). BAUMANN[3] has suggested that for the most efficient working a maximum angle of 20° between the jaws should be used; with angles greater than this there is a risk of slipping. GIESKIENG[4] suggests that the capacity of a jaw crusher increases by 3% for every 1° at which the jaws are set below 26°. GAULDIE[5] suggests, as was previously observed by HERSAM[6] and indeed noted by Gieskieng, that the capacity is related to the cotangent of the angle at which the jaws are set. Thus, if the capacity of the plant is unity when the jaws are set at 26°, according to the two investigators, the capacities at the other angles would be:

Angle (in degrees)	25	24	23	22	21	20
Capacity (Gieskieng)	1.03	1.06	1.09	1.12	1.15	1.18
Capacity (Gauldie)	1.05	1.10	1.15	1.21	1.27	1.34

Gauldie's figures would suggest an improvement nearer 6% for every 1° decrease in angle below 26°. A test[7] has shown an increase of about 2% from 27° to 20°, and then a slight falling off in capacity at 14°. HERSAM[6] observed the same effect. A figure in common use with manufacturers is 27°, although $17\frac{1}{2}°$ is quoted[7]. Gieskieng uses a nip angle factor in his formula for capacity (p. 129) such that, if θ is the angle of nip in degrees, the nip angle factor is $(26-\theta)1.03$.

Rate of crushing strokes

The rates of strokes in use vary from 100 to 350 c/min, the speed increasing inversely as the size of feed (i.e. size of equipment). However, GAULDIE[5] has shown that the speed is related to the size of product as follows. For a fragment to be less than h inches the frequency must be greater than $417/\sqrt{h}$ c/min or $665/\sqrt{h}$ if h is in centimetres. Thus, for the product to be less than 3 cm, the frequency should be greater than 400 c/min. For the size to be more than 7 cm, the frequency must be less than 250 c/min.

Throw or length of stroke of jaw movement

A large throw or stroke is said to produce large quantities of fines. Brittle materials can be broken with a small throw and materials which are inclined to be plastic or to pulverise need a large throw. Capacity is said by GIESKIENG[4] to be directly proportional to the length of stroke, except at very small strokes below about 5 mm, when output is disproportionally small. Throws are usually in the range 1–7 cm, depending upon the size of machine.

Capacity

GIESKIENG[4] suggests that capacity of jaw crushers, for straight and either smooth or corrugated jaws, is measured by

$$\text{capacity } (C) = f \cdot \rho \cdot w \cdot q \cdot l \cdot N \cdot a \cdot r$$

In this formula f is a factor dependent upon grading of feed and surface of jaw as follows:

	Smooth plates	*Corrugated plates*
Normal fines	4.14×10^{-5}	3.05×10^{-5}
Fines scalped out	3.64×10^{-5}	2.54×10^{-5}
Large pieces only	3.12×10^{-5}	2.07×10^{-5}

C is measured in small tons per hour (1 small ton = 2000 lb); ρ is the bulk density of the product in lb/ft^3; w is the width of the crushing chamber in inches; q is the open-side setting of the crusher in inches (if corrugated sides, measurement is from tip of corrugation); $l = q-t$ is the length of stroke at the lower tip of swing jaw plate; N is the number of strokes per minute; a is the nip angle factor (see p. 141)—for a curved plate, the nip angle is taken to be that formed by the lines joining the tips of the jaws and the arbitrary points where the distance between plates is $0.6\times$ feed opening (see *Figure 8.6*); and r is a factor having the form shown in *Table 8.3* (figures in parentheses are extrapolated from Gieskieng's results). When choke feeding is used, the factor r is unity.

GAULDIE[5] has analysed performance in the following way. The action of the jaws is to produce a succession of wedges (*Figure 8.6*) which are each reduced progressively to smaller sizes by compression and compaction until the first in the series is small enough to fall out of the crusher, after which the second takes its place, with the third coming on in a procession. Thus, in *Figure 8.6* the wedge bounded by the levels A and B will leave the crusher at the next backward movement of the jaw and the wedge bounded by the levels B and C will take its place to be compressed and crushed by the forward movement of the jaw and to be released and discharged, in turn, by the backward movement of the jaw. The dimensions of the fragments leaving the crusher are limited in three ways: (1) by the width of the plates, i.e. the width of the crusher feed opening (w); (2) by the maximum opening of the jaws at the discharge or product end (q); (3) by the time permitted by the frequency of opening and shutting for the material to fall freely before it is nipped again. Thus, the dimensions of the largest fragment would be:

width of feed opening (i.e. of crusher plates) $w\times$

$\times$ maximum opening of jaws at the discharge end $q\times$

$\times$ function of frequency of jaw movement cycle

These will be called the maximum width, thickness and length of the product fragment.

These ideas will be developed by reference to *Figure 8.6*, in which the movement of the jaws is first considered to be such as to maintain them in parallel. The extreme positions of the jaw are represented by a continuous line at minimum and by a broken line at maximum. The position in the throat of the jaws at which material has the same dimension as the maximum discharge opening, i.e. q, is seen to be B. If the vertical distance of B above A were h, then, for a fragment to have maximum length equal to h, the frequency

Table 8.3. VALUES OF GIESKIENG'S REALISATION FACTOR *r*

Size passing all feed / *Size of crusher receiving opening*		0.4	0.5	0.6	0.7	0.8	0.9
Truck dumping directly into converging hopper onto crusher	*single-toggle*	(0.95)	0.92	0.87	0.73	0.52	
	double-toggle	(0.93)	0.90	0.80	0.61	(0.35)	
Controlled feeding into straight-sided hopper onto crusher	*single-toggle*	(1.00)	0.99	0.97	0.95	0.90	0.76
	double-toggle	(1.00)	0.98	0.96	0.92	0.82	0.60

of backward movement of the jaws would have to allow the material to fall a distance h between being at minimum distance apart and being at maximum distance, i.e. during half the cycle time. If the frequency of the jaw movement is N c/min, then the time for one whole cycle is $60/N$ s and the greatest distance the fragment can fall freely in half a cycle is

$$\left(\frac{30}{N}\right)^2 \frac{g}{2}\,\text{ft} \quad \text{or} \quad \frac{174\,000}{N^2}\,\text{in} \quad \text{or} \quad \frac{442\,000}{N^2}\,\text{cm}$$

Thus, for a fragment to have a length h inches the frequency must be less than $417/\sqrt{h}$ c/min or, if h is in centimetres, $664/\sqrt{h}$. The value of h can also be expressed in terms of the minimum (t) and the maximum (q) size of the bottom opening and the angle between the jaws θ:

$$h = (q-t)\cot\theta$$

h can therefore be decreased by narrowing the difference between minimum and maximum openings and by increasing the angle between the jaws.

A fragment could have a greater length than h, as determined above, if its original thickness were less than t and if it were orientated to pass straight through the jaws.

The quantity issuing from the crusher in conditions such that the frequency is about the optimum $417/\sqrt{h}$ is obtained from the product of (i) area of parallelepiped bounded by parallel sides of length q and t, respectively, and a side at right angles to these of length h, so that area $= \frac{1}{2}h(q+t)$; (ii) width of plates $= w$; (iii) density of compacted mass, which can be taken to be about 0.8 times the true density of the material; and (iv) frequency of jaw movement:

$$\text{quantity per minute} = \frac{h}{2}(q+t)\cdot w\cdot\frac{417}{\sqrt{h}}\cdot 0.8\rho = 167\rho w(q+t)\sqrt{h}$$

or, more generally,

$$\frac{\Delta(q+t)\cdot w\cdot(gh)^{\frac{1}{2}}}{4\sqrt{2}}$$

where Δ is the bulk density in the section, or, in terms of the angle of the jaws to each other,

$$\frac{\Delta(q+t)\cdot w\cdot[g\cdot(q-t)\cdot\cot\theta]^{\frac{1}{2}}}{4\sqrt{2}}$$

The effect of the movement of the jaws is both (a) to crush and break and thereby to increase the voids between pieces and to decrease the bulk density and (b) to compress and compact and thereby fill the interstices between pieces and to increase the bulk density. The effect of this conflict on the bulk density of the material as it passes down and between the jaws is therefore not easy to predict, but the broad tendency must be to increase the bulk density of the material. The greater the depth of the jaws, i.e. the greater the number of wedges they can accommodate, the greater the compaction to be expected, i.e. the wider the size distribution of the product.

It is to be expected that this effect will be enhanced by the initial compaction of the feed, i.e. by the degree of choking. The effect of absolute density of the material on capacity is not found to be quite linear, as *Table 8.4* shows; but, when transposing from a low- to a high-density material, Gieskieng's equation gives conservative results.

Figure 8.7 indicates the effect of altering the settings on the capacities of some jaw crushers.

Table 8.4. EFFECT OF DENSITY OF MATERIAL ON JAW CRUSHER CAPACITY

Specific gravity (*A*)	1.1	1.9	2.7	3.0	4.4	6.2
ton/h (*B*)	100	170	380	435	640	950
B/*A*	91	90	140	145	145	153

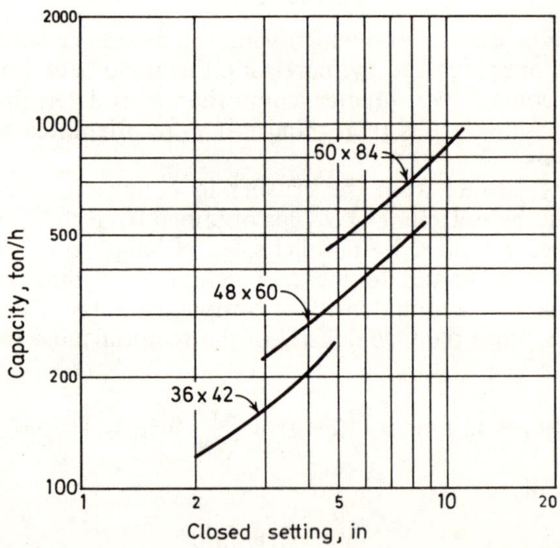

Figure 8.7. Performance of Blake jaw crusher: capacities at various settings for three different feed openings

Power Consumption

BONWETSCH[8] has suggested for economy in power consumption that two motors be used to drive the crushers; one of the two motors should be twice the power of the other and should be used for starting loads.

ROSE and ENGLISH[7] suggest that power absorbed is related to Bond's work index W_i, so that power absorbed per ton of feed is proportional to

$$W_i w\rho(q-t)^{\frac{1}{2}} \frac{q+t}{2}\left(\frac{R}{R-1}\right)^{\frac{1}{2}}$$

multiplied by a factor which, they say, is about unity for granite, quartz and trap rock, and about 0.5 for coal and coke.

JOISEL[9] maintains that the effective strength of a sphere of a heterogeneous material compressed between two 'stamps' is that of tension and that the ultimate strength is the ultimate tensile strength (and not the ultimate compressive strength, which for most materials is about ten times as large as the ultimate tensile strength). He argues that the power required in a jaw crusher is

area of 1 plate × ultimate tensile strength of material × stroke × frequency

As an example, take a Blake crusher of 24 in × 36 in opening, which for an angle of 25° means plates 56 in deep, and let frequency be optimum. Then, according to Joisel, the power required, measured in horse-power, for, say, limestone (ultimate tensile strength 0.2×10^3 lb/in^2) equals

$$\frac{36 \times 56 \times 0.2 \times 10^3}{33\,000} \times \frac{q-t}{12} \times \frac{417}{\sqrt{[(q-t)\cot 25^\circ]}} = 290\sqrt{(q-t)}\ \text{h.p.}$$

Say $q-t$, i.e. the throw, is 0.5 in; then h.p. = 200 approximately, compared with the practical 80 h.p. installed with such machines. If we use the capacity equation and assume a reduction ratio of 10 from the 24 in feed, i.e. product $t = 2.4$ in, then $q = 2.9$ in and tonnage per hour, taking $\rho = 3.2$, is:

$$\frac{167 \times 3.2 \times 62.5 \times 36 \times 53\sqrt{(2.15 \times 0.5)} \times 60}{144 \times 12 \times 2240} = 105\ \text{ton/h}$$

compared with the throughput of 70 ton/h given in *Table 8.1*.

MESSENGER[10] gives the crushing force of a jaw crusher as

$$\frac{5 \times 10^6\, W}{eN}\ \text{kgf}$$

in which $W = 0.7W_1$; W_1 is the driving motor rating in kilowatts; e is the pitman jaw stock eccentric radius; and N is the speed of eccentric shaft in rev/min.

Safety Device

Jaw crushers are protected against overload and extraneous tramp metal, etc., by the provision of such devices as (a) a torque limiter to the flywheel hub; (b) riveted toggle plate parts; (c) use of hydraulic cylinders between one end of a toggle plate and the jaw plate connection, the contents of the cylinders discharging into a reservoir if a certain pressure is exceeded and returning to the cylinder when the pressure drops back to below that pressure.

GYRATORY CRUSHERS

The gyratory type of crusher was invented about 1877 by Charles Brown and considerably developed by Philetus Gates by 1881. In its smaller or secondary form it is known as a cone crusher and sometimes as a rotary crusher.

Its essential features are a solid cone set on a revolving shaft and placed within a hollow shell which may have vertical or conical sloping sides. The angle of the apex of the solid cone may range, according to the design, from 27° for the large gyrators to about 70° for the secondary crushers. If the sides of the shell are sloping, they slope either in the opposite way to the solid cone or in the same way as the solid cone but more steeply to the vertical. Any diametrical section through them would therefore show two sets of jaws (see *Figure 8.8*). The solid cone is made to revolve eccentrically, so that any recording, at such a section, during operation would show the jaws opening and shutting like a jaw crusher. The gyratory crusher can in fact be considered as an infinite number of jaw crushers of infinitely small width.

Gyratory crushers are made to handle up to 5000 t/h and to take feeds of up to 2 m diameter.

Advantages of the machines are:

(1) They have a wide range of adjustment for product size.
(2) They will crush hard abrasive materials.
(3) They will crush wet materials provided they are not sticky.
(4) Only small headroom and space are required.
(5) Product particles have sharp angles.

However, they are not effective on soft plastic material. A useful criterion is: 'Unless it can be screened it cannot be crushed with a gyratory crusher.'

Gyratory crushers were early recognised as direct rivals to jaw crushers and in 1883[2] a contest was organised in the U.S.A. by rival protagonists, with what bases of judgement other than actual weight of rock crushed is not clear. The contest resulted in an overwhelming win for the gyratory crusher.

Gyratory crushers are for high capacity near peak operation. They occupy about two-thirds of the volume and have about two-thirds of the weight of the jaw crusher of equivalent capacity, and thus incur lower installation costs. They do not justify themselves on grounds of purchase and maintenance costs because they are a little more expensive than jaw crushers in these respects.

In its earliest design the gyratory crusher consisted of a solid cone and shaft turned out of the same piece of steel. The cone lay about midway along the shaft, leaving cylindrical portions at each end. (Nowadays the cross-section about the neighbourhood of the cone is made as large as possible while still allowing a cone to be formed and the shaft tapers away to the two ends.) The cylindrical portion at the wider end of the cone was set in a bronze eccentric sleeve which gave the tilt to the cone and was in turn set in a steel sleeve carrying a bevel gear matched with a pinion shaft at right angles which took the drive. The cylindrical portion of the other end of the cone was carried in a bearing, which in turn was carried by a two-armed spider straddling the shell. The solid cone and the shell were both lined to take the wear. Present-day machines for heavy duty do not differ essentially from this design.

The solid cone is usually called the head or the breaking head; and the shell, the bowl. The linings to the bowl are often called concaves; and the lining on the head, the mantle.

The action of the gyratory crusher, when it receives material into its top annulus, is alternately to compress and release the material gripped between the head and bowl so that it is eventually small enough to pass through the bottom annulus. As in the jaw crusher, the action can be expected to be accompanied by densification or elimination of voids, and it has been the concern of designers to regulate the densification so that the desired size distribution is achieved without unnecessary stress and wear on the machine parts.

The original form of the crusher can be represented by the sketch in *Figure 8.8a.* To make crushers of larger capacity meant making larger versions of this form. Changing the reduction ratio presented problems. For simple symmetry the angle of the cone is equal to the angle of nip

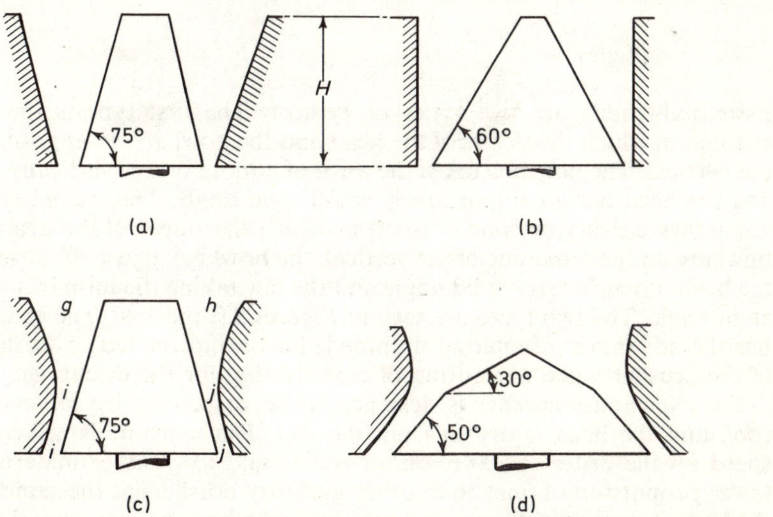

Figure 8.8. Development of gyratory crusher for secondary crushing: (a), (b) bigger reduction ratio for same height of equipment achieved by making cone base angle smaller; (c) avoiding over-densification by use of curving liners; (d) further avoidance of over-densification by increasing circumference at discharge

between the sides of the cone and the concave, i.e. about 30°. To increase this angle of the cone means making the angle of the concaves with the vertical more steep because the angle of nip cannot be increased without impairing the efficiency of the machine. Thus, to increase the reduction ratio for a machine of given over-all dimensions, the only way seems to be to increase the face of the cone, but to do this to a cone with an acute angle means making the machine higher, i.e. larger. This was the problem set to designers in adapting the gyratory crusher for the large reduction ratios required for secondary crushing. The problem was solved initially by making a cone with a large solid angle and thus increasing its length of face and then sloping the concaves at about 30° to this cone length and changing the erect V to what cowboys call the lazy V, for the position of the nipping faces (see *Figure 8.8b*).

A concomitant problem was that of densification. A point in densification was reached which was near the no-void condition beyond which the machine could compress no further and which did not necessarily coincide with the degree of size reduction required in the product. This was overcome by designing concaves whose angle relative to the head decreased, i.e. they were made to curve so as to be more nearly parallel to the head, and thus densification was controlled while the material could still be treated in the crusher (see *Figure 8.8c*). The problem was further overcome by flaring out the base of the cone so that the diameter increased, and therefore the circumferences of the annulus were increased, and densification was relieved by allowing the material to occupy a larger cross-section (*Figure 8.8d*). Thus was the cone crusher developed.

Types

Essentially there are two types of gyrator. The first type is the primary crusher, in which the slopes of the head and the bowl are on opposite sides of the vertical. The jaw thus takes the form of a more or less vertically placed V and the head has a comparatively small solid angle. The second type is the secondary crusher or cone crusher, in which the slopes of the head and the bowl are on the same side of the vertical, the bowl being just 30° or so steeper, the head having a large solid angle and the jaw taking the form of a V sloping at an angle. The two types are seen in *Figures 8.9* and *8.10*. The cone crusher has the advantage of offering unimpeded access to the feed over the mouth of the crusher while permitting of close setting for the discharge.

The Newhouse crusher is designed to be suspended by cables from the roof, and the head is driven from the top. The machine is driven at high speed (of the order of 500 rev/min) and is said to produce material with a lower proportion of fines than other gyratory crushers at the same settings. Because it is direct-driven and has no gear-box, it is lighter than other machines.

The Gervais crusher is a special type of gyratory crusher in which the head is divided into layers, each of which is separately actuated. The central

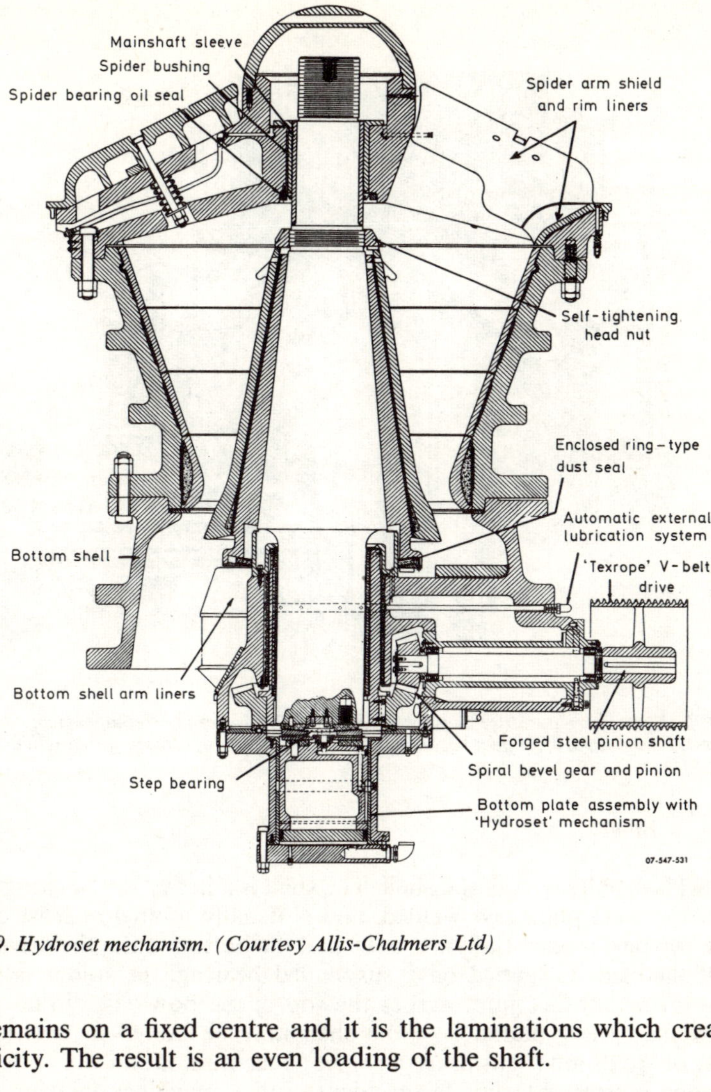

Figure 8.9. Hydroset mechanism. (Courtesy Allis-Chalmers Ltd)

shaft remains on a fixed centre and it is the laminations which create the eccentricity. The result is an even loading of the shaft.

Method of Rating

The machines are rated by: (1) the bowl diameter at the discharge point and the gape; (2) the bowl diameter at the feed point and the gape; (3) the circumference of the bowl at the feed point and the gape; (4) the maximum diameter of the head and the gape. For gyratory crushers the circumference of the bowl at the feed point is about 10 times the gape and thus the diameter at the top of the head is about equal to the gape.

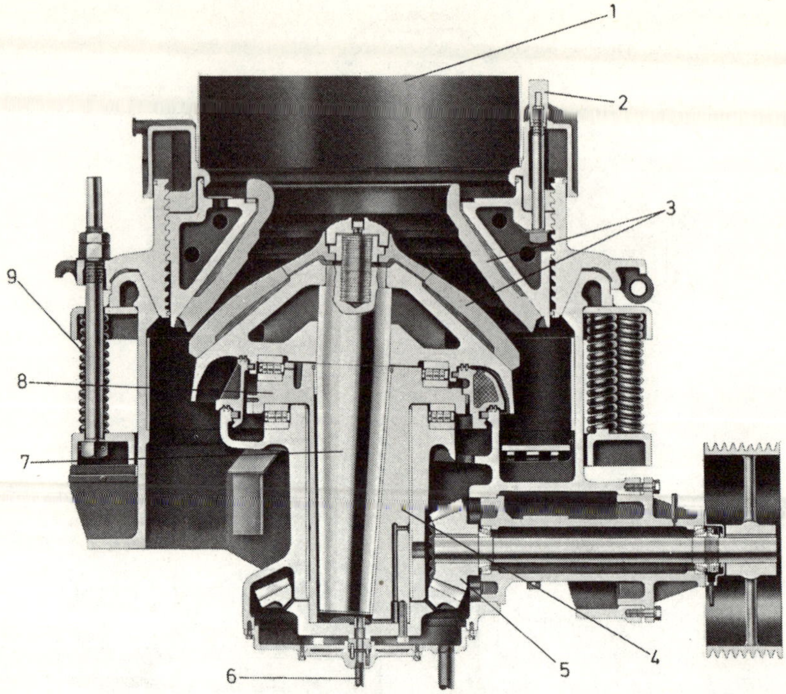

Figure 8.10. Gyrosphere gyratory crusher: 1, hopper; 2, discharge size adjustment; 3, crushing members; 4, eccentric; 5, drive gear; 6, oil supply; 7, main shaft; 8, cam; 9, spring relief. (Courtesy Pegson Ltd)

Bowl

The bowl is usually part of the shell. The shell is a heavy steel casting or may be cut from steel plate and welded, and is usually jointed at least once, so that the top part is essentially the bowl and the bottom takes the drive for the head. If the shaft is carried on a suspended bearing, the spider taking the bearing forms another joint across the top of the bowl. The joint between the bowl and the remainder of the shell may be bolted in position with springs or hydraulic cushions to give accommodation for choked or extraneous matter entering by accident; other machines incorporating a hydraulic system for head position control provide relief for the head when excessive pressures develop through that system. Again, some machines have plain bolted joints and rely upon the robust construction and some manufacturers even guarantee it against damage by tramp iron.

The bowl is protected by concaves.

Concaves

Concaves are made of manganese steel and in various sections, as has been previously discussed. For smaller crushers the concave is made in one continuous ring and held in place by bolts through lugs or other fastening devices. The larger machines have sectionalised concaves which are known as staves. These are wedge-shaped and either rest on a ring fitted between the upper and lower shell or are, again, bolted by means of lugs. The concaves are in all cases backed by Babbitt metal, zinc or, more recently, plastic-setting substances. Some manufacturers make the whole of the upper shell, containing the bowl, reversible so that the concaves can be reversed without disturbing the backing by unbolting and inverting the upper shell.

Head

The head is usually integral with the shaft, the whole, for the largest machines, being constructed from giant forgings. A mantle sits on the head and is fastened in place usually by means of threads pitched so that they are self-tightening during operation or by feathers and keys which allow the mantle to slide and tighten with agitation. On smaller machines the mantle is cast in place with Babbitt metal or zinc or with plastic-setting cements. Larger machines have a mild steel inner mantle onto which the manganese outer mantle is fastened with or without a zinc interface. Sometimes the mantle for a large machine is made in two or more rings which sit one on top of the other, the smallest ring taking the tightening device; this makes it possible for areas taking the greatest wear to be replaced without replacing other parts.

The mantles are made of manganese steel and are sometimes corrugated, which is said to improve the coarseness of the size distribution of the product. Sometimes the mantle is flared out at the bottom so as to make it bell-shaped, which is said to assist crushing of material having a tendency to choke.

Eccentricity

Eccentricity is still induced by very much the same method as that originally used for heavy-duty machines: an eccentric sleeve is fastened onto the end of the shaft and the revolution of the shaft in a bearing permitted by the suspended bearing at the other end of the shaft tilts the shaft and confers eccentricity. A long shaft is required for such an arrangement to accommodate the sleeve below the bevel gear. A development (see *Figure 8.11b*) was to fit the sleeve along the length of the shaft through the head, leaving only a short end of the shaft to be accommodated in a fixed bearing; this allowed of a shorter shaft with a saving of height and weight, on the whole machine, and thus economies in space requirements and installation costs. For machines with somewhat less arduous duties, e.g. secondary crushing, the spider was eliminated and with it a slight obstruction to the feed point (though spiders are claimed to deter overfilling). Some makers incorporate a wedge-shaped cam collar as well as the long sleeve and both the head and the main shell have roller bearings incorporated so that the cam runs between them (see

Figure 8.11c). The effect is that eccentricity is conferred both by the cam and by the sleeve, and the thrust is therefore distributed.

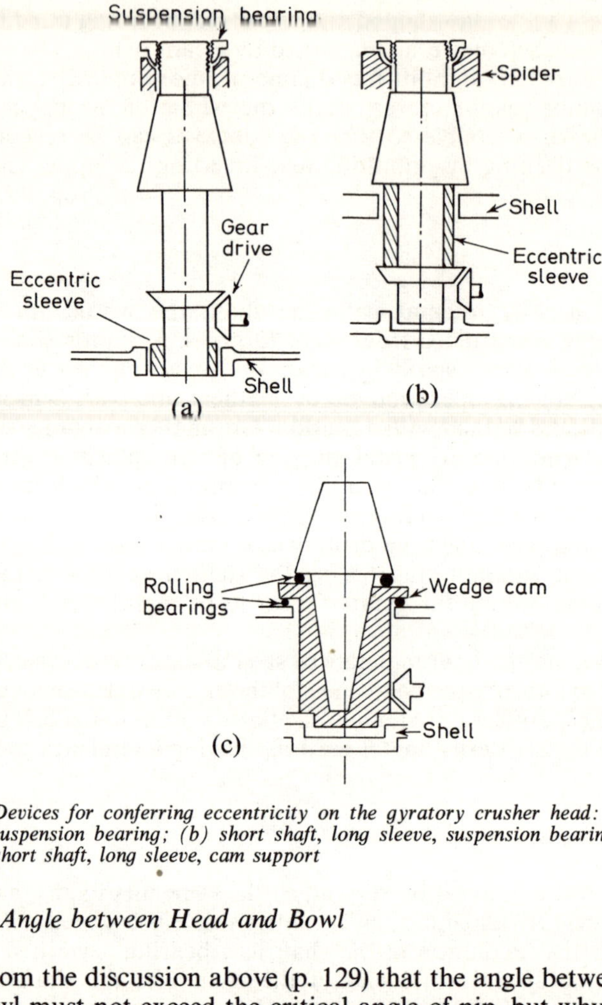

Figure 8.11. Devices for conferring eccentricity on the gyratory crusher head: (a) long shaft, short sleeve, suspension bearing; (b) short shaft, long sleeve, suspension bearing, bottom fixed bearing; (c) short shaft, long sleeve, cam support

Angle between Head and Bowl

It is clear from the discussion above (p. 129) that the angle between the head and the bowl must not exceed the critical angle of nip, but why this critical angle can be slightly larger than the jaw crusher critical angle is not known. In fact manufacturers use the range 22–30°, with most favouring 29–30° on the larger machines, in which the weight or head of feed is said to help friction.

Adjustment of Setting and Throw between Bowl and Head

The shortest distance between bowl and head at any setting is called the closed side setting and is often designated by the initials C.S.S. The distance is adjusted by moving the vertical setting of the bowl, usually a bolt or shim adjustment; or in the case of a head with a suspended bearing by adjustment of this bearing; or in the case of a machine with a hydraulic system by means of the system. The throw is altered by adjusting the eccentricity of the head, which means a substitution of sleeves.

The bowl-to-head setting should be adjusted regularly to accommodate the wear on the bowl and head linings. A setting as frequent as every 2 days may be advisable. With a hydraulic system adjustment to the position of the head is readily made by the operator at will.

Speed of Rotation of Head

The speed of rotation of the head is usually fixed for any given machine. Some cone crushers, particularly the Newhouse, which is driven from the top and can be suspended from the roof, rotate at high speed and function more like impact mills. GAULDIE[5], using arguments similar to those which he used in discussing jaw crushers, deduces that the speed of rotation for the product to have size less than h inches should be not less than

$$N = \frac{K(\sin \psi - \mu \cos \psi)}{\sqrt{h}} \text{ cycles/min}$$

where ψ is the inclination of the cone size to the horizontal, μ is the coefficient of friction of the material concerned, and K is 417 or 664, depending upon whether h is measured in inches or centimetres, respectively.

Method of Feeding Equipment

The feed should be above the minimum product size setting; fines choke the crusher and thus scalping is advisable. Gyratory crushers are sometimes operated buried in the feed. *Table 8.5*[2] suggests shovel sizes to be used in

Table 8.5. MAXIMUM SIZES OF SHOVEL DIPPERS (yd^3) FOR USE WITH DIFFERENT SIZES OF GYRATORY CRUSHER AND THICKNESSES OF ROCK STRATA (in)

Gyratory crushers size and fittings	*Maximum thickness of strata in quarry*, in 12	18	24	30	36	42	48	or more
20 in straight concaves	1½	1¼						
20 in non-choking concaves	1¼							
30 in straight concaves		2½	1¾					
30 in non-choking concaves	2½	1½						
36 in straight concaves		4½	2½	1½				
36 in non-choking concaves		3	2					
42 in straight concaves			4½	3	1½			
42 in non-choking concaves		4½	4	2½				
50 in straight concaves				4½	3	2½		
54 in straight concaves					4½	3		
60 in straight concaves						4½	3	

Table 8.5 *(Continued)* DUMPER SIZES TO FEED GYRATORY CRUSHERS WITHOUT CAUSING BRIDGING IN CRUSHER CHAMBER

Crusher size	*Car capacity*
20 in	4–5 ton
30 in	6–8 ton
36 in	8–10 ton
42 in	8–10 ton
50 in	10–12 ton
54 in	12–15 ton
60 in	15–20 ton

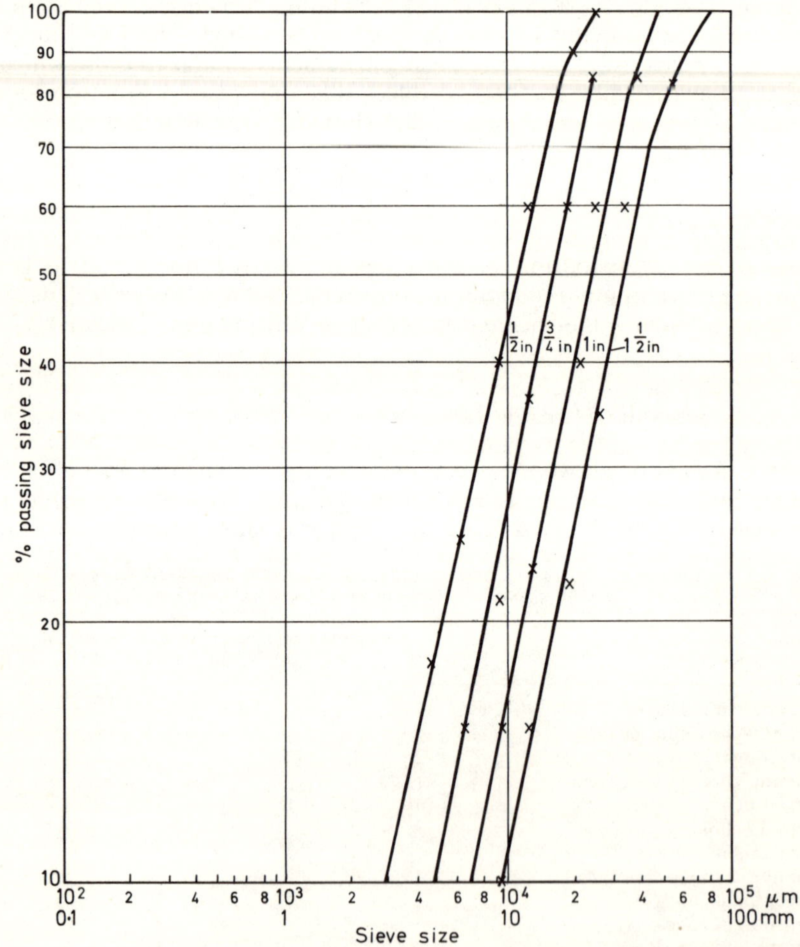

Figure 8.12. Gyratory crusher performance (for indicated closed side settings) in open circuit

conjunction with various sizes of crushers for different thicknesses of rock strata. *Table 8.5* also shows dumper truck sizes which should be used to feed these crushers so that bridging is not caused in the crusher chamber.

Secondary crushers which do not have a spider sometimes have a horizontal disc attached to the top of the head, by means of the shaft, onto which the feed drops and is distributed around the gape.

A cylindrical shell is usually fastened on the outside of the top part of the machine and feed is delivered by a belt or box conveyor or dumper to drop into this cylinder.

Size of Product

The size distribution of product for various closed side settings on a particular, but typical, machine is shown in *Figure 8.12*. MCGREW[2] states that less fines are produced by high-speed machines such as the Newhouse.

Capacity

GAULDIE[5] proposed a treatment similar to the one he proposed for jaw crushers. Instead of wedges, the elements of material in process of being crushed can be regarded as rings or helices of which a single section can be regarded as similar to that shown in *Figure 8.6*, except that it is inclined on its side at an angle ψ. Using his formula for optimum frequency and taking account of the fact that the slope of the sides of the cone head modifies the force of gravity and introduces the effect of friction on the movement of the material, we find that g, the force of gravity, in the jaw crusher relationship becomes $g(\sin\psi - \mu\cos\psi)$, where ψ is the base angle of the cone (i.e. the slope of the sides of the cone to the horizontal) and μ is the coefficient of friction between the material and the cone and bowl sides. Thus the minimum cycle frequency for the product to have size h units becomes

$$N = \frac{1}{2}\sqrt{\left[\frac{g(\sin\psi - \mu\cos\psi)}{2h}\right]}$$

and the mass of the section of this depth leaving the crusher is $2\pi r\Delta AN$, in which r is the radius of the centre of gravity of the section, Δ is its bulk density, A is the cross-sectional area and N is the frequency.

Thus if q is the maximum size of the opening at the bottom discharge between the bowl and the head and t is the minimum of this distance, A becomes $[h(q+t)\sin\psi]/2$, correcting h for a vertical measurement, and so the whole equation for optimum throughput is

$$\frac{2\pi r\cdot\Delta\cdot h\cdot\sin\psi(q+t)[g(\sin\psi-\mu\cos\psi)]^{\frac{1}{2}}}{4\sqrt{(2h)}}$$

or

$$\frac{\pi r\cdot\Delta\cdot\sin\psi(q+t)[gh(\sin\psi-\mu\cos\psi)]^{\frac{1}{2}}}{2\sqrt{2}}$$

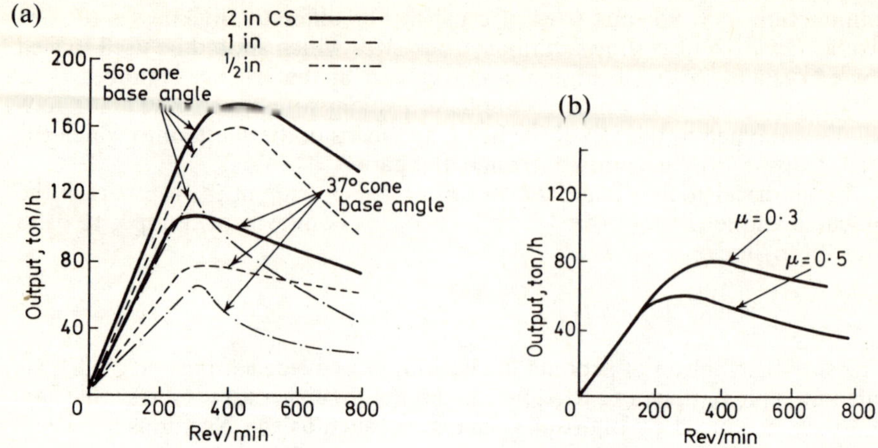

Figure 8.13. (a) Effect of closed side setting (CS) and cone base angle on capacity of 36 in *gyratory crusher. (b) Effect of coefficient of friction (μ) on output from 36* in *gyratory crusher, 37° cone angle and 1* in *CS. (After* K. GAULDIE[6]*)*

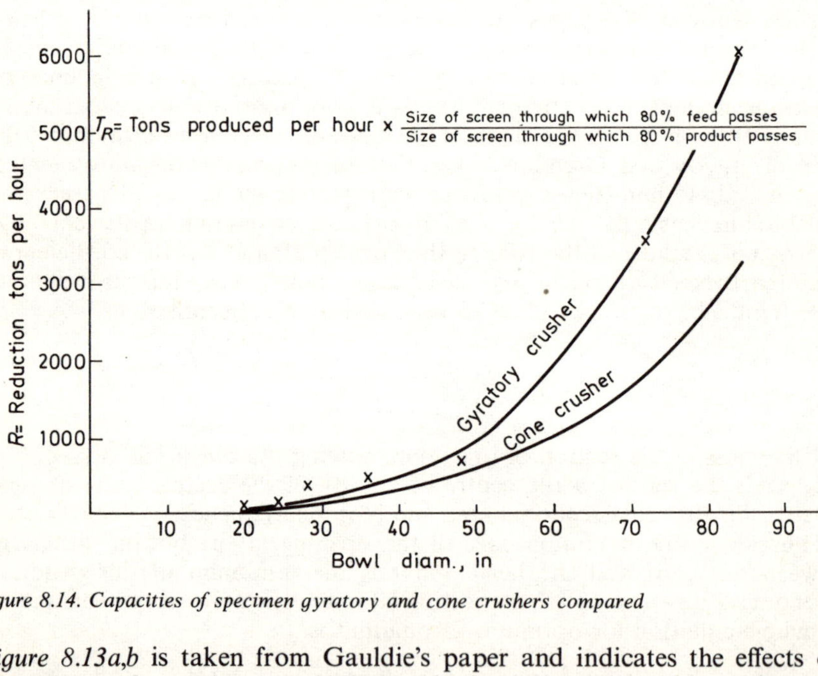

Figure 8.14. Capacities of specimen gyratory and cone crushers compared

Figure 8.13a,b is taken from Gauldie's paper and indicates the effects on throughput of cone base angle ψ, speed of rotation, close side setting and coefficient of friction μ. It is seen that maximum cone angle, largest close setting and lowest coefficient of friction at optimum speed of rotation give maximum throughput, or, broadly speaking, minimum reduction ratio to material flowing freely gives maximum capacity.

Capacities of two specimen gyratory and shorthead cone crushers are shown in *Figure 8.14* for various bowl diameters. The capacity is expressed in the form of the quotient

$$\frac{\text{ton/h}}{\text{reduction ratio}}$$

Power Consumption

JOISEL[9] suggests that the same kind of calculation can be used for gyratory crushers as for jaw crushers (see p. 145), except that double the surface of the cone crusher head is taken.

TAGGART[1] suggests that idling power consumption is about one-third of the peak load.

ROLL CRUSHERS

The roll crusher was probably first used, in the form of three rolls, to crush grain and beans and later to crush sugar-canes. It is said to have been invented in 1449 by Pietro Speciale, Prefect of Sicily, who used a wooden type with vertical axes. George Sitwell of Renishaw, an ancestor of the well-known literary family, devised in 1652 rolls in the form of wooden cores covered with iron plates. Rolls were first made to be used in horizontal form by John Smeaton in the eighteenth century. Initially they were driven by water-wheels but they were in use in the steam-driven form in Cornish mines in 1806[11] and are probably the first comminuting equipment to be power-driven. Thomas Alva Edison became involved in comminution when he developed 6 ft diameter rolls with knobbed plates.

Roll crushers are of many types (*Figure 8.15*) and some, particularly the single-roll form, can resemble impact breakers. Their essential features are that they consist of at least one cylinder rotating on its principal axis and nipping material particles between its surface and another surface to compress and break them. They do not rely upon kinetic energy to do this and are relatively slow-moving compared with the impact machines.

Roll crushers may consist of one, two, three, four, six or possibly more cylinders, which are usually horizontal and revolve either towards each other or towards a plate. If they consist of more than one roll, the diameters and speeds of the rolls may differ. The distance between the rolls is usually adjustable and in some types the rolls are at an angle to each other. The rolls may be gear-driven, but this type provides limited adjustment of distance between rolls, and they are now more usually driven by V-belt drives from separate motors. A single-roll machine uses a breaker plate as its other nipping surface. A multi-roll machine uses rolls in pairs so that material can be successively passed between pairs. Alternatively, as in the triple form, the rolls are arranged so that their centres form a triangle with one roll at the top; material is then fed under the top roll and over the other ones. Conical rolls or rolls with axes at an angle to one another have been devised to give a particular particle-size distribution.

Depending upon their form, roll crushers are also known as crushing rolls, roller crushers, Cornish rolls, giant rolls and slugger rolls. In the

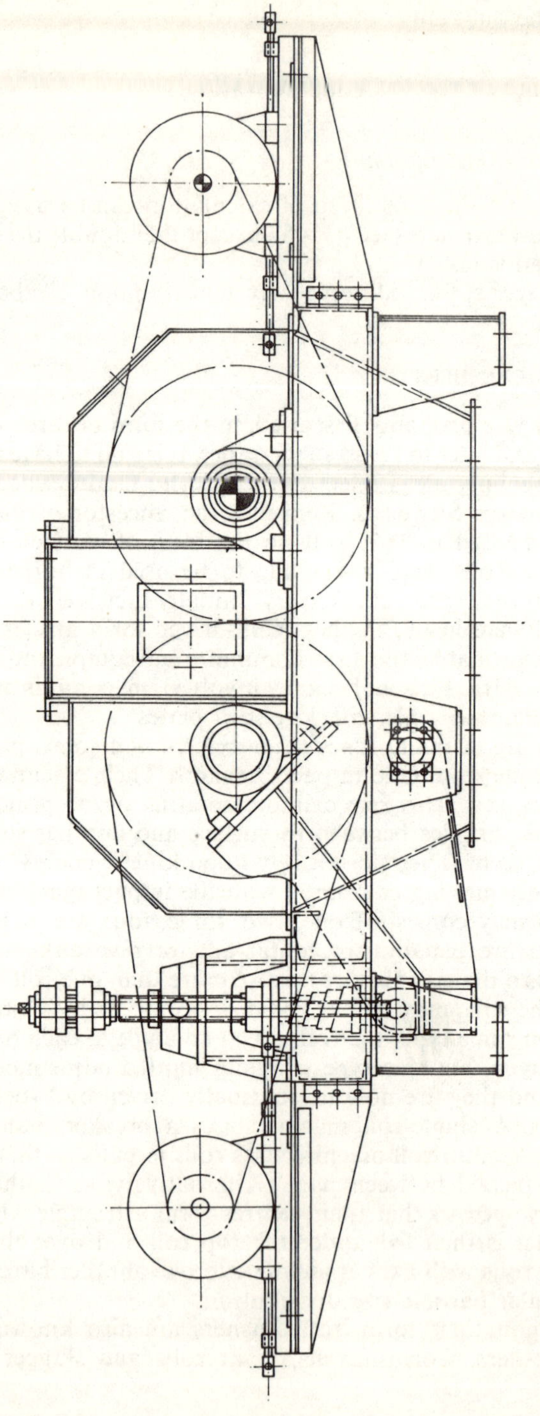

Figure 8.15. General arrangement of roll crusher. (Courtesy Aulman and Beckschulte)

single-roll form the roll crusher is known as a roller and breast mill, a kibbler or a sledging roll, depending upon details of the roll design.

Machines having more than two rolls are nowadays rare, but multi-roll systems have been devised in the past which have allowed the production of material within size ranges with a top-to-bottom size ratio of 4 or less and such that 85% of the product was within this range[11]. In a system proposed by W. F. Carey, four rolls were used, but adjustments to obtain a smooth and continuous flow were not considered economic.

The action of the roll crusher is very similar to that of the jaw crusher. Instead of the jaws moving in a horizontal direction relative to each other, the surfaces of the cylinders (or, in the case of a single roll, the surfaces of the cylinder and the breaker plate) move in combined horizontal and vertical movements relative to each other, so that the material being crushed is continuously and steadily induced into the smaller part of the mouth by the circular movement of the cylinders and the friction between the material and the cylinder surfaces. The roll crusher does not have the periodicity of the jaw crusher and its parts do not have to suffer the oscillating stresses or power demands. The roll crusher's power demands are much lower than those of both the jaw and gyratory crushers.

Many types of roll crusher, particularly the single-roll, have projections on the roll surface and in addition to the compressive action also have a piercing action.

Roll crushers incorporate a relief mechanism so that tramp and extraneous materials do minimum damage to the rolls. The mechanism is generally built on mechanical-spring or hydraulic principles and allows one roll or the hinged breaker plate to move when excessive pressures are created.

Advantages

Advantages of the roll crusher are:

(1) It will handle wet, sticky or frozen feeds, or material which easily packs.
(2) The product can be controlled so that there is a high yield within a narrow size range: within a top-to-bottom size ratio of 4, 85% yields are possible.
(3) The power consumption of the machine is low compared with that of many other machines (see p. 60).
(4) Rolls can produce a product in an intermediate range of about minus 1 cm to about 3 mm more economically than most other machines.
(5) The surface of the rolls can be heated or cooled by the circulation of fluid within the interior of the rolls.

Disadvantages

Disadvantages of the machine are:

(1) Only a relatively small size reduction is achievable by practical roll sizes, a ratio of 4 being usually accepted (but see p. 94).

(2) Materials having a foliated structure will break into leaves unless rolls with lugs or intermeshing teeth are used.

Construction of Rolls

The prime requirements for roll surfaces besides hardness and abrasion-resistance are that they be grindable, to maintain smoothness, and weldable,

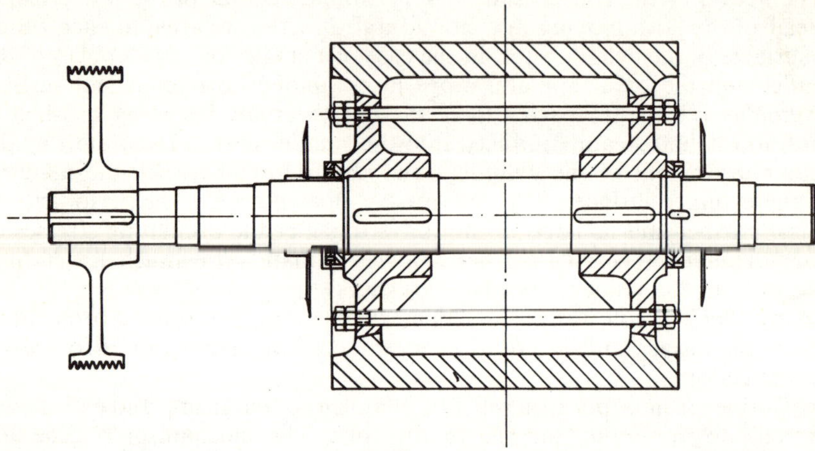

Figure 8.16. Construction of roll for roll crusher. (Courtesy Aulman and Beckschulte)

to facilitate repairs. Materials fulfilling these requirements are such alloy steels as those with manganese, high carbon, chrome manganese, chrome molybdenum. Ni-hard or chilled iron. Sometimes one roll is made of a different material from another; for example, one roll may be rubber-covered and the other exposed steel, as in the case of rolls for crushing asbestos, such a combination being found to give fibres which are not shortened.

The rolls for smooth roll crushers are made up from tyres which may be keyed onto a shaft made with a sufficiently large diameter or, with a shaft of smaller diameter, held together with hub cones at the extreme ends of the assembled set of tyres and clamped together with tie-bars running through the cones and bolted on the outside (see *Figure 8.16*). The cones are keyed onto the shaft. With some of the more massive single rolls the cylinders may be welded or they may have octagonal cores onto which wear-plates carrying the knobs may be splined. Each roll carries a pulley on its shaft for multi-V rope drive from a separate motor for each roll. The shafts are made of high carbon steel.

The roll surface may be smooth, corrugated, grooved, beaded; have intermeshing teeth, fingers or lugs (see *Figure 8.17*); or be fluted, waffled (like waffle irons) or serrated. The non-smooth surfaces are for the coarser products and may be used in a series which terminates in a crusher with smooth surfaces for the finest product. Grooved rolls are sometimes used to eliminate hard and coarser materials from softer and finer ones, e.g. stones in

Figure 8.17. Toothed rolls. (Courtesy Aulman and Beckschulte)

clay. The grooves are helical and carry the hard pieces along to the end of the roller, where they may be discharged. If material has a disposition to be foliated and cube-shaped particles are required, rolls with intermeshing teeth should be used for the bigger sizes and waffled or serrated rolls for the smaller sizes. The particle shapes of the product from these surfaces are approximately as follows:

Smooth	*Intermeshing teeth*	*Beaded or corrugated*
Cubic	Cubic	Tabular

It was Edison who conceived the idea of having on the knobbed surface two additional rows of larger knobs set diametrically opposite each other to catch the occasional recalcitrant big piece.

On some machines the distance between the rolls is adjusted by inserting shims between the fixed and sliding bearings of the rolls. On others it is adjusted by means of bearings for one roller being set upon an arm which is pivoted and controlled by a spring-tensioned square-threaded rod (see *Figure 8.15*) or by a similar device using hydraulic pressure in place of the springs. Many designs permit of adjustment during operation. The length of rolls has been found by experience to be best kept short. It is unusual for the

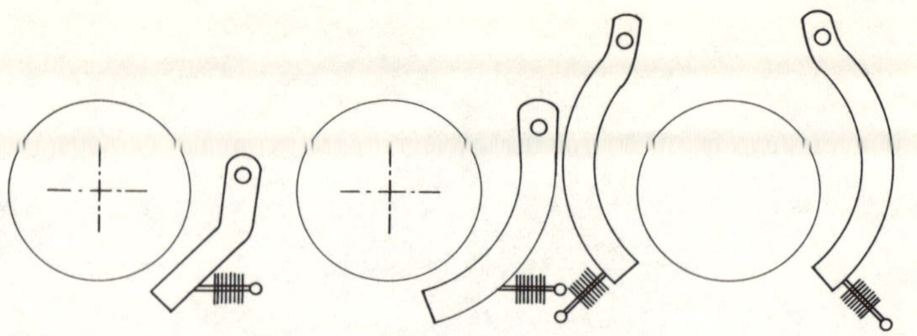

Figure 8.18. Types of breaker plate for single-roll crushers

length to be much more than the diameter of the roll, as experience has shown that short rolls offer best control of feed and cause least maintenance difficulty.

Breaker Plates

Breaker plates for single-roll crushers can take the form of a simple plate parallel to a tangent to some chosen point on the surface of the roll (see *Figure 8.18*) or can be in the form of a curved plate arranged eccentrically with regard to the roll, which may itself move eccentrically. They are made of manganese or chilled iron.

Frame and Housing

The housing of the double-roll crusher is of relatively light construction. The frame work is powerful and sufficiently robust to take the large pressures between the rolls and to carry the feed hopper which distributes the feed across the rolls. With single-roll machines, particularly of the sledging type, the housing has to be able to take the flying pieces from the piercing action. It is, in general, substantially built and of ribbed construction.

Feeding and Roll Wear

The rolls are usually fed from flat belts. Care is taken to spread the material evenly on these belts and make it fall uniformly into the hopper above the rolls. The hopper runs the length of the rolls and has a distributing plate at the bottom; cheek plates in the hopper form a seal on the rolls, so that all material is directed into the wedge space between the rolls. Rolls invariably wear more in the centre than at the edges and it is important that the distribution of feed across the rolls be as even as can be managed. Some machines incorporate a device by which one roll can be moved laterally parallel to the axis and relative to the other roll and thus be adjusted for preferential wear, particularly wear that leaves flanges at the ends of the rolls which prevent proper adjustment of the roll distance; corrugations in

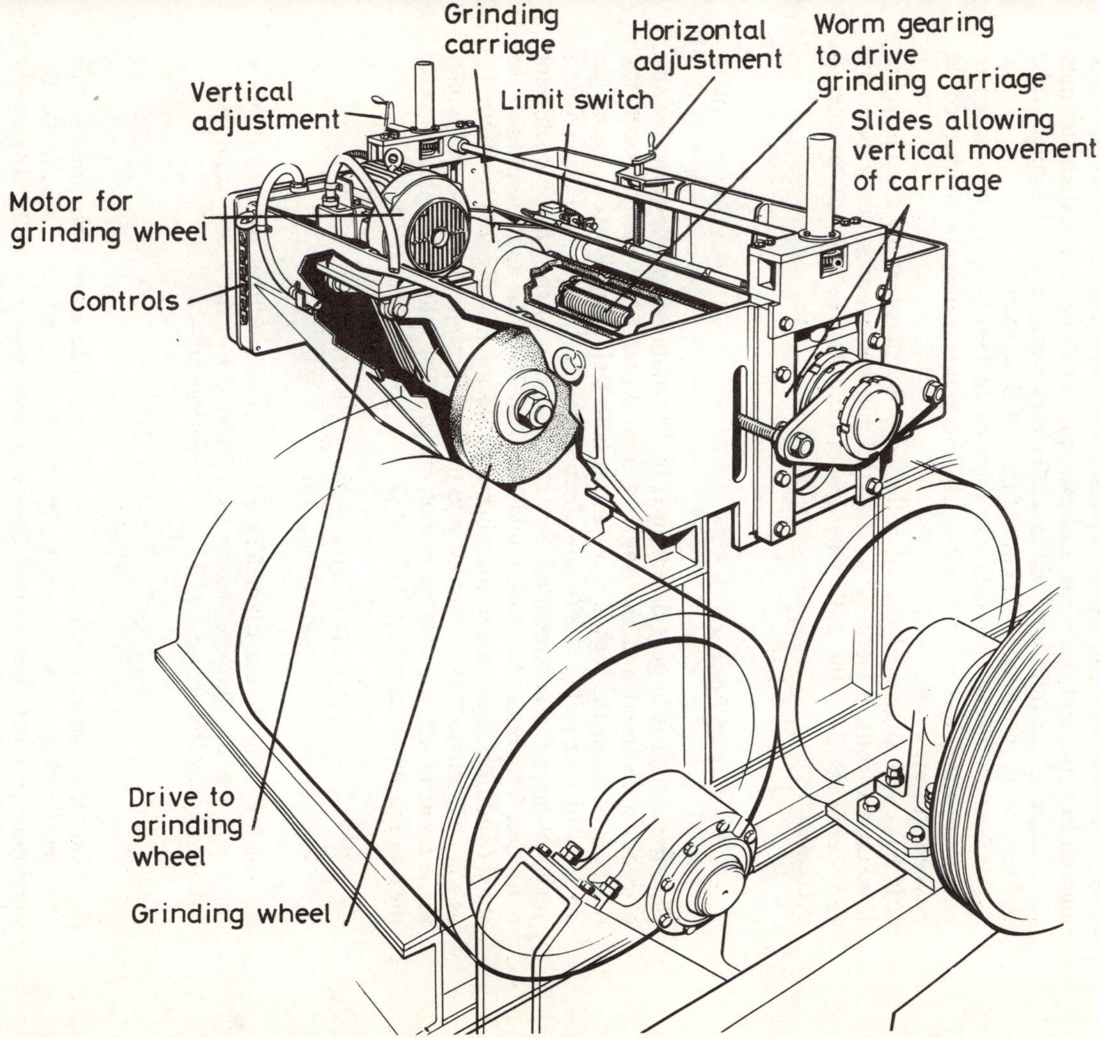

Figure 8.19. Grinding attachment for roll crushers for in situ *surface maintenance. (Courtesy Aulman and Beckschulte)*

the main body of the roller can also be prevented by such a device. Such an arrangement means having rolls somewhat longer than the capacity demands.

Rolls have to be ground occasionally to eliminate high areas. Some grinding machines, designed to be fitted to rolls (see *Figure 8.19*), can run the width of the rolls and utilize their rotation to grind the faces *in situ*.

Wear on the rolls can be quite considerable and they are designed thick enough to withstand this wear. For example, a roll 100 cm in diameter may have a tyre 15 cm thick and will take 10 cm of wear, i.e. the diameter of the roll can be worn down to 80 cm before it is discarded.

For best control of product size the rate of feed should correspond to about one-third of the rate at which the theoretical ribbon of product could pass between the rolls.

It has been suggested that if rollers are fed at speed to the clearance between the rolls, preferably so that the material reaches the rolls at the peripheral speed of the rolls, energy consumption by the roll crusher is decreased.

Size of Feed

The size of feed is largely determined by the diameter of the rolls, the size of product required and the angle of nip, although for sledging rolls the distance between the lugs is more important. Relationships concerning the angle of nip for roll crushers have been developed above (p. 129). Various angles of nip are recommended. Strictly according to the typical coefficient of friction for stones, the angle could be about 30°. This angle applies to the static condition and could apply when the rolls are moving slowly. However, as will be discussed later, practice is to use high peripheral speeds and angles in the range 11–25° are used by manufacturers. For rolls of unequal diameter the relationship is (case 1, p. 132)

$$\frac{(d_1+2a)(d_2+2a)}{(d_1+b)(d_2+b)} = \frac{1+\cos\theta}{2}$$

in which d_1 and d_2 are diameters of the rolls, $2a$ is the distance between them and b is the diameter of the feed particle. For the more usual case of equal roll diameters (case 2, p. 132), the relationship simplifies to

$$\frac{d+2a}{d+b} = \cos(\theta/2)$$

To take the range of angles used by manufacturers, this means that $(d+2a)/(d+b)$ can vary from 0.976 to 0.995. Now $2a$ is nominally the maximum product size, so that the reduction ratio R is $b/2a$. Taking the mean of the values in the above equation, say 0.985 (corresponding to $\theta = 20°$), and solving for b,

$$b = 0.015d + 2.031a \quad \text{or} \quad \frac{b}{2a} = R = \frac{0.015d}{2a} + 1.016$$

or, approximately,

$$R-1=\frac{0.015d}{2a}=\frac{d}{67\times 2a}$$

For an angle of nip of 20°, to obtain a reduction ratio greater than 2, the diameter of the roll must be more than 67 times the product size required.

It is clear that for smooth surface rolls the reduction ratio will be small unless very large rolls are used. *Table 8.6* gives the factor by which the product size has to be multiplied to obtain the roll diameter at which a particular reduction ratio is obtained.

Table 8.6

Reduction ratio (R)	2	2.5	8.5	10
Diameter of roll/diameter of product particle ($d/2a$)	67	100	500	600

Thus, taking the range in which roll crushers are said to be particularly economic, 1 cm to 3 mm, the diameter of the roll for this angle of nip (20°) must be 156 times the diameter of the product particle, or 469 mm.

The diameter of rolls decreases with wear and this should be taken into account in deciding upon roll size for a particular feed size. A reasonable assumption is that the rolls will wear down to three-quarters of their original diameter, i.e. one-quarter will be lost as wear.

For single sledging rolls the relationships are different from multi-roll machines and much larger pieces can be handled relative to the roll diameter. Examples given by MCGREW[2] based on limestone are:

Roll diameter, cm	60	90	150
Feed size, cm	35	60	90

These figures correspond with the thickness and length of the pieces and pieces of much larger width, limited only by the width of the machine, can be handled.

While sledging mills can handle best soft to medium-hard rocks having compressive strengths below 15 000 lb/in^2 (1000 kg/mm^2), the rock being preferably friable and not very abrasive, wet, clayey and sticky materials are also handled superbly. Materials like fuller's earth cling to the rolls and breaker plate and are not well handled.

Size of Product

The nominal size of the product is the setting between the rolls, but this only controls one dimension and a crude factor is that a proportion of about one-third of the product will have at least one dimension three times as large as this setting. The lowest setting for rollers is about 1 mm and, for free crushing conditions, this should be reckoned the lower size limit; smaller sizes can be obtained by choke feeding, which will also increase throughput but will lead to less precision of size of product. Typical size distributions are shown in *Figure 8.20*.

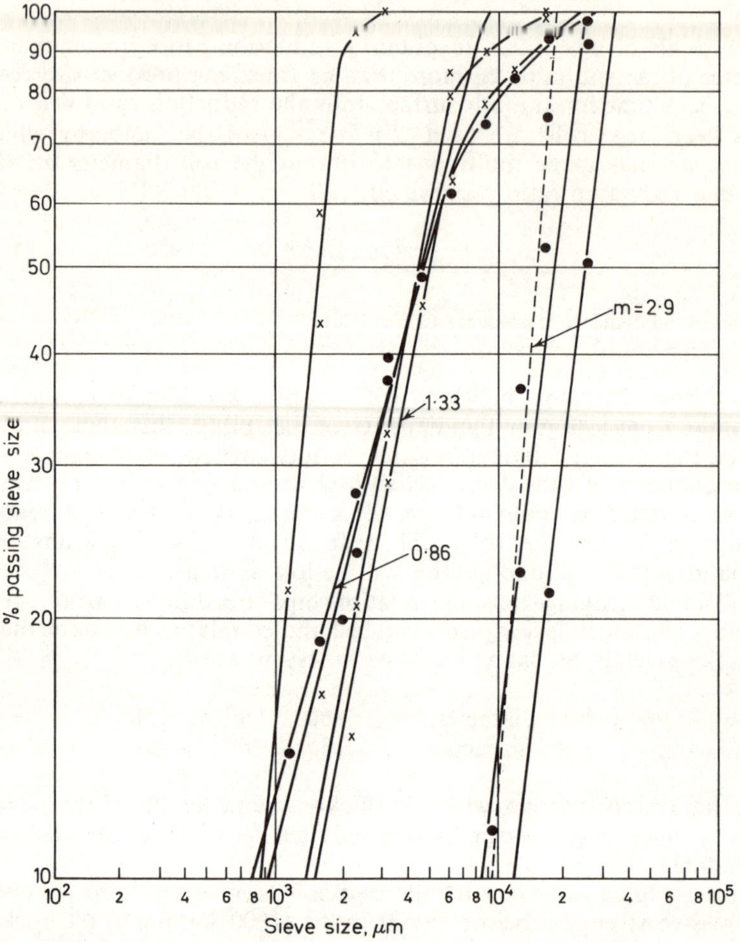

Figure 8.20. Product size distributions from roll crushers (Gaudin m-values indicated)

Speed of Rolls

The speed of the rolls is usually considered in terms of peripheral speed. The larger the angle of nip used the slower the peripheral speed which still allows a particle to be gripped. A very approximate relationship for an angle of nip of about 30° is

$$v = 1.8D - 40x + 128$$

in which v is the peripheral speed in metres per minute, D is the diameter of the roll and x is the sieve size through which 80% of feed passes, both in centimetres. For smaller angles of nip the speed can be much larger, e.g. for angles of about 17° speeds actually used are nearly five times the figure calculated by the above formula. The range of speeds used is from 60 to 1200 m/min.

The effect of using larger rollers than is demanded by the required reduction ratio is to decrease the angle of nip. If the ratio between the minimum roll diameter required and the diameter actually used is α, the lengths of arc of the rolls in contact with material of a given feed size are in the ratio of $\sqrt{\alpha}$; and if the peripheral speeds of the rolls are the same, the same quantity of material is displaced through the same opening between the rolls, but the rate of reduction is $1/\sqrt{\alpha}$ times as slow, which means that crushing is more gradual and more likely to produce fines. This can be counteracted by increasing the peripheral speed of the rolls, which also has the benefit of increasing capacity.

The peripheral speed is limited by that at which slipping occurs. Slipping causes not only inefficient crushing but also intense wear. Sledging rolls are moved much faster than smooth rolls and speeds of over 1000 m/mm are used on larger sledging machines.

A slight speed differential of the order of 10–20% between a pair of rolls is sometimes used, especially for plastic materials in the treatment of which a rubbing action is advantageous. Other machines use entirely different speeds for the rolls, of ratio of the order of 1–5, but differences of speed between rolls favour fines production.

Force between Rolls

Forces used per centimetre of length of roll vary from 90 kg to 6000 kg, depending upon the material to be crushed.

Capacity

An estimate of capacity is obtained by considering the material as passing through the rolls as a ribbon and then applying a factor for voids between the particles. Thus capacity is equal to the product,

$$\pi N d l 2 a \rho f$$

in which N is the speed of the rolls in revolutions per unit time; d is the diameter of the rolls; l is the width of the rolls; $2a$ is the distance of setting between rolls; ρ is the true density of material; and f is the packing factor, usually taken to be about 0.3.

Power Consumption

Power consumption varies with the reduction ratio (see p. 62 for equitable comparison with other equipment). The following figures are taken from

Bond and Wang's chart (*Figure 4.3*) and conform with several practical results obtained from manufacturers in the U.K.:

Reduction ratio	2	4	6	8	10
Power range, kWh/t	0.4–1.5	0.5–2	0.6–3	0.7–3	0.8–4

COMPARISON OF JAW, GYRATORY AND ROLL CRUSHERS

A comparison of jaw, gyratory and roll crushers may be made on the basis of capacity for the same size of product from the same feed and the interrelationships given above. Jaw crushers are, for the same reduction ratio, of smaller capacity than gyratory crushers, while roll crushers are in these circumstances of vastly larger capacity than either. If typical dimensions of these machines are taken for the feed size b and the smaller open bottom setting t, then the maximum capacities of the machines at their optimum frequencies will be:

jaw crusher $\Delta(q+t)\cdot w(g\cdot h)^{\frac{1}{2}}/4\sqrt{2}$
gyratory crusher $\Delta(q+t)\cdot \sin\psi\pi r\cdot[gh(\sin\psi-\mu\cos\psi)]^{\frac{1}{2}}/2\sqrt{2}$
roll crusher $\rho\pi N\cdot d\cdot l\cdot 2a\cdot f$

The frequencies of the machines are as given earlier:

jaw crusher $\frac{1}{2}(g/2h)^{\frac{1}{2}}$
gyratory crusher $\frac{1}{2}[g(\sin\psi-\mu\cos\psi)/2h]^{\frac{1}{2}}$
roll crusher N

while the minimum discharge settings are:

jaw crusher t
gyratory crusher t
roll crusher $2a$

If the coefficient of friction of the material (μ) is taken to be about 0.27 and the angle of nip (θ) of the roller crusher is taken to be 30°, the diameter of the roll (d) for a feed size b is obtained from

$$\frac{d+t}{d+b} = \cos(\theta/2) = 0.966$$

and $d = 28.4b-29.4t$. The width of the roll (l) is usually equal to the diameter.

The width of the jaw crusher (w) is usually twice the gape, which in turn is about 1.1 times the maximum feed. Thus width $w = 2.2b$. A typical throw is 4 cm.

The circumference of the gyratory bowl at the gape is about 10 times the gape or $11b$. If the cone base angle (ψ) is 75° and the angle between the jaws is 30° for a throw of 4 cm, the average radius of all material in the crusher is easily calculated from the radius of bowl at the gape ($11b/2\pi = 1.75b$) and shown to be $1.2b-2$ cm.

If the frequencies of the machines are made equal and the minimum discharge settings are also made equal, in the determination of the ratio of the capacities of the machine, these factors may be eliminated from the above expressions for the respective machines and these become:

jaw crusher $$\frac{\Delta(q+t)\cdot w\cdot h}{2}$$

Since $q = t+4$ and $w = 2.2b$, this equals
$\Delta(2t+4)\cdot 2.2b\cdot h/2$ or $\Delta(t+2)\cdot 2.2b\cdot h$

gyratory crusher $\Delta(q+t)\sin\psi\cdot\pi rh$
Since $q = t+4$, $r = 1.2b-2$ and $\sin\psi = 0.97$, this becomes
$\Delta(t+2)\cdot 6.1\cdot(1.2b-2)h$

roll crusher $\rho\pi(28.4b-29.4t)^2\cdot t\cdot f$
if the bulk density Δ is taken to be about 0.8ρ and $f = 0.3$, the expression becomes
$\Delta 950\cdot(b-1.04t)^2\cdot t$

Now if a further assumption be made that the length of the product particle (h) is equal to its thickness (t) and the reduction ratio (b/t) is equal to 5, these three ratios reduce to:

jaw crusher $(t+2)\cdot 11t^2$
gyratory crusher $(t+2)\cdot 6.1(6t-2)t$
roll crusher $950(3.96t)^2\cdot t$

and are approximately in the ratio as 1:3.3:1355.

The enormous factor for roll crushers arises from the relatively large diameter of roll necessary to take a given size of feed and from the continuity of crushing from the revolution of the rolls compared with the intermittent production from the jaw crusher and the gyratory crusher. The jaw crusher produces in one revolution a wedge piece of volume $\frac{1}{2}(q+t)\cdot w\cdot h$ or, with the simplifications used above,

$$(t+2)(2.2b)\cdot h = (t+2)(11t)\cdot t \approx 11t^3$$

whereas the roller produces a layer t thick round its whole circumference in one revolution $= \pi\cdot d\cdot l\cdot t$, which, with factor for looseness of packing of this layer of 0·3 and the simplifications above, becomes

$$\pi\cdot 0.3\cdot(113t)^2 \approx 13\,000t^3$$

i.e. of the order of 1300 times the capacity as shown above.

TAGGART[1] states the rule that, if the hourly tonnage required is less than $0.115\times$ the square of the gape in inches, then a jaw crusher should be used.

The above calculation showed that, to utilise fully a gyratory crusher instead of multiple jaw crushers, the capacity required needed to be about three times that obtained from a single jaw crusher of equal gape. These rules relate to utilisation of capacity and not necessarily to the over-all costs.

Table 8.7 summarises some of the properties of the three kinds of equipment. Power costs of operating the equipment are compared with other comminuting equipment in Chapter 4 (p. 64).

DISC CRUSHERS

The disc crusher was originally developed as an answer to the problem of secondary crushing with a gyratory crusher, before the cone crusher came into being.

The essential feature of the disc crusher (see *Figure 8.21*) is two concave

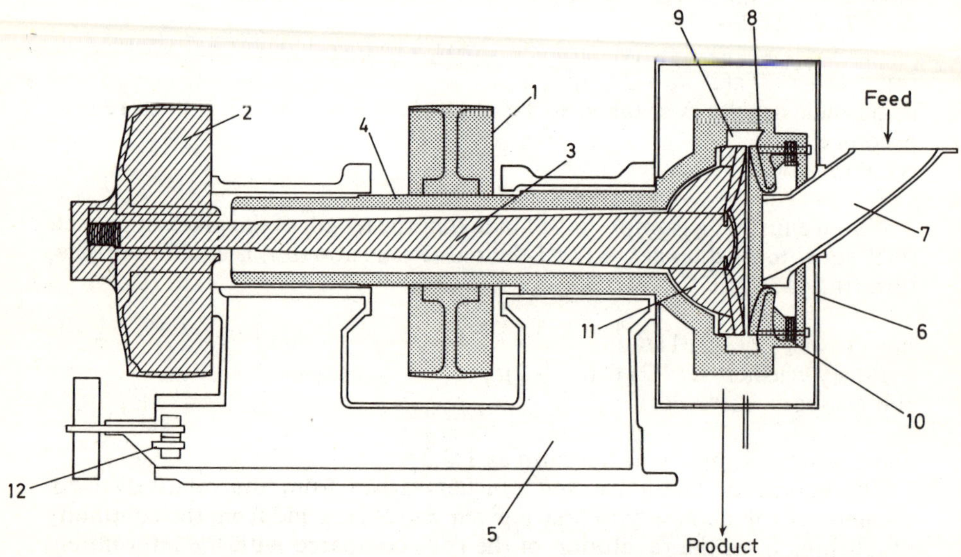

Figure 8.21. Disc mill: 1, driving pulley; 2, eccentric pulley; 3, inner shaft; 4, outer shaft; 5, main frame; 6, hood; 7, spout; 8, outer disc; 9, inner disc; 10, adjusting shims; 11, spherical bearing

surfaces facing each other, being rotated in opposite directions, one disc having a gyratory movement imparted to it so that the edges of the discs approach and recede from each other during a cycle. The material is fed to the centre of the discs, where the distance between the discs is maximum. From here the material is thrown out radially, by centrifugal force, towards the circumference in locations which correspond with the particle sizes. The particles are compressed and broken and thrown out to positions of greater radius until the particles are able to pass through the distance between the discs at the circumference.

Table 8.7. COMPARISON OF JAW, GYRATORY AND ROLL CRUSHERS BASED UPON SAME FEED AND SAME SIZE OF PRODUCT

Per unit quantity of product	*Jaw*	*Gyratory*	*Roll*
Capital cost including installation	—	Lowest	Highest
Weight (proportional to installation)	Heaviest	—	Lightest
Power consumption	Highest	—	Lowest: balanced mechanism. Low starting peak
Idling load % of total	45	30	30
Headroom required	—	Highest (maintenance to lift out head)	—
Lubrication	Least simple	—	Most simple
Robustness	—	—	Most robust
Type of material processed	Blocks, massive rocks. Clayey, sticky, spongey rocks can be handled. Disposition to bridge. Product tends to be tabular	Slabs and stratified stone handled. Product tends to be cubical	Any kind of material. Shape can be controlled by type of roll surface
Feeding	—	Simplest: can be buried in feed, if necessary, and always fed from two sides	Most fastidious: requires very even feed across rolls
Wear and maintenance	Retains adjustment even when worn. Wear is localised and intense. Requires crane with two-way travel	Maintenance costs highest	Easy to maintain *in situ*. Maintenance costs lowest
Safety guarding	—	Least extensive	—

Advantages

(1) The crusher will accept material of any degree of wetness.
(2) It will accept material of any association of particle sizes, such as gravel boulders and sand, and reduce them all to a given size.
(3) The machine cannot be choked; it cleans itself.
(4) The product is cubic in shape and does not possess strain, so that it is admirable for aggregate for concrete and road surfacing.
(5) The crusher produces very little fines.

One form of construction of the mill is shown in *Figure 8.21.* The outer disc is bolted to a hemispherical flange on a hollow shaft which carries a pulley at about the centre of its length; the inner disc is set on a spherical bearing, which in turn is set on a tapered shaft. The spherical bearing sits in the cup formed by the hemispherical flange of the hollow shaft. The tapered shaft passes through the hollow shaft and is set eccentrically in a bearing in another pulley. The hollow shaft is fitted with oil as a seal. The outer disc has a central hole through which a spout passes sufficiently close to seal the disc against egress of material but loose enough to allow the disc to be rotated freely without turning the spout. The two pulleys are driven so that they move in the same direction, but one disc has a horizontal pulsating movement as well. There is no grinding action between the discs. (Consequently the wear on them is small.) Feed is introduced through the spout and product is thrown out at the periphery into a surrounding casing, from whence it can be directed on to a conveyor belt, etc.

In some disc crushers the discs rotate in the horizontal plane; in others, in the vertical plane.

Method of Rating

The usual method of rating is according to the diameter of the discs.

Size of Feed

Crushers are made which can take feed up to 25 cm.

Size of Product

Crushers are designed for both primary and secondary crushing and the product can be as small as 80% passing 5 mm. Very little fines are produced.

Discs

The discs are usually made of manganese steel. The surfaces of the discs may be plain or may include lugs. The lugged discs are used for the coarser sizes. The minimum distance between the discs can be adjusted for product size, but the distance cannot usually be decreased below 5 mm.

PAN MILLS

Pan mills are also known as edgerunners, chasers and mullers; a special design with more than two rollers is known as a Chilean mill. Pan mills with perforated bottoms are sometimes known as grate mills.

The essential features of this kind of mill (see *Figure 8.22*) are rollers or

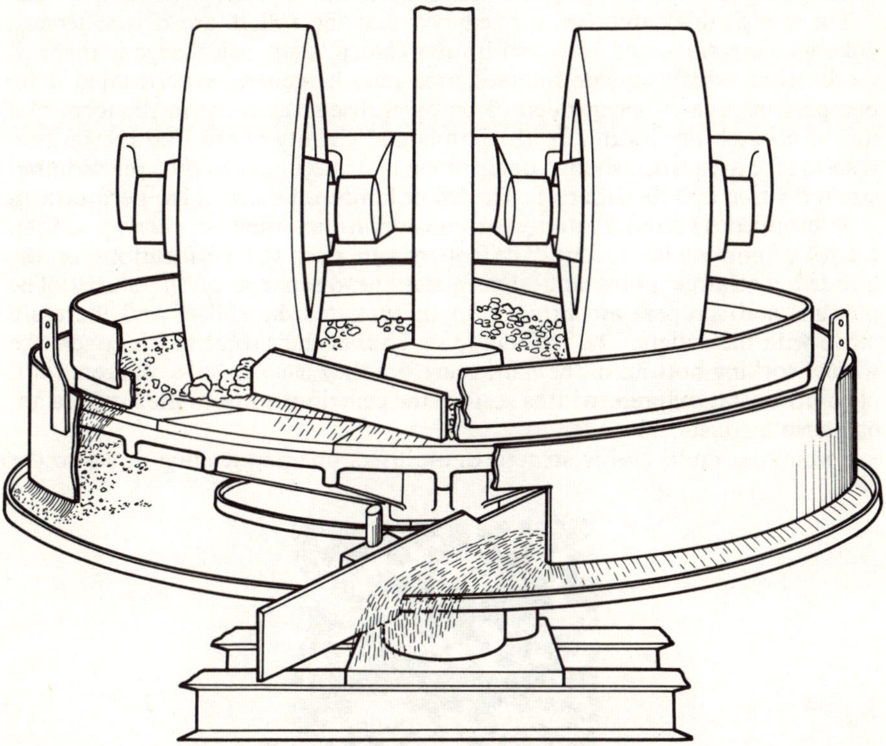

Figure 8.22. Pan mill with rim discharge. (Courtesy Bradley and Craven Ltd)

mullers which turn on a horizontal axis and which rest either under gravity or with the assistance of an impressed force on a bed of material in a pan. Ploughs are set directing material back to the centre of the pan.

The milling may be wet or dry but the design of the mill alters according to the method to be used.

The dry type has a rotating pan and the bottom either is perforated or has rim discharge. Rim discharge uses a slot, the size of which is controlled by an adjustable steel slat, in the side of the pan; the slot is usually set at about

1 cm and material is impelled through the slot by the centrifugal action of the rollers. In the dry form of pan mill the perforated plates are set at an angle of 15–20° to the horizontal, again to take advantage of the centrifugal forces induced by the rollers and so that particles of sufficiently small size are shot through the perforations. In the dry form, too, the rollers are equidistant from the centre of the pan and revolve on solid plates which form a circle in the centre of the pan. The ploughs and scrapers are fitted to a bar which forms part of the main structure of the mill. The rim discharge can be adapted for batch processing by substituting a bottom discharge door for the run opening, or indeed by merely shutting the rim opening and using a plough device when discharge is required which utilises the rotation of the pan to direct material up it and over the rim into an appropriate collecting device.

The wet form of mill has a fixed bed and the rollers are driven from a yoke which rotates and is driven from a central shaft. Discharge is through the bottom, which outside the feed area may be generally perforated or in one part may be either gridded or be open discharge or be in the form of a door for batch processing. In the immediate vicinity of the feed the bottom is solid; the area to be solid is determined by experiment to give the required size reduction and distribution. The lower limit to the size of the perforations is determined by strength and wear considerations: they are rarely less than 1 cm or more than 2.5 cm. The rollers run over the perforations or the gridded discharge point and thus assist the discharge of the slurry. The ploughs and scrapers are fitted onto the yoke of the rollers and therefore move with the rollers. The rollers are set so that their total width covers the whole working bottom of the pan; therefore they usually have different radii of rotation and different widths so that the centrifugal forces they induce are balanced against each other.

Sometimes multi-storey structures are used, one pan feeding down to the next.

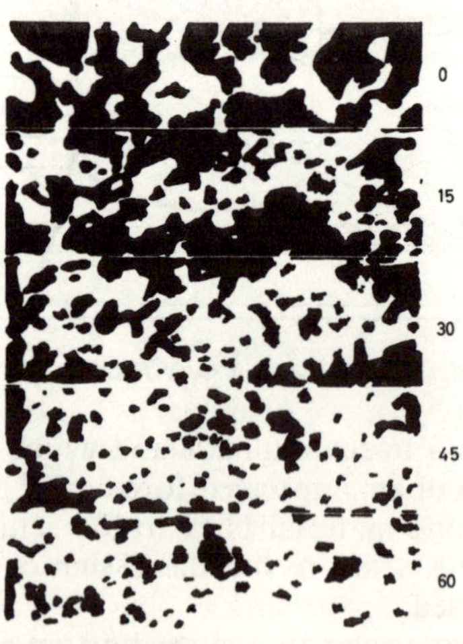

Figure 8.23. Progressive elimination of voids (black areas) in course of pan milling dolomite. Time of milling indicated on right-hand side in minutes. (After A. H. B. CROSS, D. F. MARSHALL *and* R. J. SARGANT[12]*)*

Normally the rollers rest on the bottom of the pan, but they can be made so that the distance between their rest position and the bottom of the pan is variable, one device for this being a spring-loaded bolt arrangement. The pan can be open (available for inspection) or closed (to control the dust and fume).

The pan bed is made of manganese steel or cast-iron.

Feeding

Feeding may be by means of a screw feeder or a chute. If the pan is elevated, then an elevator or a conveyor may be interposed. A chute is usually used to direct material between the ploughs, which in turn are bringing material back from the edges to the centre of the pan. By feeding in this position, the material acquires an impetus directing it under the rollers and inducing circulation. For wet grinding the slurry is fed in front of the rollers at the point where the solid plates begin.

The depth of material in the pan is controlled by rate of feed.

Size of Feed

Pan mills can be used as primary crushers, since they can be designed to break material as large as 15 cm.

Size of Product

The fineness of the product is largely determined by the depth of the bed in the pan and the roller weight.

Under the pan is placed a collecting dish for the product emerging through the perforations or the rim discharge. In the dish is set a shaft on which are blades which both agitate the contents and have a positive movement towards a discharge point.

The principal advantage of the machine, particularly the solid-bottom form, is its ability to produce a mixture of sizes which jointly constitute a material of very small porosity or voidage, i.e. high bulk density. This ability the machine possesses in a high degree, so that, when a mixture was synthesised from seven different fractions proportioned so as to give minimum porosity, a bulk density was not achieved as great as that obtained in a pan mixer[12]. This is due to the sliding and compressive action of the rollers, which induces the particles to marry in with one another and thus minimise voids. The effect is shown in *Figure 8.23*, which is copied from a contact print of a negative sensitised plate which has received transmitted light through a bed of material at various stages during a course of pan milling. The black areas are those transmitting the light, i.e. the voids. Another effect which augments the quality of high bulk density is the production of rounded particles in the machine, even at the coarse end of the size range. *Figure 8.24* illustrates this

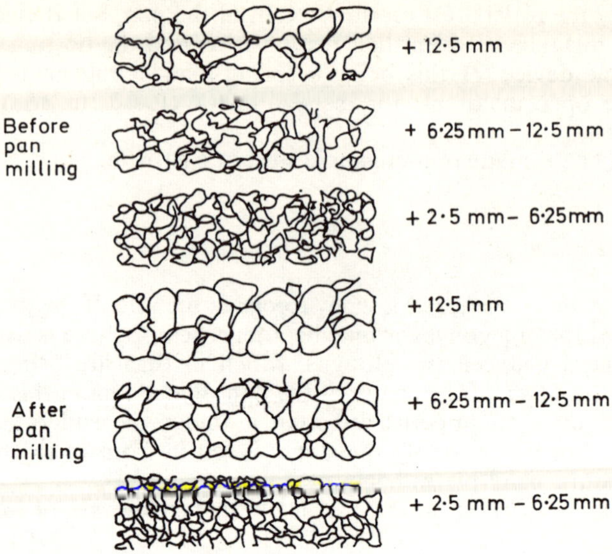

Figure 8.24. Rounding of particles during pan milling: screen ranges before and after milling. (After A. H. B. CROSS, D. F. MARSHALL *and* R. J. SARGANT[12]*)*

effect. This rounding generally increases the bulk density, so that particles in a given size range will increase in bulk density during processing.

The mechanism of increasing bulk density appears to be of three kinds: (1) an increase in the number of different sizes, particularly fine sizes; (2) selective crushing of softer constituents; (3) making the shapes of particles more spherical.

High bulk density is of considerable benefit in such industries as brick-making, in which a compact mass is the objective; mixtures of high bulk density show less disposition to segregate into different sizes during transport and storage than do less bulk-dense products.

The action of the pan mill has been analysed[13] and shown to be partly rolling (compressive) and partly sliding. The areas in which these actions take place are related to the distance of the rollers from the centre of the pan (R) and the width of the rolls (B), so that

$$\frac{\text{area of rolling action}}{\text{area of sliding action}} = \frac{4R}{B}$$

Thus, to maximise the sliding, i.e. the grinding action, R is made as small as possible and B as large as possible. To maximise the compressive action, the converse actions should be taken. Sliding is encouraged by putting the centre lines of the rollers out of line with the centre of the dish and maintaining balance on the driving mechanism by putting the rollers out of centre on opposite sides of the pan centre. Skidding is thereby induced.

Rollers

The rollers may be caused to move by the horizontal movement of the yoke set on a central vertical shaft or by the rotation of the pan and the friction between them and the pan. The rollers are best independently suspended so that they can take up the contour of the material passing beneath them, ideally if they are suspended from opposite sides of the pan so that material is rotated under them in the same sense as that in which they are rotating; however, this has the operating disadvantage that both sides of the pan are obstructed, which creates difficulties in feeding and inspection as well as maintenance.

Typical sizes of rollers are:

Dish diameter, m	1.5	2.1	2.4	2.9	3.4
Roller diameter, m	1.0	1.1	1.3	1.5	1.8
Roll width, cm	40	33	40	40	56

Thus roller diameter is often about half the pan diameter. The rollers are usually made of cast-iron, are hollow and have chilled cast-iron or manganese steel tyres which may be replaceable. The tyres may be split and fastened with countersunk bolts; they may be wedged in place; or they may be caulked on with a mixture of iron chippings and ammonium chloride. The first type of fixing is best for tyres which need frequent replacement. Occasionally, rollers are made of stone, sometimes plain and sometimes fitted with steel tyres. The rollers may consist of a hub connected to the rim by a pair of discs or, with the heavier type, by a double set of ribs.

The pressure of the rolls, or the crushing load, has been investigated[13] and the effectiveness has been shown to be related to the shrinkage or compressive properties of the material. The investigators state that the pressure (P) is related to the compressibility of the mass (y) by $P = Cy^{\mu}$, where C and μ are constants related to the mass and the height of the material in the pan. Using rut formation technology and applying it to the weight of the roll (Q), its diameter (D) and its width (B), they deduce

$$P = C\left[\frac{Q}{BC\sqrt{(D)}[1-0.33\mu]}\right]^{\mu/\mu+0.5}$$

It was possible to measure P by a device based upon calibrated copper probes whose degree of compression corresponded with a pressure and which were fitted in the rim of the roller. Agreement with the formula was obtained to about 15%.

It is seen that the effectiveness of the rollers is related to the expression $Q/B\sqrt{D}$; B, in particular, should therefore be as small as possible. A crude factor used is 120 kg per centimetre of roll width as an effective roll weight for most materials. The weight of the rolls can be augmented by inserting springed bolts between the shaft and the yoke, by taking supports from the yoke to a box container (in some mining operations heavy minerals were placed in this box) or by compressed air or by hydraulic means. The size of the roll should take account of maintenance costs. The wider the roll, the greater the amount of maintenance.

A somewhat surprising report[14] was that, when material was damp and soft, there was an optimum pressure beyond which crushing decreased.

Ploughs and Scrapers

The wearing life of ploughs is said to be much increased if they are faced with ceramic. Scrapers are usually fastened in positions on the shaft carrying the rollers for the wet pan but fixed on a stationary part for the dry. It is important that these be kept in a good state of adjustment and set close to the pan bed to prevent build-up in the pans and to sustain output.

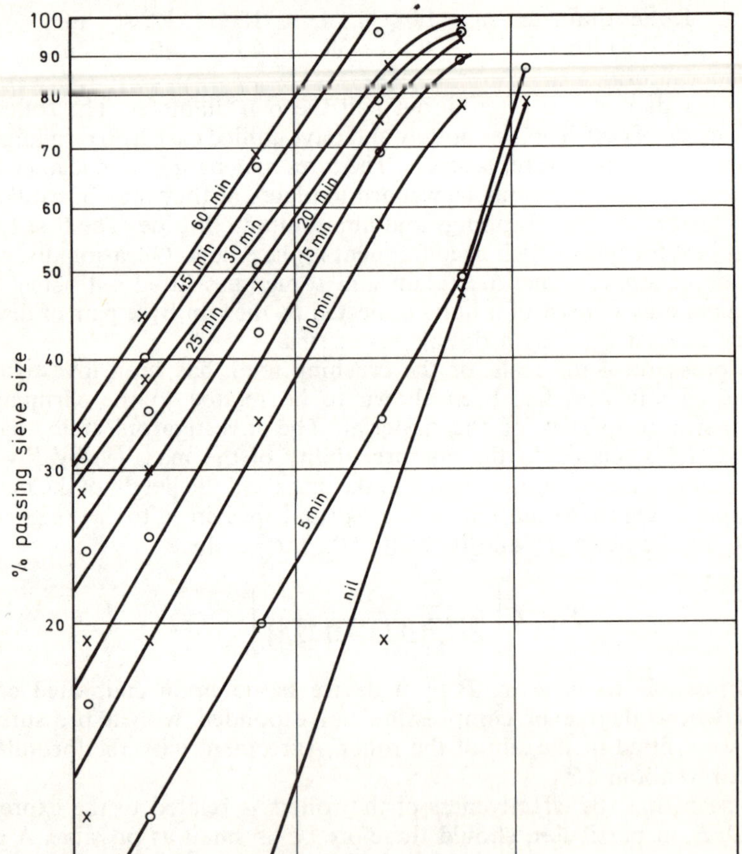

Figure 8.25. Pan milling: effect of time of milling on size distribution

Time of Grinding

There is an optimum time of grinding for maximum bulk density, not only from the point of power consumption but also from that of bulk density itself. For example, during the pan milling of dolomite the density changed with time as follows[12]:

Time, min	0	5	10	15	20	25	30	45	60
Bulk density, g/cm^3	1.95	2.36	2.31	2.48	2.49	2.52	2.46	2.46	2.32

If the screen analyses of these products are plotted in the form of log (screen size) versus log (% passing screen size), a series of straight lines is obtained (see *Figure 8.25*), which after the first minutes of grinding become parallel to one another, illustrative of the Gaudin law (see p. 70).

Calculations have been made[13] of the mixing factor based upon the number of times the area covered by the rollers in a given time covers the area of the pan; the investigators add a factor for the mixing due to the ploughs or scrapers. They thus compare the number of mixings in one pan with those of another, taking into account the pressure exerted by the rollers.

Rotation of Mill or of Rollers

The following relationship between crushing load and number of rotations was found in one mill to give maximum bulk density for a given charge in the least time:

Total crushing load, ton	7.2	5.4	4.2	3.1	2.5
Number of revolutions	100	200	300	400	500
Crushing load × revolutions	720	1080	1320	1240	1250

From this it appears that for some of the range optimum revolutions are inversely proportional to the crushing load. Rate of rotation is important; the effective angle of nip (see p. 129) is decreased as the speed of the roller increases. A rough guide is:

Feed size, cm	–1.25	1.25–4	4–7.5	7.5–15
rev/min	40	36–40	32–36	24–26

Crushing Load and Grinding Time

In a detailed investigation with dolomite[12], using a series of pans 1.5–2.75 m in diameter, with roller diameters of 0.9–1.8 m and widths of 30–35 cm, with different crushing loads, the optimum number of grinding revolutions of the rollers (i.e. the minimum number of revolutions to reach maximum density) was found to be related to the charge loading per unit area. (See *Table 8.8*.) If log (optimum revolutions) is plotted against crushing load, straight lines

Table 8.8. OPTIMUM NUMBER OF REVOLUTIONS OF A PAN MILL TO SECURE MAXIMUM BULK DENSITY FOR VARIOUS CRUSHING LOADS AND CHARGE LOADINGS

Charge loading,	*Total crushing load,* tonnes									
N/m^2 *of pan surface*	*3*	*4*	*5*	*6*	*7*	*8*	*9*	*10*	*11*	*12*
960 (20 lb/ft²)	570	460	380	320	280	240	210	190	180	160
580 (12 lb/ft²)	470	370	290	240	190	160	140	120	115	115
380 (8 lb/ft²)	320	260	200	160	130	90	80	70	70	70

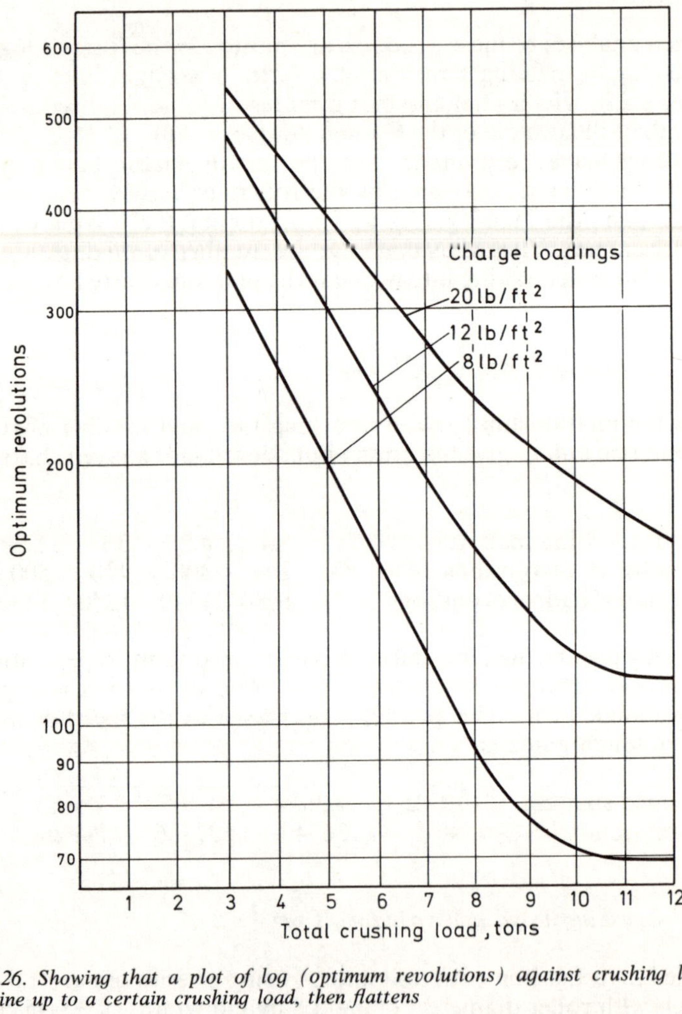

Figure 8.26. Showing that a plot of log (optimum revolutions) against crushing load gives a straight line up to a certain crushing load, then flattens

are obtained, as shown in *Figure 8.26*, up to crushing loads of about 8 ton for this particular roller, of width 10 in, but at higher loads the lines flatten out. This indicates that the crushing load is having less effect and indeed that the minimum number of revolutions to secure maximum bulk density for a given loading of material in the pan (charge loading) is being approached. What is even more interesting is that this flattening begins at about the same crushing load for each of the charge loadings.

Charge Weight

In the same set of investigations[12], the optimum grinding revolutions were found to be related linearly with charge weight (*Table 8.9*).

Table 8.9. PRODUCT OF NUMBER OF REVOLUTIONS OF PAN MILL TO ACHIEVE OPTIMUM BULK DENSITY AND TOTAL CRUSHING LOAD

Charge loading, N/m^2 A	*Total crushing load,* tonnes								*Average of revolutions* × *crushing load* B	B/A
	3	*4*	*5*	*6*	*7*	*8*	*9*	*10*		
960	1 710	1 840	1 900	1 920	1 960	1 920	1 890	1 900	1 880	2.0
580	1 410	1 480	1 450	1 440	1 330	1 280	1 260	1 300	1 370	2.4
380	960	1 040	1 000	960	910	720	720	700	880	2.3

RING ROLL AND RING BALL MILLS

Ring roll and ring ball mills were developed to prepare material of intermediate hardness, which had already been reduced below 1 cm, to finenesses of the order of 100 μm for the process industries. They were designed in an attempt to decrease the cost of similar size reduction in ball mills while retaining their high availability and freedom from breakdown. Many of them began as pulverisers of coal for fuel in power stations.

The distinguishing feature of this group of mills is that of a number of rolls or balls pressed by mechanical or gravitational means onto a ring of more than five times their diameter. Either the ring or the other parts are driven, or they are both driven, and the action is of rolling and thereby nipping material and comminuting it. These mills always incorporate fast-moving arms or parts which act like blades and are therefore disposed to create air flows which can be utilised to move the material and to separate fractions of it. The principle usually is to use the flows of air to lift out of the vicinity of the comminuting parts those particles small enough to be entrained, and the rates of flow of the air are adjusted so that the size range of the entrained particles coincides with that required for the product. The air flow can also be used to separate constituents in the feed which it is not desired to grind, e.g. pyrites in coal. Once separation at grain boundaries has been effected in the mill, the heavier particle is not supported by the air-stream and can be made to fall back through the air annulus at the base of the mill and be collected. Thus closed-circuit grinding is almost always the rule with this type of mill, which means that dust is closely confined.

These mills will not handle moist or sticky materials unless the circulating air is heated and drying thereby effected *in situ*. The ready adaptation of these types of machine for *in situ* drying is an important advantage: in some of them gases can be handled at about 350°C and material entraining as much as 25% moisture can be dried down to less than 1%. Inert gases can in the same way be introduced into the fluid circulation system for grinding materials that cannot tolerate air. Until recently mills in this group were not suitable for materials with hardnesses much above Mohs 4 (Vickers 150), quartz and cement clinker, for example, causing such wear as to make the cost of operation compared with that of ball mills prohibitive. More recently, however, alloyed steels have been developed and used which increase the range of hardnesses which can be handled economically in these mills.

Table 8.10 summarises the features of some ring roll and ring ball mills.

Table 8.10. TYPES OF RING ROLL AND RING BALL MILLS

Name	*Roll or Ball*		*Ring*
	No.		
Lupulco (Loesche)	2–3	Tensioned by spring to ring. Not driven	Surface horizontal. Axis vertical. Driven
Raymond	3–6	Axles swing freely from spider on vertical shaft. When rotated axles swing out like governors to touch ring	Surface vertical. Axis vertical. Stationary
Bradley Poltte	3	Spring-loaded against ring. Rotated	Surface horizontal. Axis vertical. Stationary. Screen sometimes incorporated in housing
Sturtevant	3	Located on spider on a door which can be opened for inspection and roll mainten- ance. Spring-loaded. Not driven	Surface horizontal. Axis horizontal. Rotated
Babcock and Wilcox	8–14	Single horizontal circle of balls	Surface horizontal. Axis vertical. Rotated
Fuller		Double layer of balls, each layer of balls in horizontal circle	Two surfaces horizontal. Axis vertical. Rotated
Maxecon	3	Tensioned by spring to ring. One driven, other two rotated by rotating ring	Surface horizontal. Axis horizontal. Rotates
Griffin	1	Axle suspended from driving head	Surface vertical. Axis vertical. Stationary

The rolls are usually arranged so that they can move away from the ring and break contact so that material can enter, but the loading between the rolls and the ring is such that this can only be done by overcoming a considerable force. This force may be gravitational, or in the form of centrifugal force, tensioned springs or hydraulic pressure. Frequently a setting device is incorporated which ensures that the clearance does not decrease below a certain minimum.

RAYMOND MILL

The Raymond mill consists of a spindle on which is set a spider onto which the axles of the rollers are pivoted (see *Figure 8.27*). At a lower level, corresponding with the level of the rolls, is set a ring with axis coincident with the axis of the spindle. The spindle is driven from underneath the machine. A duct brings air to a level just below the grinding ring. Material is fed from the side through a star feeder and is moved into the grinding zone by ploughs suitably deployed and fixed to the spindle. It is usual to fit a size separator on the top of the machine which can be of the so-called whizzer design, which consists of a bladed wheel. The number of blades in this wheel, and its speed, can be altered, depending upon the air flow required through the machine: it is clearly far less interrupting and more economic to operate if the control can be on the whizzer speed once the range has been set against the number of blades installed. The whizzer works on the principle that the heavier particles will be thrown by the whizzer to the outer perimeter, i.e.

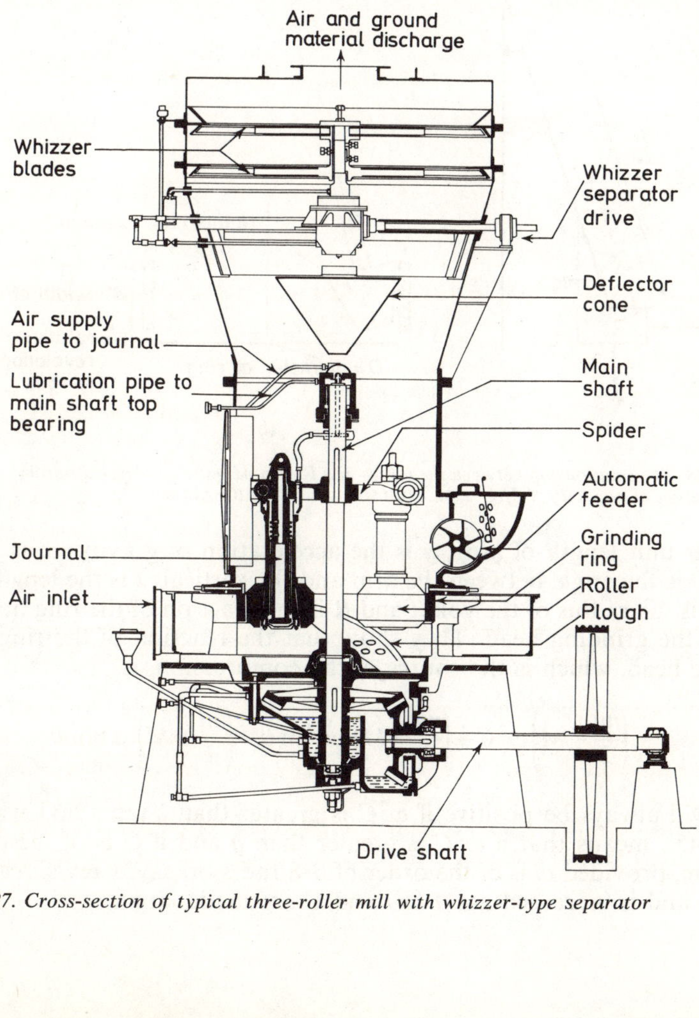

Figure 8.27. Cross-section of typical three-roller mill with whizzer-type separator

the mill wall, where the upward air velocity is least, and will fall back to the grinding zone. The material entrained in the air-stream leaving the top of the mill is first passed into a cyclone and then through a bag filter to remove the finest particles.

The feed rate is often controlled automatically by relying upon the flow of fluid from the top of the mill to exert suction in a pipe at right angles to the flow which is connected to bellows, which in turn move in accordance with the pressure in the pipe and move the star feeder proportionately. Thus, if the flow is small, more feed is added until the pressure builds up above that corresponding with the capacity of the mill, when the feed is shut off.

The parts of the mill are fabricated in different metals. The grinding ring is usually forged steel and the rollers are made of high carbon steel with deep-chilled faces.

COULSON and RICHARDSON[15] analyse the action of the Raymond mill as shown in *Figure 8.28a*, in which M is the mass of the grinding head; m is the

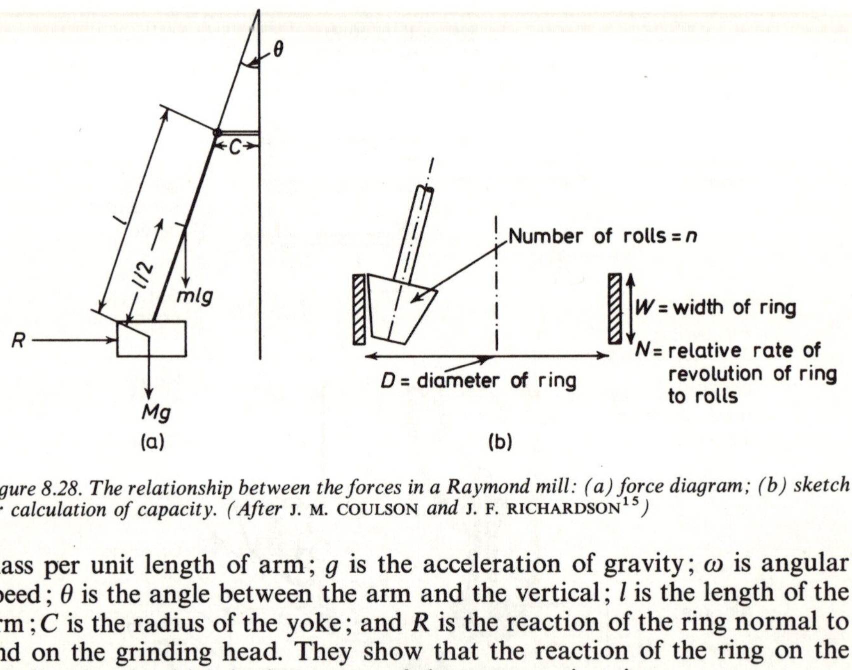

Figure 8.28. The relationship between the forces in a Raymond mill: (a) force diagram; (b) sketch for calculation of capacity. (After J. M. COULSON *and* J. F. RICHARDSON[15]*)*

mass per unit length of arm; g is the acceleration of gravity; ω is angular speed; θ is the angle between the arm and the vertical; l is the length of the arm; C is the radius of the yoke; and R is the reaction of the ring normal to and on the grinding head. They show that the reaction of the ring on the grinding head, which is the source of the compression, is

$$\left(\frac{ml}{2}+M\right)\omega^2 C+\left(\frac{ml}{3}+M\right)\omega^2 \sin\theta-\left(\frac{ml}{2}+M\right) g\tan\theta$$

which will always be positive if $\omega^2 C$ is greater than $g\tan\theta$ and which, up to $\theta = 45°$, means that if $\omega^2 C$ is greater than g and if C is of the order of 15–30 cm, provided ω is of the order of 6–8 rad/s or, say, 1 rev/s, reaction is positive and is increased as $Mml\omega$ and C are made larger.

The size of feed to the mill is usually below 2 cm, though there is no fundamental reason why rolls should not be designed for larger sizes.

The product from Raymond mills can be of very small particle size, down to about 35 μm, and mixtures containing 80% through 60 μm are quite usual. For examples of the size distribution of the products from a Raymond mill at various rates see *Figure 8.29*. Of all the ring roll mills, the Raymond type handles sticky material best; not so well as the ring ball type, according to some investigators.

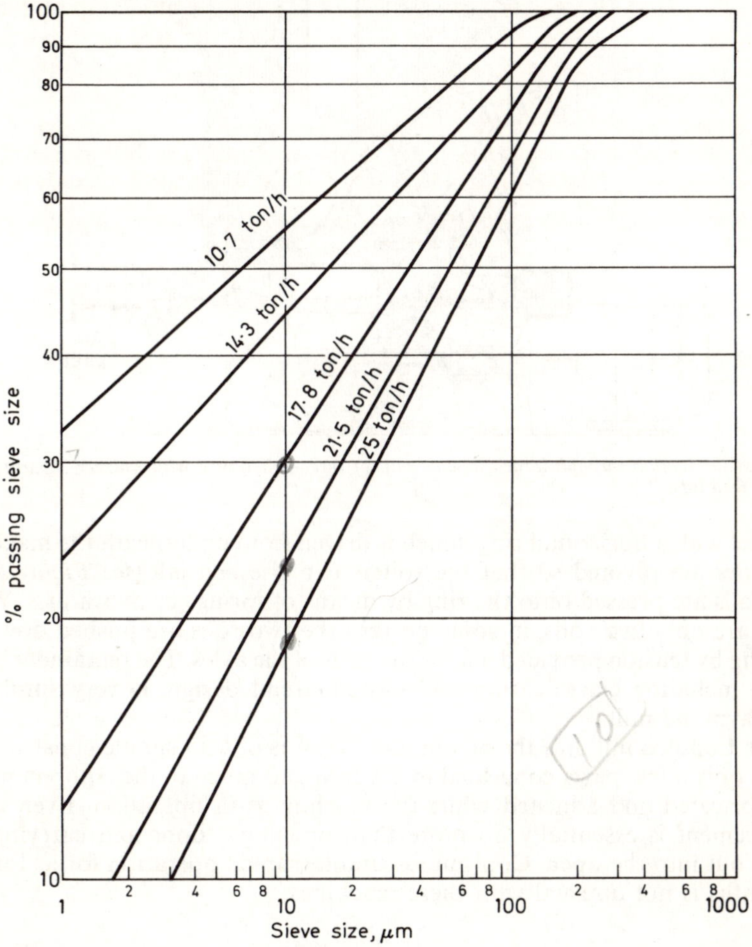

Figure 8.29. Raymond mill: effect of throughput on size distribution

LOPULCO OR LOESCHE MILL

The Lopulco or Loesche mill consists of two or more rollers shaped like cone frusta and having axles which are set at angles to the horizontal corresponding to (half) the solid angle of their cones so that their faces are

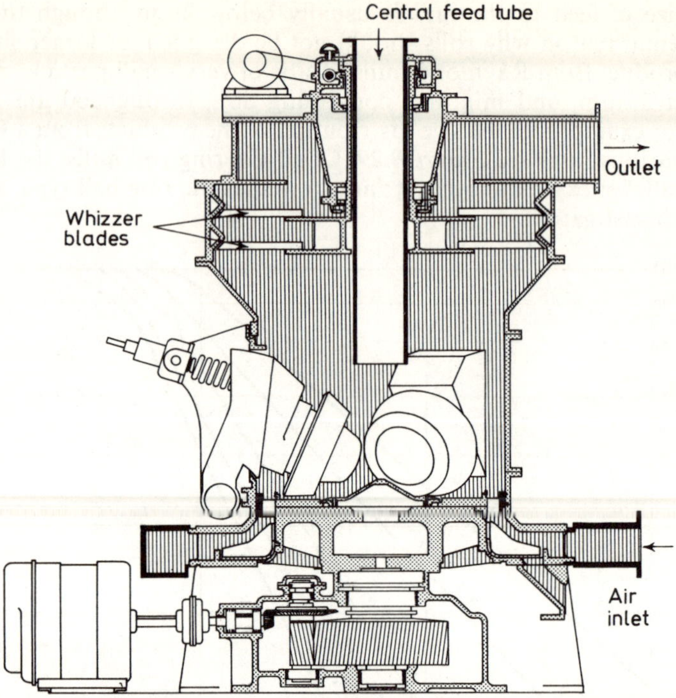

Figure 8.30. Section through large Lopulco mill of three-roller type with whizzer separator and central feed tube

parallel with a horizontal ring which is driven from underneath the machine; the axles are pivoted so that the rollers can rise and fall (see *Figure 8.30*). The rolls are pressed onto the ring by means of springs or hydraulics. When there are only two rolls, in some designs the two rolls are pushed down on the ring by tension provided across the ends of the axles. The remainder of the plant, including classification and closed-circuit design, is very similar to the Raymond mill.

The Lopulco mill has the advantage that it is of very simple construction. It has only a few parts concealed in the mill and much of the equipment can be lubricated and adjusted while the machine is in operation. Even roller replacement is essentially no more than opening a door and carrying the roller out into the open. Continuous uninterrupted operation for as long as 6 months is not unusual with these machines.

BABCOCK AND WILCOX RING BALL MILL

The Babcock and Wilcox mill is sometimes known as the E-type mill. It was developed in the U.S.A. by the company which employed Hardgrove (see p. 58). It is no coincidence that the Hardgrove grindability apparatus bears

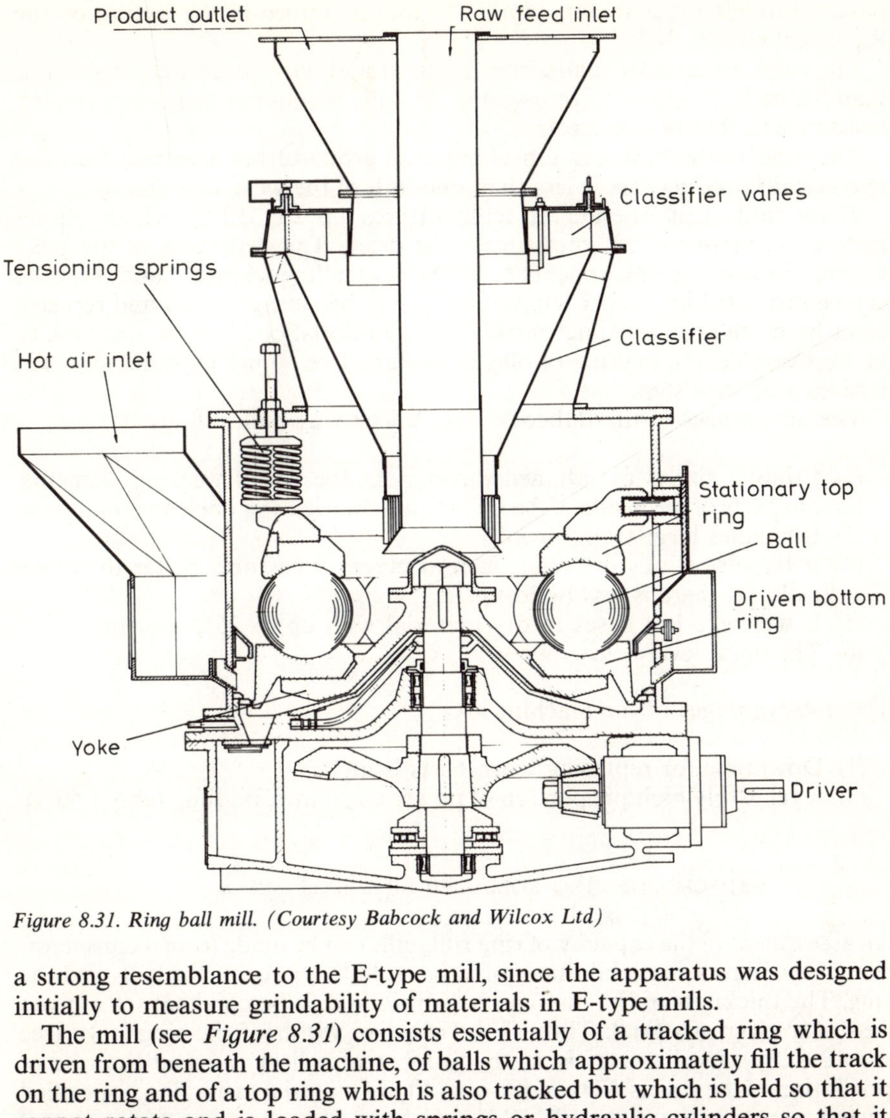

Figure 8.31. Ring ball mill. (Courtesy Babcock and Wilcox Ltd)

a strong resemblance to the E-type mill, since the apparatus was designed initially to measure grindability of materials in E-type mills.

The mill (see *Figure 8.31*) consists essentially of a tracked ring which is driven from beneath the machine, of balls which approximately fill the track on the ring and of a top ring which is also tracked but which is held so that it cannot rotate and is loaded with springs or hydraulic cylinders so that it exerts a high pressure on the balls. This pressure, however, can be overcome by a high enough resistance so that the top ring can move vertically.

The balls are hollow and usually made of cast steel; ten has been found[16] to be about the optimum number for a compromise between length of life and consumption of idling power.

The feed axis is coincident with the vertical axis of the machine. Material is taken between the top and bottom rings and is thrown out radially by the rotating bottom ring to the balls. Air enters on the sides of the bottom ring and entrains solid; the fluid containing the suspended particles of solid then passes through a system very similar to that described in the section on the Raymond mill (p. 183).

The rings are usually made from Ni-hard and have more wear resistance than the balls. Ni-hard is not used for the balls because of its inferior impact resistance to that of cast steel.

The wear of the balls per ton of material ground from a certain feed size to a certain product size is less than one-fifth of the wear in a ball mill.

These mills can operate at temperatures up to 350°C, which allows material containing 25% moisture to be dried. The deflection of the yoke driving the bottom ring appears to be the controlling feature[16] and the yoke can be protected for higher temperatures (e.g., by strengthening and replacement by more heat-resisting parts) if the additional cost can be justified, as in, for example, the making of plaster of Paris from lump gypsum, in which it saves a process step.

The advantages of the Babcock and Wilcox ring ball mill are[16]:

(1) Product size is maintained throughout the life of wearing elements.
(2) No parts of the inside of the mill require lubrication or clean conditions.
(3) It handles large tramp iron.
(4) It handles sticky material, such as clayey limestone, better than ring rolls, i.e. there is less build-up.
(5) It will take hot gases to dry material with up to 25% moisture.
(6) The noise level is low.

The disadvantages of the machine are:

(1) Downtime for replacing worn parts is high.
(2) A relatively high proportion of power consumed is idling (about 60%).

CAPACITY OF RING ROLL AND RING BALL MILLS

An assessment of the capacity of ring roll mills can be made from a consideration of the film of material that would be included between the rolls and the ring. The thickness of the film can be considered to be constant for a given feed of a given material. Thus, if the effective width of the ring is W (see *Figure 8.28b*), its diameter is D, the number of rollers is n and the rate of revolution of the ring is N, then the quantity of material ground to a specified degree is proportional to $\pi D \cdot W N n$. *Table 8.11* gives some capacities of Raymond mills and sets the value given by this formula beside them: they are seen to be nearly proportional in this instance.

For ring ball mills an approximate relationship is:

$$\text{capacity} \propto d^2 n \qquad \text{or} \qquad \frac{C}{d^2 n} = \text{a constant}$$

in which C is the capacity, d is the diameter of the balls and n is their number.

Table 8.12 gives some capacities of E-type mills and sets the value given by this formula beside them. The pressure on the balls has to be sufficient to

Table 8.11. RELATIONSHIP FOR CAPACITY OF RAYMOND MILL

Code No.	*D*, in	*n*	*W*, in	*N*, rev/min	*Capacity in* lb/h *Florida phosphate rock, 60%–74* μm (*A*)	$\pi DWNn$ (*B*)	*B/A*
5048	50	4	8	435	12×10^3	24×10^5	2×10^2
5448	54	4	8	435	15	27	1.8
6058	60	5	8	435	21	33	1.6
6669	66	6	9	600	34	67	2
73612	73	6	12	600	50	99	2

For meaning of symbols see *Figure 8.28.*

Table 8.12. E-TYPE RING BALL MILL: RELATIONSHIP BETWEEN BALL NUMBER AND DIAMETER AND CAPACITY (COURTESY BABCOCK AND WILCOX LTD)

Maker's identification	*Loading per ball,* kg	*Type of loading*	*Ball diameter* (*d*), cm	*Number of balls,* *n*	*Capacity coal* (*V*), tonnes/h	$\pi d^2 n/V$
E26	550	Springs	23.5	8	2.6	5350
E35	550	Springs	23.5	11	4.3	4450
E44	550	Springs	23.5	14	6.0	4050
E50/47	2300	Springs	42.5	8	8.5	5300
E56/50	2300	Springs	42.5	9	9.5	5360
E64/60	2300	Springs	42.5	11	12.0	5200
E70/62	3600	Springs	53.5	9	14.0	5490
7E	2700	Pneumatic cylinders	53.5	9	17.0	4750
8.2E	2600	Pneumatic cylinders	53.5	12	22	4950
8.5E	4000	Pneumatic cylinders	65.5	10	27	5000
9.1E	5500	Pneumatic cylinders	77	9	33	5100
10E	5500	Pneumatic cylinders	77	10	40	4650
11E	5500	Pneumatic cylinders	77	11	49	4200
11.8E	5500	Pneumatic cylinders	77	12	58	3850
12E	8000	Pneumatic cylinders	92	10	67	4000
14E	11000	Pneumatic cylinders	107	10	93	3900

break the material, i.e. pressure per unit of surface of the ball in contact with the material has to exceed a certain value. The amount by which the certain value is exceeded does not appear to make much difference to the capacity.

POWER CONSUMPTION OF RING ROLL AND RING BALL MILLS

Nearly all mills of the ring roll or ring ball type include air separation equipment, and the total power consumption is approximately double that actually consumed by the mill; in other words, the power consumed by the fan and auxiliaries is about equal to that consumed by the mill itself. The justification for the additional power consumption is discussed in the section on closed-circuit grinding in Chapter 5 (p. 93).

The over-all power consumption of ring roll mills is somewhat less than that of corresponding ball mills operating under the same conditions of feed and product size. A typical over-all figure for this type of mill grinding limestone from 2.5 cm to 80% through 76 μm is 18 kWh/t (but see p. 60). Other typical data are shown in *Table 8.13*.

Table 8.13. PERFORMANCES OF RING ROLL MILLS

(*a*) *Data on Sturtevant mills fed with material 80% passing 500* μm, *product 80% passing 147* μm. *Density of both rocks = 3.2; see p. 63 for definition of factor (surface created)*/h.p.

	Algerian phosphate rock			*Florida phosphate rock*		
Gross h.p.	*Output*, ton/h	h.p. $ton^{-1} h^{-1}$ *average*	(*Surface created*)/h.p.	*Output*, ton/h	h.p. $ton^{-1} h^{-1}$ *average*	(*Surface created*)/h.p.
25	3–4	7.2	1.7	2–2.25	11.7	1.0
45	5–7	7.5		3.5–4	12	
90	10–14	7.5		7–8	12	

(*b*) *Data on Raymond with various materials*

Material	*Specific gravity*	*Feed size*, μm	*Product size*, μm	h.p./ton
Coal	1.3	−3 000	90% 147	19/20
Limestone	2.35	−3 000	99% 147	
			90% 74	24
Barytes	4.5	−10 000	99% 53	52

REFERENCES

1. TAGGART, A. F., *Handbook of Mineral Dressing*, 1953. New York; Wiley
2. MCGREW, B., *Rock Prod.* onwards from 1950 (June)
3. BAUMANN, V. A., *Mech. Constr. Moscow*, **11**(7) (1954), 21
4. GIESKIENG, D. H., *Trans. Am. Inst. Min. Metall. Eng.*, **184** (1949), 239; **187** (1950), 568
5. GAULDIE, K., *Engineering, Lond.*, **176** (1953), 456; **177** (1954), 557
6. HERSAM, E. A., *Trans. AIME*, **68** (1923), 463
7. ROSE, H. E. and ENGLISH, J. E., *Trans. IMM*, C32 (1969)
8. BONWETSCH, A., *Quarry Mgrs' J.*, **30** (1947), 586
9. JOISEL, A., *Dechema Zerkleinern Symposion, 1962*, 49
10. MESSENGER, J. F., *Design and Application of Crushers*, Discussion Instn Mech. Eng., 1972
11. MILLER, W. T. W. and BADGER, G., *Instn Mech. Eng. Yorkshire Meeting, 1938*
12. CROSS, A. H. B., MARSHALL, D. F. and SARGANT, R. J., *Iron and Steel Institute Report on Development of Monolithic Dolomite Linings*, 1946
13. MAKAROV, V. I. and LARIN, A. P., *Ogneupory̆*, **9** (1968), 18
14. KOEHLER, W. A., *Trans. Am. Inst. Chem. Eng.*, **40** (1944), 575
15. COULSON, J. M. and RICHARDSON, J. F., *Chemical Engineering*, 1955. London; Pergamon
16. MCKENZIE, E. C., *S. Afr. Mech. Eng.*, **22** (1972), 62

9

Impacting Machines

Many types of mills can be classified as impact mills: they include hammer, cage, centrifugal, drum and chain, prall, swing, sledge, stamp, vibration and pin mills.

The distinctive feature of an impact mill is that it has a fast-moving part which transfers some of its kinetic energy to the material particles by coming in contact with them, so that internal stresses are created in the particles. These stresses are often large enough to cause the particles to shatter but are supplemented by allowing the struck particle to impact upon an anvil of some kind.

Many mills in this class resemble roll crushers in their broad design; but impactor rotors move much faster and nipping, as a breaking process, is a comparatively minor feature. This difference in mechanism results in important differences in the finished product. In roll crushing the slow action of nipping appears to leave a residual strain in the material which may manifest itself much later in cracks and severance. Because the action of impact is upon the suspended particle and stress waves are free to concentrate at areas of weakness, it appears to produce immediate effects, with little after-effect, so that subsequent cracking and breaking is much rarer. This stress-free condition is particularly valuable in stone used for brick-making, for building and for road-making, in which integrity depends upon binding agents covering all surfaces; in such processes as vertical kilning; and for such purposes as packing in scrubbing towers, in which disintegration would lead to blockages and unequal distribution of fluids.

A type of mill which relies upon the impact of a jet of fluid to transfer momentum to a particle is the fluid-energy mill, but this is not classed as an impact mill, because it is thought to depend more upon the subsequent abrasive action between the fast-moving particles than upon the inducement of internal stresses within the particle. Such mills are therefore classed as attrition mills.

Advantageous features of the impact machine are:

(1) The energy required relative to new surface created is smaller than for most comminuting equipment.
(2) Usually comparatively little space is required and underground installation of some of the total volume is sometimes possible.
(3) Worn parts are usually easily replaced.

(4) Most materials can be broken in it (albeit some not economically, owing to excessive wear), fibrous material such as bones, wood, leather and asbestos included.
(5) The product from minerals tends to be cubic in shape.
(6) Strain-free particles are produced that are not susceptible to after-cracking and breaking.

Disadvantageous features are:

(1) Very hard materials cause intense wear on the moving parts. Metallurgical developments with alloyed cast-irons are permitting harder materials to be processed economically.
(2) Some of the material that the impact machine can handle without intense wear (particularly fibrous material) builds up, over long periods, on the parts.
(3) Product size is often not easily adjustable, because parts such as pulleys have to be substituted, and in some equipment clearances between rotor and impact plate cannot be altered during operation. This disadvantage has been eliminated in some designs.

MECHANISM OF IMPACT

When two bodies collide, i.e. impact, they compress each other until they have the same velocity and remain in this state until restitution of the compression begins; then the bodies push each other apart and go their separate ways. If one of the bodies is held in position, stationary or otherwise, then the other body has to conform with this position during a small interval. The restitution of compression takes time, albeit a very short time, to complete, and during that time a body possesses strain energy which may

Table 9.1. CLASSIFICATION OF IMPACT MILLS

IMPACT MILLS

ROTARY		OTHER
Nature of rotor	*Additional breaking device*	
Impactors set round periphery of rotor disc:		Stamp, Vibration
Fixed hammers	Breaking plate	
Pivoted or swing hammers	Breaking bars or grid	
Short chains	Drum	
Rotor in form of drum:	Cage	
Rotor in form of cage:	Concentric cages	
Pieces set over surface of drum or disc		
Lugs		
Rods (pins)		

lead to fracture. Only the restitution which takes place while the bodies are in contact is likely to have much effect on the kinetic energies of the bodies; thus the faster the bodies move away from each other, the more energy is likely to be available for subsequent fracture. If the bodies are travelling in the same direction before impact, the glancing blow may be more effective for fracturing than direct collision. The glancing blow may also be more effective for the reason that the faster the rate of impact, the stronger the bodies become (see p. 34), though this effect is only logarithmic in its proportions. It is a curious fact, to which reference has already been made, that the faster the rate of application of the force, the more quickly does it manifest itself in fracture as distinct from remaining stored as strain energy.

Impact mills can be broadly divided into types with impactors which rotate and those with other motions. Rotary machines can be further subdivided. A classification is shown in *Table 9.1*.

ROTARY HAMMER MACHINES

Rotary machines are those with rotating drums or discs as the source of the energy to be imparted to the material particles. This (kinetic) energy is usually transmitted through certain parts, which may be attached to the rotor; or may be bars forming the interrupted surface of the drum, like a cage; or may be flails. These hammers, as they may be called collectively, can be rigid or pivoted. If they are pivoted, they are more limited as hammers but more accommodating to chance large lumps and extraneous pieces of metal. Pivoted hammers are therefore more usual for smaller machines or those built to handle softer materials.

The rotors can be mounted on shafts, horizontal or vertical and sometimes designed so that they can be reversed to even out wear. The breaking devices can be stationary; co-rotating, either differentially or in synchronism with the rotor; or contra-rotating, at either the same or different speeds. The breaking devices can be smooth, corrugated, straight, curved, slotted or solid. Traditionally they are made of Hadfield's (13% manganese) steel. More recently, chromium carbide nodulised cast-iron has been used, which does not depend upon work-hardening for its effect and is therefore instantly abrasion-resistant. The hammers are usually of the same material as the breaker plates. Hammers are made in various shapes, including those of simple rectangles, hammer ends and flare ends. In general, single-rotor machines are used for coarse crushing and augmented or double-rotor machines for secondary crushing. The substitution of double-rotor for single-rotor machines has been found to prevent build-up and blocking in the handling of clays.

The swing-hammer machine (see *Figure 9.1*) is designed for material pieces weighing from 1 to 15 kg. The pivot is provided to allow a hammer, on encountering the occasional excessively large or hard lump or piece of extraneous metal, to move out of its path. The mill is provided not with a free exit, but one that is grilled or perforated, so that material not broken to the size required is retained and swept up again by the rotor for further impacting.

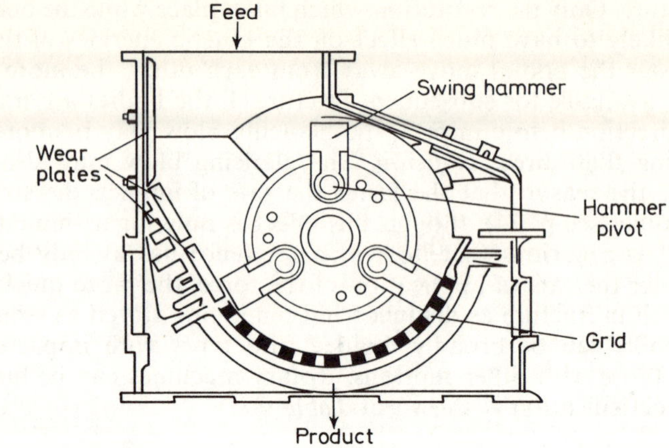

Figure 9.1. Swing-hammer mill

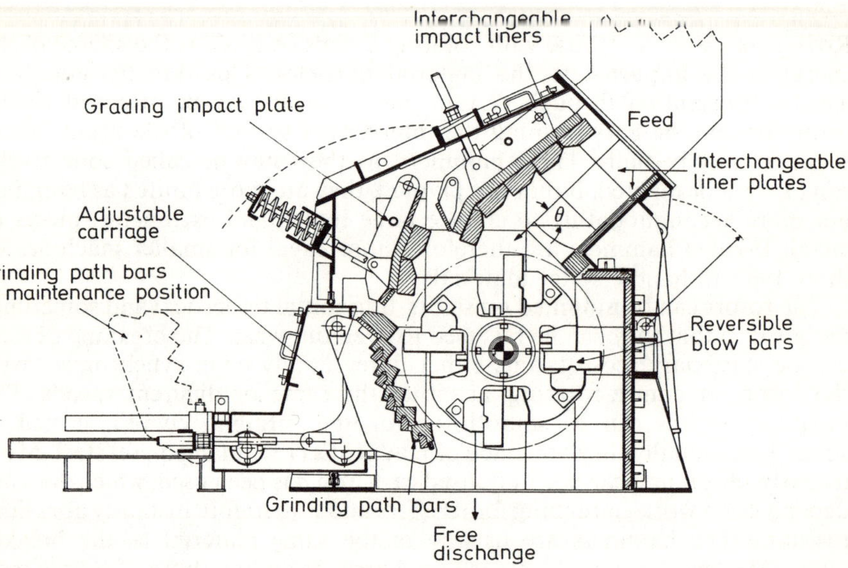

Figure 9.2. Holmes–Hazemag impact crusher with grinding path

This type of machine is designed to give to the pieces fed to it velocities of the order of the velocity of the rotor, and fracture takes place either through the severity of the blow received from the hammer or through subsequent impact with the casing or grid. Attrition can play a part in particles meeting others set on the grid.

The fixed-hammer machine (see *Figure 9.2*) is produced for an altogether larger duty. It is produced for pieces of the order of 100–5000 kg and the feed chutes are designed so that the material enters tangentially to the rotor hammers to receive a glancing impulse to send it spinning towards the breaker plate. The velocity imparted to the piece is deliberately restricted to a fraction of the velocity of the rotor, to avoid what would otherwise be an enormous strain upon the rotor bearings. The fractured pieces which are able to pass through the clearance between the rotor and the breaker plate are, in some machines, made to enter a chamber created by another breaker plate and having a somewhat smaller clearance than the first one before entering a third classification made by what is called a grinding path, which is so designed (see *Figure 9.2*) that flakiness is reduced and the particle leaves the system with nearer to cubic dimension ratios. Machines are available which have double rotors, one being of the fixed-hammer type and the other a swing-hammer type. Such machines can reduce from pieces of over 5000 kg down to pieces all less than 40 g.

Thus, the rotary impact mill affords control of the product size, the facility to control product shape and the energy-saving aspect of the removal of the particle, once it has reached the size required, from further size reduction.

A type of machine developed as a macerator for waste claims to handle anything. It consists of a cylinder standing with axis vertical, on the bottom of which is a plate disc with tungsten carbide lugs set round the face; steel knives project from the cylinder sides as radii so that their tips nearly touch the lugs. The waste is added with water jets which hold the waste onto the

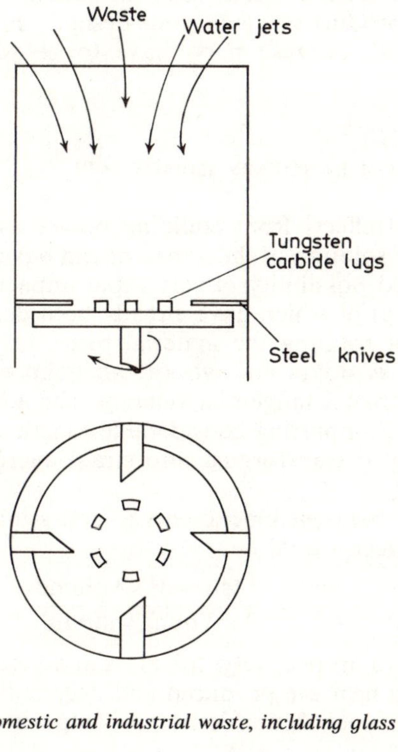

Figure 9.3. Macerator for domestic and industrial waste, including glass

plate until it is small enough to pass out through the annulus between the disc and the cylinder (see *Figure 9.3*).

Particular advantages claimed[15] for rotary impact mills are:

(1) A higher proportion of in-grade product size is possible with little intermediate grade and only a small amount of fines owing to the classifying action of the clearances.
(2) The even balance of the rotor does not demand expensive foundations.
(3) Because of the enclosed construction dust can easily be removed by suction and outside cooling applied with water jets.
(4) The breaking effect on a particle projected against a plate tends to be along grain boundaries and thus there is some quality separation in breaking a mineral of mixed composition, which can be useful in mineral processing.
(5) Weathered material, the outside of the pieces of which have been softened, but of which the cores remain hard, is particularly easily broken.
(6) Very large reduction ratio.

Disadvantages are:

(1) The machine can act like a fan because of the fast movement of the impactor and entrains material in the subsequent air flow.
(2) The smaller product size is in some designs, though not in others, not easily adjusted, because parts have to be substituted to decrease clearances.

MECHANISM IN ROTARY IMPACT MILL

Besides the general effects from colliding bodies as described above, the additional factor of rotation of the impactor can have two important effects. One is the increased possibility of very rapid impact through the glancing blow, the advantages of which have already been described; and the other is the possibility of rotating the material piece, i.e. imparting a spinning action to it, which as MOYES and WOODHEAD point out[1], creates less strain on the rotor drive than a tangential velocity. The advantage of the rotor is that it is a means of imparting considerable kinetic energy to the material piece, which energy is transformed into strain energy by impact with the breaker plates.

The relationship between kinetic energy and strain energy, if the transformation were perfect, would be:

$$\frac{mv^2}{2} = \frac{(\text{stress})^2 \times \text{volume}}{2 \times \text{Young's modulus}}$$

in which m and v are, respectively, the mass and velocity of the particle. In practice, sound and heat are produced and decrease the strain energy. The blow bar must be of such dimensions and composition that it is capable of absorbing the strain energy from the reaction of imparting kinetic energy to the material piece.

Moyes and Woodhead suggest that effectively the initial collision is between the projecting part of the blow bar (hammer) and the material piece and that, for most effect, the volume of the projecting part of the blow bar should be adequate to absorb this strain energy and be travelling fast enough to break the material piece. This is most important in considering the number of hammers to be fitted to the rotor at the design stage: the effectiveness of impact is likely to be lost when trying to increase the selection function (for meaning of this term see p. 75).

ROTOR SPEED

Rotor speed is a very important factor because it has a square-law effect on the kinetic energy of the rotor, the source of energy to the particle. The selection function increases with speed of rotation[2], as would be expected, because the striking rate also rises. The most important factor is peripheral speed, since it is this that influences the particle impact. Peripheral speed reaches optimum for the selection function. This is sometimes stated to be about 300–500 m/s, to cover most materials, although there were signs[2] of a maximum being reached for fertiliser at about 50 m/s.

A relationship between the energy to break a particle and the equivalent kinetic energy is

$$\frac{mv^2}{2} = \frac{(\text{compressive strength})^2}{2 \times \text{Young's modulus}} \times \text{volume}$$

The mass m is proportional to the particle size (x) cubed and multiplied by its density (ρ).

Thus

$$\frac{\rho x^3 \cdot v^2}{2} = \frac{\sigma^2 \cdot x^3}{2E}$$

where σ is the ultimate strength, or

$$v = \sqrt{(\sigma^2/\rho E)} \qquad \text{or} \qquad 5 \text{ m/s}$$

for material such as limestone. This is about one-tenth of the speed normally used for the rotor, which by design is greatly in excess of that estimated to be transferred to the particle. Thus the theoretical result is not greatly different from that obtained in practice.

JOISEL[3] approached the question of speed by assuming that the work done in impact is the product of the force necessary to break the particle and the average distance moved by the particle from moment of impact to collapse of the particle. Thus the force is tensile strength (he argues that this is the determining strength rather than the compressive strength) multiplied by the area of a cross-section of the particle, i.e. σx^2 in the symbolism used above. The average distance moved by the particle from impact to collapse is the distance from the periphery of the particle to the centre of gravity of the particle before collapse from which has been subtracted the similar distance after collapse, i.e. the difference between the radii of the particles before and after impact. Thus, if R is the reduction ratio,

$$\text{work done} = x^2\sigma\left(\frac{x}{2}-\frac{x}{2R}\right)$$

and thus equals the kinetic energy $\frac{1}{2}\rho x^3 \cdot v^2$ or

$$v = \sqrt{\left[\frac{\sigma}{\rho}\left(1-\frac{1}{R}\right)\right]}$$

or as maximum $=\sqrt{(\sigma/\rho)}$. For a material such as limestone this would be about 50 m/s.

Size of product is affected by speed of rotation. The faster the rotor the smaller the size of the product; or, expressed in terms of the breakage function (see p. 75), for a given size the breakage function increases. For example, GARSIDE and WILDSMITH[2] found that the breakage function for a given feed size (5.2 mm) and for a given product size (0.6 mm) altered with peripheral speed as follows:

Peripheral speed, m/s	15	30	45
Breakage function (B)	0.26	0.43	0.63

Harder materials require higher speeds than softer materials to obtain the same breakage functions.

Reduction ratios (as defined in Chapter 2, p. 10) as high as 40 to 1 are achievable with these machines and ratios of from 10 to 20 are common.

NUMBER OF IMPACTORS OR HAMMERS AND THEIR SHAPE

The selection function is increased by the number of hammers provided they are individually large enough to effect breakage at the speeds used. The distribution function is not much affected by the number of hammers. The value of swing hammers is in accommodating extraneous excessively large and hard pieces. The shape of hammers appears to be important in certain applications, e.g. in municipal refuse destruction, when the originally rectangular bars wear and become round shouldered, bite is reduced and the bars have to be replaced if efficiency of comminution is to be maintained.

The spacing of hammers must be such as to balance the rotor and correspond with the rotor speed so that each piece in the feed is likely to be hit by a hammer before it can penetrate more than a few millimetres into the cylinder of rotation of the hammer tips.

NATURE AND DESIGN OF ROTOR

Various designs of rotors or impactors are shown in *Figure 9.4*. Broadly, the appropriate blow-bar designs can be classified for fixed-hammer machines

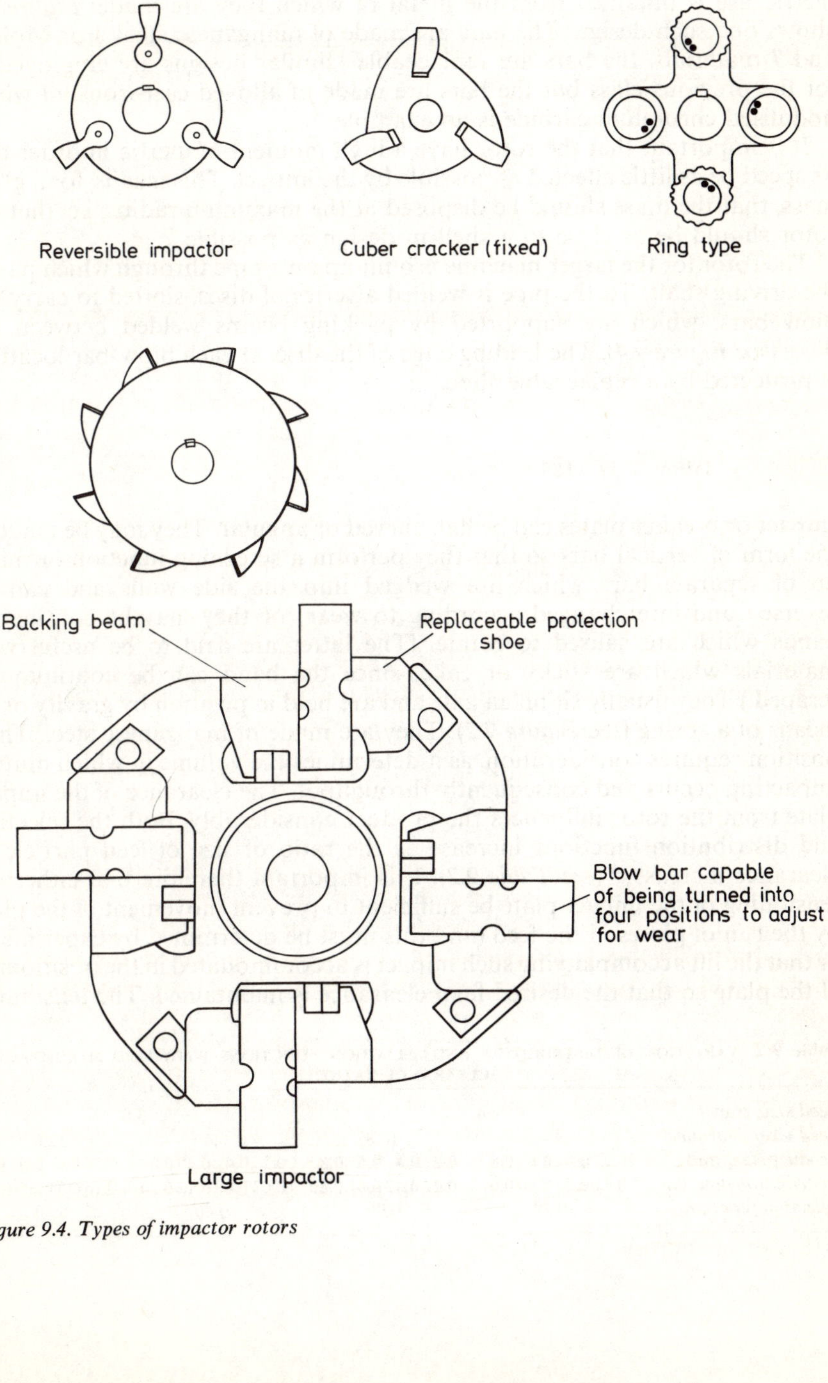

Figure 9.4. Types of impactor rotors

as follows according to the hardness of the material to be handled. For Mohs 1 and 2 materials, the blow bars are usually non-replaceable and they usually last as long as the machine; they are made of manganese steel. For Mohs 3 and 4 materials, the blow bars are replaceable and various ingenious designs have been developed for turning the bars so that maximum volumetric use is obtained from the metal of which they are made. *Figure 9.4* shows one such design. The bars are made of manganese steel. For Mohs 5 and 7 materials, the bars are replaceable; similar designs are employed as for the previous class but the bars are made of alloyed cast-irons, of which nodulised chromium carbide is an example.

It is important that the rotor have a high moment of inertia in order that its speed be as little affected as possible by the impact. This means, for a given mass, that the mass should be disposed at the maximum radius, i.e. that the rotor should be as close to a shell in design as possible.

The rotor for the larger machine is built up on a pipe through which passes the driving shaft. To the pipe is welded a series of discs, slotted to carry the blow bars, which are supported by packing beams welded between the discs (see *Figure 9.4*). The leading edge of the disc, at each blow-bar location, is protected by a replaceable shoe.

IMPACT PLATES

Impact or breaker plates can be flat, curved or angular. They may be made in the form of vertical bars so that they perform a screening function or made up of separate bars which are wedged into the side walls and can be reversed and interchanged according to wear; or they may be continuous bands which are caused to rotate. (The latter are said to be useful with materials which are sticky or cake, since the band can be continuously scraped.) They usually sit on an axle and are held in position by gravity or by means of a spring (see *Figure 9.2*). They are made of manganese steel. Their position requires consideration, as it determines the volume in which mutual impacting occurs and consequently throughput. The clearance of the impact plate from the rotor influences the product considerably. Both the selection and distribution functions increase as the ratio of size of feed particle to clearance increases[2] (see *Table 9.2*). It is important therefore that either the tensioning of the impact plate be sufficient to prevent movement of the plate by the run of pieces in the feed (and this must be determined by experiment) or that the lift accompanying such impact is accommodated in the positioning of the plate so that the desired final clearance is maintained. The tensioning

Table 9.2. VARIATION OF DISTRIBUTION AND SELECTION FUNCTIONS WITH FEED SIZE/IMPACTOR CLEARANCE RATIO

Feed size, mm	4.7								3.0							
Feed size/clearance	9.4				18.8				3.0				12.0			
Product size, mm	0.2	0.4	0.6	0.8	0.2	0.4	0.6	0.8	0.2	0.4	0.6	0.8	0.2	0.4	0.6	0.8
Distribution function	0.11	0.23	0.40	0.61	0.21	0.42	0.61	0.80	0.03	0.08	0.12	0.20	0.16	0.31	0.50	0.66
Selection function	0.79				0.90				0.46				0.80			

of the impact plate can also act as a safety device for tramp metal. In the fixed-plate design a shear pin is sometimes incorporated. On some machines adjustment of the position of the impact plates is possible from outside the casing. The tensioning should, however, be small enough to allow extraneous and excessively hard oddments to be released from the system to avoid damaging the rotor.

The incorporation of two pinned impact plates allows the clearance at the front of the rotor to be adjusted on one and at the rear of the rotor on the other so that, if there is any asymmetry in the flow of the particles, the effects can be adjusted.

The individual wear parts on the impact plates and on the walls of the mill should be of such a volume as to permit absorption of the energy of impact from the flying pieces without breaking. Some manufacturers make a feature of ease of replacement of wall wear plates, which are made in the form of square tiles and which can be moved from one location or orientation to another according to the degree of wear they suffer.

SIZE OF PRODUCT

The rotary impact crusher design is so adaptable and has such large potential reduction ratios that any size of product is possible down to about 100 μm. Machines can therefore be designed for almost any range of comminution in the primary and secondary categories, and very often what are for other machines two operations can be concentrated into one in the rotary impact mill. A property of impact mills in comminuting mixtures of materials of different hardnesses to very nearly the same degree of fineness is frequently of great value in industry. Two examples may be quoted from the brick-making industry. A clay contained pyrites and when formed into earthenware pipes exhibited unsightly brown areas, so-called iron spots. The clay was put through a rotary impactor and the pyrites was reduced almost to the fineness of the clay. When formed into pipes and fired, the colour was uniform. Another example is of clay used for brick-making and containing lumps of limestone. When the clay was fired and the resulting bricks placed in a store, the bricks broke out into a most repulsive excrescence due to the burnt lime hydrating. After using a rotary impact mill and reducing the whole of the material to below 400 μm, combination of the lime with the clay body silica took place on firing. The bricks displayed no visible faults in use.

SHAPE OF PRODUCT PARTICLE

Many uses of stone require that the shape of the stone be as close to cubic as possible. The rotary impactor is naturally more disposed than, say, a jaw crusher to produce cubes and can also be adapted fairly readily to the consistent production of this shape. What is called a 'grinding-path' is fitted near the end of the path of stone. It consists (see *Figure 9.2*) of an impact plate having a staircase of parallel beams at right angles to the direction of the flight of the stone which collect the slabs and orientate them with respect to the rotor so that they are sheared across their longest axis.

Increased throughput tends to allow more slabs to pass, i.e. for flakiness index (see p. 85) to increase. The larger the size of the product, the more evident this tendency becomes (see *Figure 9.5*).

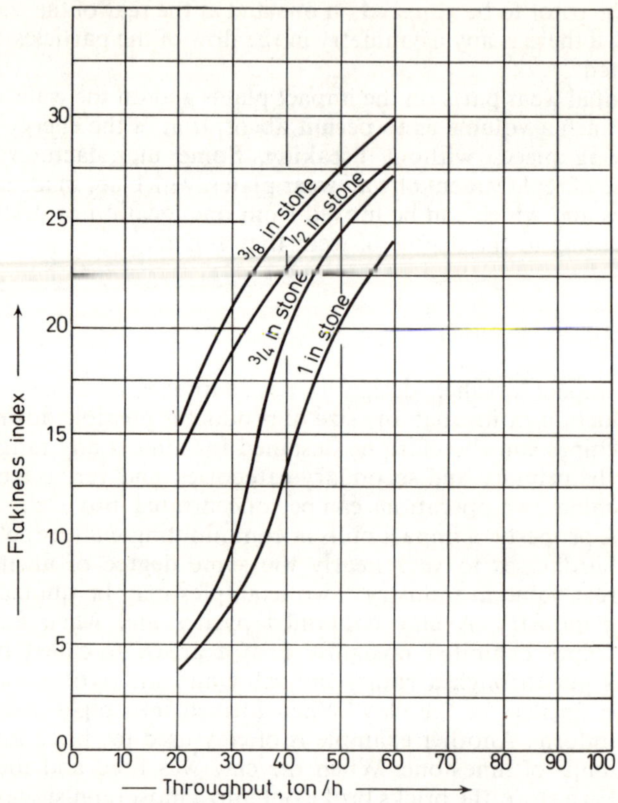

Figure 9.5. Flakiness index versus rotary impactor throughput. Material is gritstone. (After A. J. MOYES *and* P. WOODHEAD[1]*)*

METHOD OF FEEDING

As rotary impact machines usually have little protection against tramp metal, except accommodation in the breaker plates and their general robust structure, a magnetic separator should be installed beside which the feed passes before it enters the mill. Feeding may be by rubber belt, box conveyor, apron feeder or grizzly, depending upon the size and quantity being fed, and then into a chute. Rotary impact mills should not be choke-fed because the feed needs to have kinetic energy on its own account to enter the rotor circle. With smaller or swing-hammer machines this energy is usually imparted by allowing the material to fall through a distance before being led into the mill.

The minimum distance for this fall is given by $H = PnN/240$, where H is the fall distance in millimetres; P is the horizontal penetration required in millimetres into the rotor circle; N is the number of revolutions of the rotor per minute; and n is the number of hammers.

As previously mentioned, MOYES and WOODHEAD[1] have shown that there is an advantage in giving spinning velocity to the feed to decrease strain on the rotor drive. If the feed could be given such motion on entry to the mill, this would therefore be an advantage. The feeding of slabs to the impact mill should be avoided because slabs tend to hit the blow bars 'end on' and not to receive the glancing blow for which the rotor is designed. Slabs can damage the hammers.

The angle of feeding is also important (θ in *Figure 9.2*), as it determines the residence time of the material in the crusher. Also, if it is too small, eddies can be created near the entrance which increase blow-bar wear and retard flow of material into the mill. The sharp angle between the chute and the body of the mill at the entrance can be a source of trouble due to build-up unless this angle is rounded off.

To avoid bridging, which may occur in large-scale operation, the impact plates are sometimes equipped with a hydraulic lift mechanism.

An interesting development, which has been applied to this kind of crusher but which is clearly not exclusive to it, is to put the crusher on crawlers or on hydraulic walking mechanisms which take the crusher right to the quarry face so that it can be fed directly by the shovel and thus process material in more inaccessible quarry locations as well as save double handling.

AIR FLOW IN MACHINES

The high rotation speed of the rotor creates a fan-like condition in the mill body in which there are induced large air-flow rates. Unless the mill installation is properly designed and erected, these air flows can cause much trouble by entraining dust and interfering with feeding and product removal. The technique used is to make air intake and discharge take place through the exit of the machine by sitting the machine on a discharge chute long enough to calm some of the turbulence prior to material dropping out for collection and to have a chain curtain or double gate at the feed inlet to inhibit air flow.

MAINTENANCE

Substantial maintenance is likely to be required for these machines, handling lumps weighing several tons at high throughput and run of quarry conditions; but modern designs make provision for this to be carried out with the minimum of shut-down. The casing is hinged in convenient positions and sometimes hydraulic rams are provided to open it in the minimum of time.

Bearing design for such machines is clearly crucial for maintenance time, and modern machines incorporate roller bearings for the shaft with block shock absorption pads.

Wear rates for the blow bars depend very much on the material and the speed of rotation, but 20 g per ton of the material ground is common for a manganese steel, rising to 50 g for an abrasive material containing, for example, silica.

CAPACITY

Size distribution does not appear to be very sensitive to volume of throughput (see *Figures 9.6* and *9.7*) until a certain maximum loading is reached. When breaker plates are used, the tendency will be for them to lift more, and thus for the size of product to increase on that account, unless breaker-plate clearances are adjusted to take account of this.

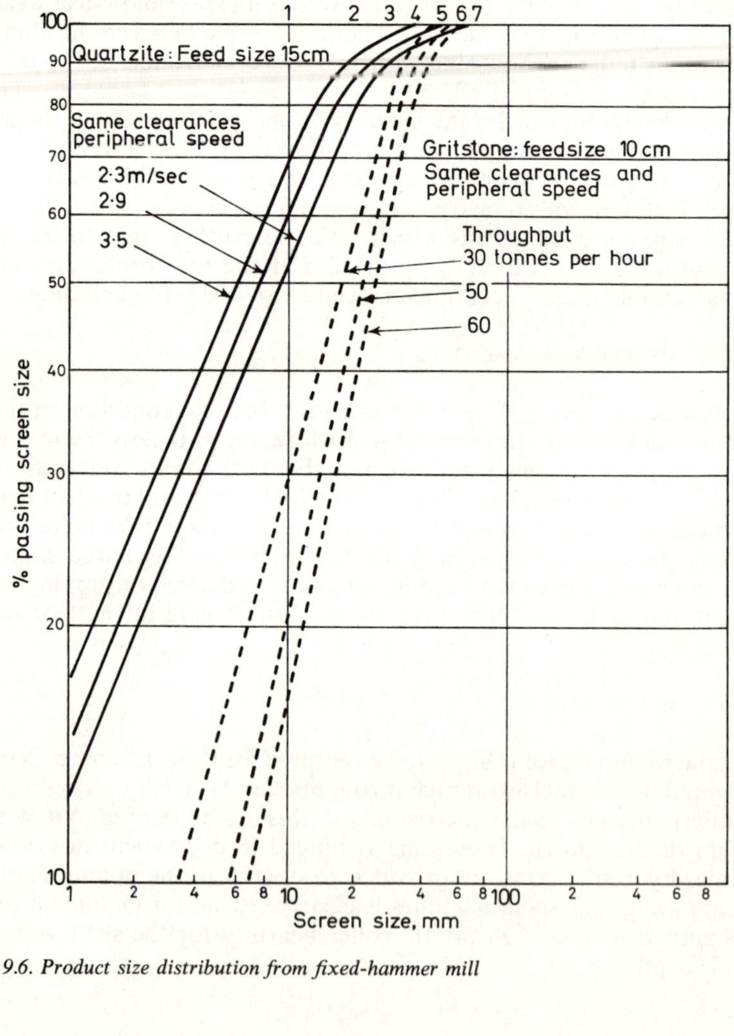

Figure 9.6. Product size distribution from fixed-hammer mill

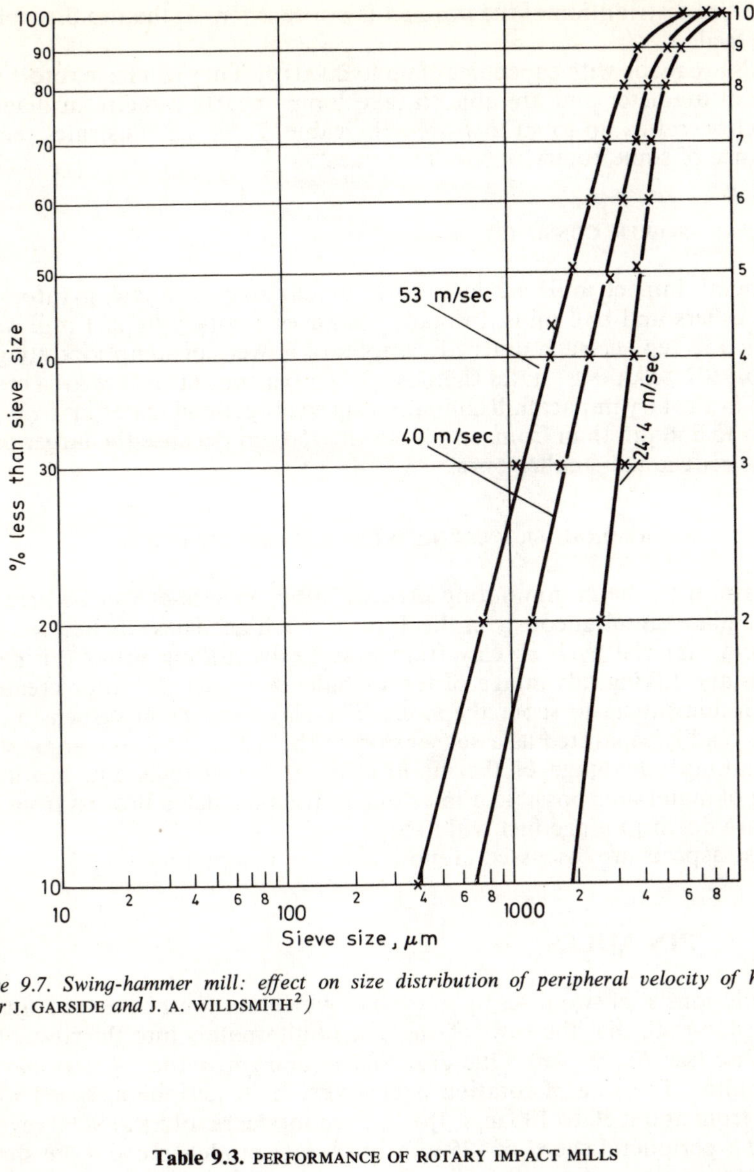

Figure 9.7. Swing-hammer mill: effect on size distribution of peripheral velocity of hammer. (After J. GARSIDE *and* J. A. WILDSMITH[2]*)*

Table 9.3. PERFORMANCE OF ROTARY IMPACT MILLS

Size of rotor drum, cm		*Output*, t/h		*Power consumption*, kWh	
Length	*Diameter*	3 mm *screen*	2.5 cm *screen*	*Total*	*Per tonne*
40	45	0.75–1.0	2–3	5–8	2.7–6.7
50	30	1.5–2.5	4–6	9–11	4.4–6.0
60	45	3–4	8–10	15–20	5.0
200	150			45	

The size distribution of the product is not sensitive to the size distribution of the feed.

Mills are made with capacities of up to 2000 t/h. They can have rotors up to 1.5 m in diameter and are able to take lumps nearly a metre in diameter. Reduction ratios up to 40 to 1 are achievable. *Table 9.3* illustrates the performance of some rotary impact machines[4].

POWER CONSUMPTION

In general, impact mills are less power-demanding than jaw, gyratory and roll crushers and ball mills. In many instances a rotary impact mill can be designed to replace such mills[5]. Examples of power consumptions are given in *Table 9.3*. FARRANT[6] states that less power per unit of surface area created is used in a rotary impact mill comminuting easily ground materials from, say, 20 mm to 750 μm than from, say, 3 mm to 750 μm because the larger pieces help to reduce the smaller ones.

ADDITIONAL PROCESSING WITH ROTARY IMPACTOR

In addition to the comminuting process, other processes can be arranged to take place simultaneously in this type of machine. These include washing adhering material such as clay from a stone by adding water during the milling and taking advantage of the turbulence which the rotor creates in such circumstances to scour the stone. The clay remains in suspension and can be readily separated as a suspension in the water from the larger stone.

By taking advantage of the air flow in such machines and heating it, drying of material is possible. Limestone is dried in such a process from 20% moisture down to a product with 0.5%.

These aspects are discussed in more detail in Chapter 6 (p. 111).

PIN MILLS

Pin mills consist of two discs upon each of which concentric rows of pins are set, of such radii that the rows of one disc fit alternately into the rows of the other disc (see *Figure 9.8*). One disc rotates and one either is stationary or also rotates. The rate of rotation is at a very high peripheral speed which ranges from about 80 to 130 m/s; the 12.5 cm model revolves at 9000 rev/min or with a peripheral speed of 125 m/s, but larger models move more slowly, both rotationally and peripherally. The machine is fed through the centre of the stationary disc either by gravity or, if the material is of a suitable density and particle size, by use of the very considerable suction induced by the rotating disc. A general arrangement of equipment for suction feeding is shown in *Figure 9.8*. The pins are usually square in cross-section because this shape has been found to give a narrower size range than circular cross-section, a result which finds an echo in the need of some users for square-shouldered hammers in hammer mills.

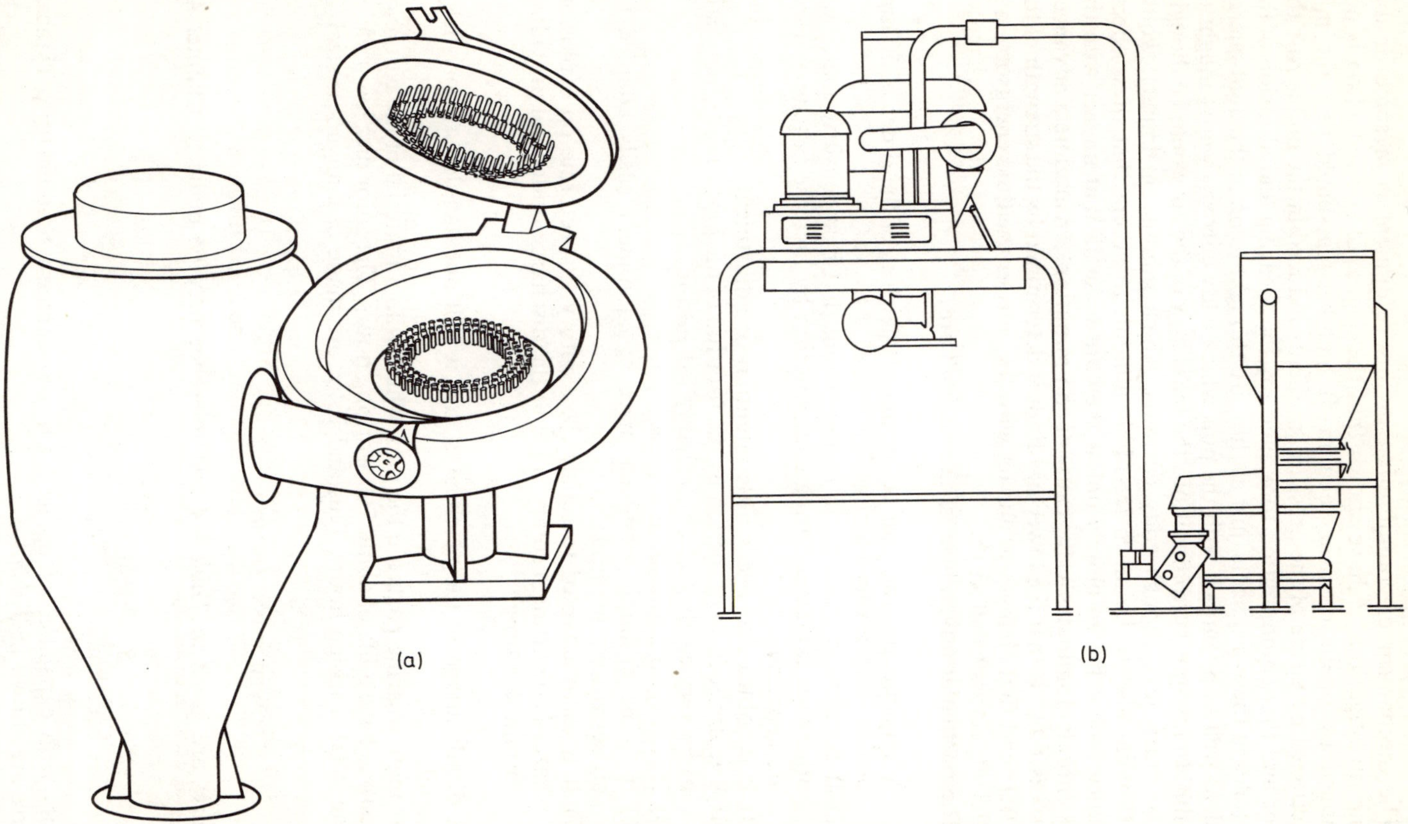

Figure 9.8. Pin mill showing connection with cyclone and general arrangement for suction feed. (Courtesy Kek Ltd)

The mechanism of comminution in this equipment is in some doubt: some investigations[7,8] have shown that the degree of size reduction is not changed by operating in a vacuum. It had been assumed that the mill functioned by virtue of the pressure gradients induced in the mill across the rotor and that comminution took place by the shearing and abrasion of the particles in these gradients, just as they are thought to do in the fluid-energy mill, as well as by impact with the pins which eddy currents were thought to inhibit. In low-pressure conditions the mass flow of the air would be thought to be much less, though eddy currents, which inhibit pin impact, would be thought also to be much less. It is difficult to accept that these effects balance each other so closely unless they are so small that impact with the pins entirely dominates as the cause of comminution. An argument advanced by MARSHALL[8] is that the lower size limit is determined by the strength of the particle and that changes in impact with the pins or the fluid energy are not sufficient factors to affect this.

The particular advantages of the pin mill are:

(1) A very fine product can be produced which is of the order of 50 μm, or less, with a very narrow size range.
(2) It entrains so much air in its movement that heat-sensitive materials, otherwise damaged during comminution by increase in temperature, are kept cool.
(3) It is robust and needs little attention or adjustment.
(4) The production of suction by the rotation of the pins enables materials to be very easily and controllably fed into it.
(5) It is easy to maintain and clean.
(6) It forms a simple, compact piece of equipment easily adapted to air-separation methods.
(7) It is easily adapted to satisfy the Factory Inspectorate on explosion hazards by fitting a rotary valve to the discharge outlet and providing certain shielding.

A disadvantage of the machine is that it will not handle fatty or sticky materials.

Its main wearing parts are the pins, which with a material of the nature of limestone have a life of about 700 working hours. The rotor disc has a life of about 4000 working hours. Pins and rotor are made of manganese steel.

POWER CONSUMPTION

Figures are given in *Table 9.4* and reference may be made to Chapter 4, p. 60.

CAPACITY

Plants with capacities of up to 1 t/h are standard manufacture. Typical figures are shown in *Table 9.4*.

Table 9.4. PERFORMANCE OF PIN MILLS ON LIMESTONE. FEED SIZE, MINUS 3 mm; PRODUCT SIZE, 96% THROUGH 50 μm

Diameter of disc, cm	*Speed* rev/min	m/cm	*Capacity*, kg/h	*Power consumption*, kWh/h	kWh/t
12.5	19 000	125	60	2.2	37
17.8	9 000	84	360	5.6	16
27.5	6 000	86	540	7.5	14
33	5 000	86	720	15.0	21

STAMP MILLS

The stamp mill dates from the Middle Ages, when it was used by German miners and was made of wood. Its chief attraction was the fact that amalgamation of gold ores could take place simultaneously with size reduction. Stamp mills owe their continued existence mainly to the uneconomic cost of replacing them.

The mill consists essentially of a stamp or block which is alternately lifted and dropped under gravity onto material placed on an anvil. A device called a mortar is used to place material on the anvil; it is a housing unit into which a slurry of material flows and overflows a weir through a plate with perforations of the selected size (see *Figure 9.9*). Material, broken to about 5 cm, usually in rolls, is led into A together with a plentiful supply of water. The

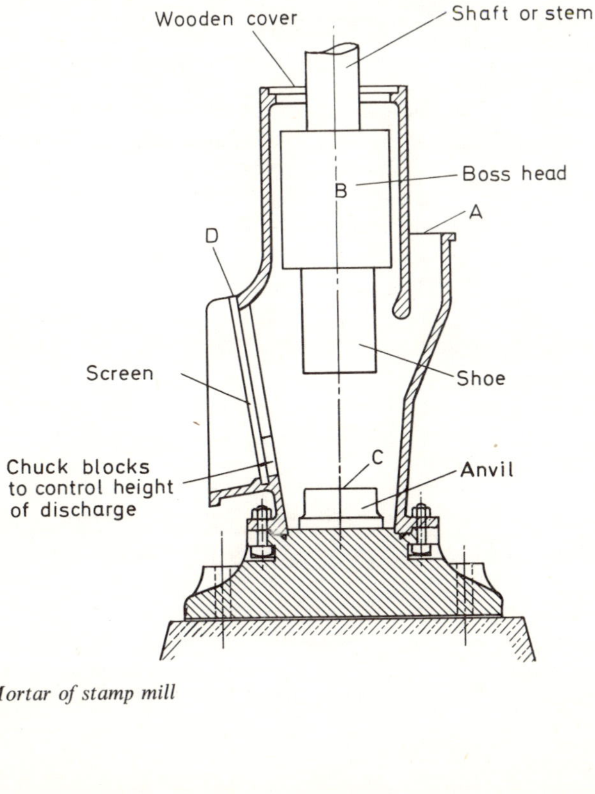

Figure 9.9. Mortar of stamp mill

pestle shaft (B) passes through the top of the mortar. By the movement of the shaft, the pestle drops periodically on the anvil (C) and is lifted up to drop again. Material which has been broken fine enough is carried by the flow of the water through the perforated plate (D). Three of these plates sit on three sides of the mortar and the height of discharge can be regulated so that a sufficient depth of material covers the anvil. The mills are used in batteries of five to ten.

The pestle and the anvil wear and are replaceable.

Specifications of a typical operation still in use in some mines are:

Weight of stamp	600 kg
Set height of drop	20 cm
Drops per minute	98
Duty per stamp per 24 h	6 t
Mortar height of discharge	7 cm
Screen size of mortar plate	0.7 mm
Ratio water to ore, by weight	8
Ore feed +3.5 cm	55%
−2 cm	30%
Product +0.3 mm	30%

PESTLES

Pestles up to 1 t in weight are used, but the weight should be carefully determined to give optimum breaking rate. Energy consumed is very nearly directly the energy required to lift the pestle and there is clearly no point in lifting unnecessary weight.

Pestle shoes wear and break and it is usual to compensate for the decreased weight by adding split cast-iron discs to the shaft. The flatness of the pestle shoe has an important effect on capacity and shoes should be turned flat regularly.

DROP

The drop is usually less than 30 cm and the frequency about 100/min. The height has a considerable influence on the size of product.

WATER USAGE

Water usage is in the range of 5–10 times by weight of the ore processed. No advantage was found in using quantities of water much outside this range[9].

SIZE OF FEED

Feed may be up to 10 cm depending upon mining practice; but the larger the pieces, the shorter the distance the stamp is able to drop and thus the less its effect. The usual practice is to feed less than 5 cm. No advantage is found from crushing feed much below this size[9], because smaller material distributes the impact and less comminution results.

SIZE OF PRODUCT

Discharge sizes are up to 2 mm but can be as small as 0.07 mm. The smaller the product, the greater the cushioning effect on the pestle owing to the thicker deposit in the mortar; impact gives place to abrasion. Ball mills are used for further reduction in, for example, gold mines.

CAPACITY

The capacity of a mill is insensitive to both feed and product size. The effect on auriferous quartz is typical with $\frac{1}{2}$ t stamps:

Feed size, mm	40	80	100
Product size, mm	1.4	1.7	4.7
Capacity, t/h	0.3	0.3	0.25

POWER

A figure often given is 1 h.p. per 180 kg of weight of stamp. This corresponds very approximately to the theory of power being the energy required to lift up the stamp, or

$$\text{power} = \text{weight of stamp} \times \text{distance lifted} \times \text{frequency}$$

Say that weight is 2200 lb (1 t), distance is 10 in (25 cm) and frequency is 100/min. Then

$$\text{horse power} = \frac{2200 \times 10 \times 100}{33\,000 \times 12} = 5.5 \text{ h.p.}$$

1 t is 5.5 times as large as 180 kg.

Figures which deviate much from these do so as a rule because material is not being flushed efficiently from the mortar.

VIBRATION MILLS

The first vibration mill was produced by the Sieb-technik company in Germany about 1936—no doubt an off-shoot of their experience with vibratory screens. The elucidation of the behaviour of cylindrical vibration mills is largely due to the investigations of ROSE and his co-workers[10]. The principle is of loose bodies or media contained within a shell which is caused to vibrate. In contrast to tumbling mills, the media in vibration mills move only a few millimetres and their general path is a complex spinning helix. Thus there is some shearing as well as impacting action on the material between the media. The mills can be run batchwise or continuously and either dry, i.e. with material less than 1% moisture, or wet, i.e. containing about 50% moisture. Intermediate moistures are likely to cause plugging. Mills can be used in a series so that successively more reduction is achieved. Broadly speaking, there are two types of vibration mill, one with a torus-shaped shell and the other with a shell of cylindrical shape (*Figure 9.10*).

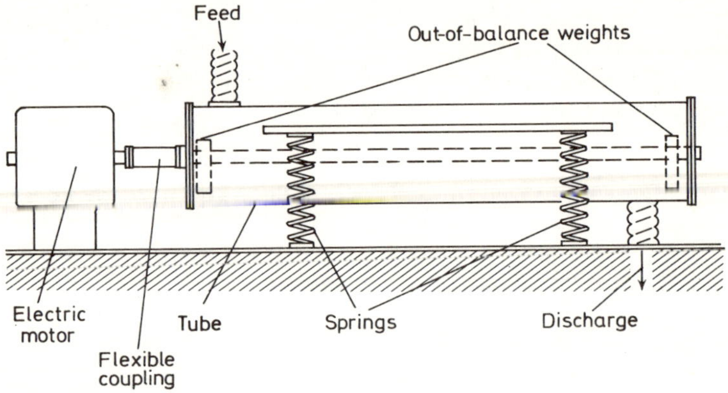

Figure 9.10. Diagrammatic sketch of tubular or cylindrical vibration mill

ADVANTAGES OF VIBRATION MILL

The advantages of the mill compared with a tumbling mill of equivalent capacity are:

(1) The power consumption for the same degree of comminution is only about 10%.
(2) The weight of the equipment is only about 20% and capital tied up in media is only about 14%.
(3) The floor space occupied is only about 30%.
(4) The volume is only from 5 to 30%.
(5) The replacement of liners and media is quicker, and wear is only about 90%.
(6) The size distribution of the product is narrower, especially for small ratios of ball diameter to mill cross-sectional diameter.
(7) Because the shell does not rotate but only moves a small distance, it can easily be attached to ducting and venting devices and made dust-tight. It can easily be supplied with a special atmosphere or made to work under vacuum.
(8) It can deal with a material such as mica, which defies a ball mill.
(9) It tends to produce particles with jagged shapes compared with rounded shapes from a ball mill.

(10) It can deal with thixotropic substances because vibration from the mill can be transmitted through a feed hopper and they can thus be made to flow.
(11) It can easily be jacketed for cold or heat without complications of joints to allow rotation.

A final advantage, shared with the tumbling mill is:

(12) It can easily be fitted with sonic feeding control.

Advantages 2, 3 and 4 decrease cost of installation and facilitate placing on upper floors of buildings, on tops of bunkers, etc.

DISADVANTAGES OF VIBRATION MILL

The disadvantages are:

(1) The vibration mill cannot reduce particles larger than a certain critical diameter determined by the ball diameter and amplitude of vibration.
(2) It is unsuitable for heat-sensitive materials.
(3) It will not grind tough materials (see p. 44).
(4) A mill of greater capacity than 5 ton/h involves great engineering problems.
(5) Vibrations make equipment prone to disintegration.
(6) The mill product size is very sensitive to feed-size variation.
(7) The mill's capacity is very sensitive to the moisture content of the feed and sticky, clayey materials tend to plug. A general rule for dry grinding in this type of mill is to keep moisture below 1%; but where more moist materials can be handled, the effect on capacity is considerable. As an example:

Moisture content, %	0	1	5.5
Capacity, kg/h	570	280	170

TORUS MILL

The shape of the shell of the torus mill can best be visualised as a hollow ring which has been cut in two across its largest diameter; each of the halves constitutes a torus. In fact in this mill the torus has a U-shaped section, so that it may be considered to be formed out of the torus described, but with vertical sides added to the semicircular base. The media and the material to be ground sit in the U-shaped section, which is usually constructed of steel and lined with abrasion-resistant rubber, but may be of unlined mild, stainless or manganese steel. The shell may be jacketed for heating or cooling.

In the middle of and centred on the axis of the ring is the mechanism which is the source of the vibration to the ring. It is essentially an electric motor with its shaft extended at both ends; an arm with a weight attached is fitted upon each of the shaft ends. The degree of vibration can be adjusted by altering the relative angle between the two arms. The motor is carried on a flange bracketed from the inner diameter of the torus (see *Figure 9.11*). The outer diameter of the torus is bracketed to another flange carried on springs which in turn are carried on the mill base. In this way a three-dimensional vibration is given to the torus.

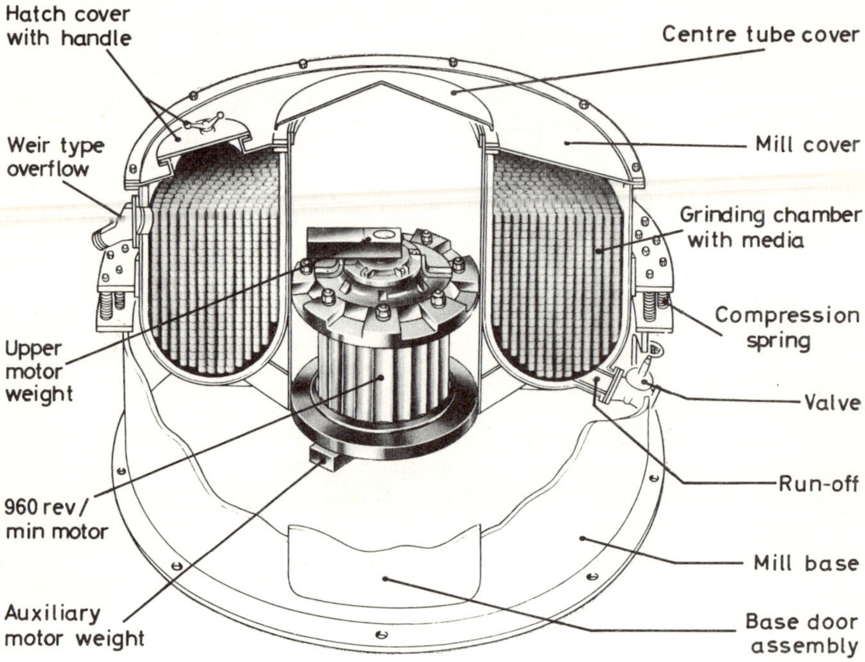

Figure 9.11. Toroidal types of vibration mill. (Courtesy Podmore and Sons Ltd)

Mechanism

According to the manufacturers of this type of mill, the objective is to place material in thin layers between the media so that particles cannot avoid the force of the impact of the media by redeploying themselves in the bed. By doing this, the crushing intensity can be kept to a minimum and the strain and possibility of damage to the equipment is decreased. It is also claimed that, for the production of the smallest particles, the impacting faces of the media should be flat—in fact, polished flat—to achieve a relatively large undeviating plane of contact. In this way the units of the media may be small because the surfaces of contact, being relatively large, transmit large forces from the vibrations.

Media

The media are usually made of corundum or zircon, though steel and tungsten carbide have been used. The shape is usually cylindrical, with height and diameter being equal, though, again, spheres and random shapes have been used. Density does not seem to affect rate of grinding in this type of mill as it does in the cylindrical type.

Cylinders of equal height and diameter are said[10] to produce less than 3% as much new surface in a given time as spheres of the same diameter. Others[11] claim that such cylinders are necessary to produce the finest products and that when used with the main axis vertical they offer flat surfaces for impacting, provide adequate space (25% voids) between them horizontally to enable materials to be circulated, and are small enough to ensure that very thin films of material are interposed between the flat surfaces. Spheres, it is said, provide too much voidage (37 %) and because of their point contact are most likely, through greater intensity of force compared with planar or distributed force, to suffer indentation and therefore jeopardise the possibilities of close enough contact to produce the very finest product. Moreover, spheres are less likely to attack preferentially the large particles (because point contact is unselective) than the planar surfaces, which are restricted and obstructed by the largest particles and have to break down the largest particles before the smaller ones can be attacked. Size distribution of the products from spherical media is wider in range than from cylindrical media.

The torus type of vibration mill is therefore closely packed with small ceramic cylinders vibrated with sufficient amplitude in three dimensions just not to disrupt the alignment of the cylinders assembled with their main axes vertical.

Media wear in grinding a material such as limestone from 250 to 53 μm would contaminate the material with, say, corundum to the extent of 0.006%. The losses amount to 10^{-6} parts per part of media by weight per hour.

Vibration

The general arrangement of the equipment to produce vibration has been described above. The motor usually rotates at 16 Hz and has a hollow shaft to facilitate cooling.

The amplitudes in the three dimensions are found by experiment to be best when they do not exceed about 6 mm in the two horizontal axes and about 5 mm in the vertical. There is an optimum frequency of vibration which is the product of the number of impacts and the effectiveness in size reduction of each impact. Frequencies that are too high do not permit sufficient force to be exerted between the surfaces of the media. The optimum frequency appears to be related to the extreme diameter of the shell and is about 25 Hz for a diameter of 1.8 m. It is important that the frequency used allows the media to spin, otherwise flats develop on the cylindrical surfaces, packing density of media increases locally and circulation of the material is inhibited.

The movement of the media is so slight that there is no danger of the media being lifted out of the mill. The mill cover need be only lightly constructed.

Feed Size

The maximum size of feed is related to the amplitude of the vibration and the size of the media. A size of about 6 mm is the largest that these machines will take; but for efficient comminution the feed size should not be more than about 0.25 mm.

For wet grinding the viscosity should not exceed 400 mNs/m^2. The very fine grinding of which this mill is capable tends to emphasise the frictional behaviour and to reveal thixotropic properties of the slurries, and care should be taken to control the liquid–solid ratios and to maintain proper circulation in the mill.

Product

Product sizes of 80% passing 2–3 μm are relatively easily obtained from feed sizes of about 250 μm. Sometimes dispersants such as sodium silicate, sodium hexametaphosphate (Calgon), Lissapol and Lissatan are necessary to obtain these very small particle sizes.

Very little mixing takes place in the torus mill, and it cannot, therefore, be used for simultaneous comminution and mixing.

CYLINDRICAL MILL

The cylindrical mill is essentially a tube inside of which are loose balls and the material to be ground. To it is imparted a vibration composed of a horizontal component at right angles to the axis of the tube and a vertical component; sometimes the tube is also given a swing about the centre of the tube in a vertical plane.

The mills can be run batchwise or continuously and either dry or wet.

Shell

The shell may be circular or U-shaped in section. The circular cross-section is said to aid the helical movement of the media and to produce a greater rate of grinding than the U-shape. The length and diameter do not appear to have any influence other than a directly proportional capacity effect.

The shell can be lined with mild steel, stainless steel, manganese steel or rubber; it has been suggested that rubber decreases the rate of grinding[10]. Linings for iron-free products are made of stainless steel. In some cylindrical machines an interior tube runs the length of the axis, making the space for the media and the material annular, which is said to improve the imparting of vibration to the media. The axes of the cylindrical form can be horizontal or vertical.

Media

The media can be made of mild steel, manganese steel, porcelain, corundum, tungsten carbide, zircon, agate or rubber: for iron-free products one of the ceramics is used. In shape they can be spheres or rods of length nearly equal to that of the shell or small cylinders. Balls make point contact, whereas cylinders make line and plane contact. Rods behave like spheres of equal diameter. Short cylinders of equal height and diameter are said[10] to produce less than 3% as much new surface in a given time as spheres of the same diameter; other investigators[11,13] claim that such cylinders are necessary to produce the finest products (see p. 215).

Wear of balls is about one-fifth of that of cylinders of the same volume and material and about one-third of that of long rods. Tests made with plastics-lined and exposed mild steel shells showed that wear of balls was about five times the wear on the shell walls[12].

The rate of grinding in cylindrical mills is proportional to the ball density and increased ball density allows of decreased shell volume for a given capacity of mill; thus there is an optimum capital case for expenditure on dense balls versus expenditure on the volume of the shell. A counteracting influence to that of ball density is the greater tendency of coating to occur with balls of higher density; thus, with materials which coat easily, e.g. cement and marble, a further compromise is necessary.

Vibration

Vibration is usually effected by out-of-balance weights at either end of a shaft which may constitute the axis of the mill cylinder and be driven by a flexible coupling from an electric motor or may constitute the rotary shaft in an electric motor so suspended that its vibrations are transmitted to the tube of the mill. Other means of vibration, such as electromagnetic methods, have been used, but not on a commercial scale.

The frequency of vibration is usually about 1500 c/min, because this is usually the standard rate of rotation of electric motors in the U.K. Frequencies below 1000 c/min are not sufficient to separate the balls from the shell lining unless the amplitude is high. The theoretical relationship is that $\omega^2\alpha$ must be greater than g, which would mean that for amplitudes (α) of more than 1 mm, lower frequencies (ω in radians) than 1000 c/min would be tolerable. Russian workers are using vibrations of 3000 c/min. ROSE and SULLIVAN[10] state that the vibrations should be as close as possible to the resonance of the system to obtain the highest rate of grinding for the lowest power consumption.

The amplitude of the vibration can have two variables—size and shape.

The size of the vertical component of the vibration must be at least equal to the size of particle it is wished to comminute. In practice, vertical amplitudes in the range 3–20 mm have been used.

The path of the vibration is important. Those paths which have equal horizontal (β) and vertical (γ) components, i.e. those which are circular, are usually the most effective. SMITH[12] found that, for the comminution of iron oxide in oil, eccentricity of vibration gave improved rates of grinding. The vertical component appears to contribute more to the rate of grinding. The relationship has not yet been ascertained but *Figure 9.12* indicates a few points determined by Rose; lines have been interpolated through these results of Rose, to indicate a tendency. The dominant effect of the vertical component is thus shown in the figure, which also shows how the dominance decreases as the ratio γ/β increases. Rose suggests that, for small differences of distance from those used in his experiments, the ratio of the components (β/γ) can be used to determine the resultant.

Broadly speaking, the rate of grinding increases with the amplitude up to a value equal to twice the diameter of the largest particle in the feed. However, for mechanical reasons there are advantages in keeping the amplitude small and, since the rate of grinding is proportional to the product of the amplitude and the frequency (to a common exponent), in keeping the frequency as high as possible. The reasons given by Rose for this choice are:

(1) The flexibility of the coupling from the motor to the mill shaft can be lower.
(2) The spring design is less difficult.

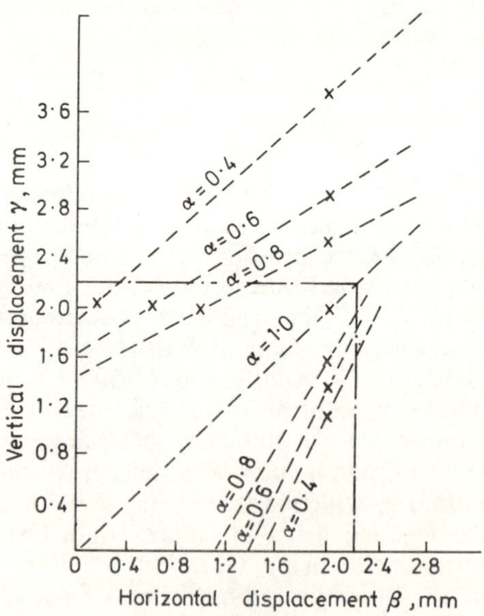

Figure 9.12. Effect (α) on rate of grinding in vibration mill of vertical (γ) and horizontal (β) displacement. Crosses define points actually determined by H. E. ROSE *and* R. M. E. SULLIVAN[10]*; broken joining lines are suppositious and intended to indicate trends*

(3) The smaller the amplitude, the shorter the period in which the powder and the balls can reach equilibrium against the mill shell and the smaller the proportion of time in which no milling occurs.

Flow of Fluids through the Mill

SMITH[12] observed that the balls in a cylindrical mill followed a helical path in which they had a frequency about one-hundredth of that of the machine and that this rate of gyration could be decreased by introducing a flow of air up through the bottom duct, which is normally the outlet. The gyration could be halted completely by a suitable rate of air flow. He found, however, that by pumping slurry through the mill in this direction, i.e. against the normal direction of the ball spiral, improved rates of grinding were obtained. Smith explained the result as being due to residence time being increased by this direction of flow because the pumping action of the ball spiral was being opposed.

Size of Feed

The subject of feed size is discussed more fully in subsequent sections (see p. 221). It can be said here that feed size must be related to size of media. If the latter is too small, build-up and plugging results; conversely, the size of feed should be carefully controlled, by, for example, the use of a rod mill to process the feed. The rate of feeding must also be carefully controlled to obtain a uniform product; provision of a surge hopper above the mill is desirable.

Size of Product

Very fine product sizes are possible, certainly down to 1 μm. The size distribution of products from a vibration mill are shown in *Figure 9.13*. Oversize is invariably present in the mill product unless closed-circuit grinding is used.

Relationship of Balls to Mill

For ball filling from about 0 to between 30 and 50%, depending upon the size of the mill, the rate of grinding is constant. If the filling is increased outside this range, the rate increases steadily in such a way as to suggest that at 100% filling the rate would reach twice the constant rate in the 0 to about 50% filling range. In fact at about 80–90% filling (see *Figure 9.14*) the rate of increase begins to fall off. Rose explains this behaviour as being due to the fact that in a mill of circular cross-section, up to the point at which it is about half-filled, the balls are projected upwards by the vibration and fall freely; beyond this value of filling, the top side of the circle begins to restrict the lift

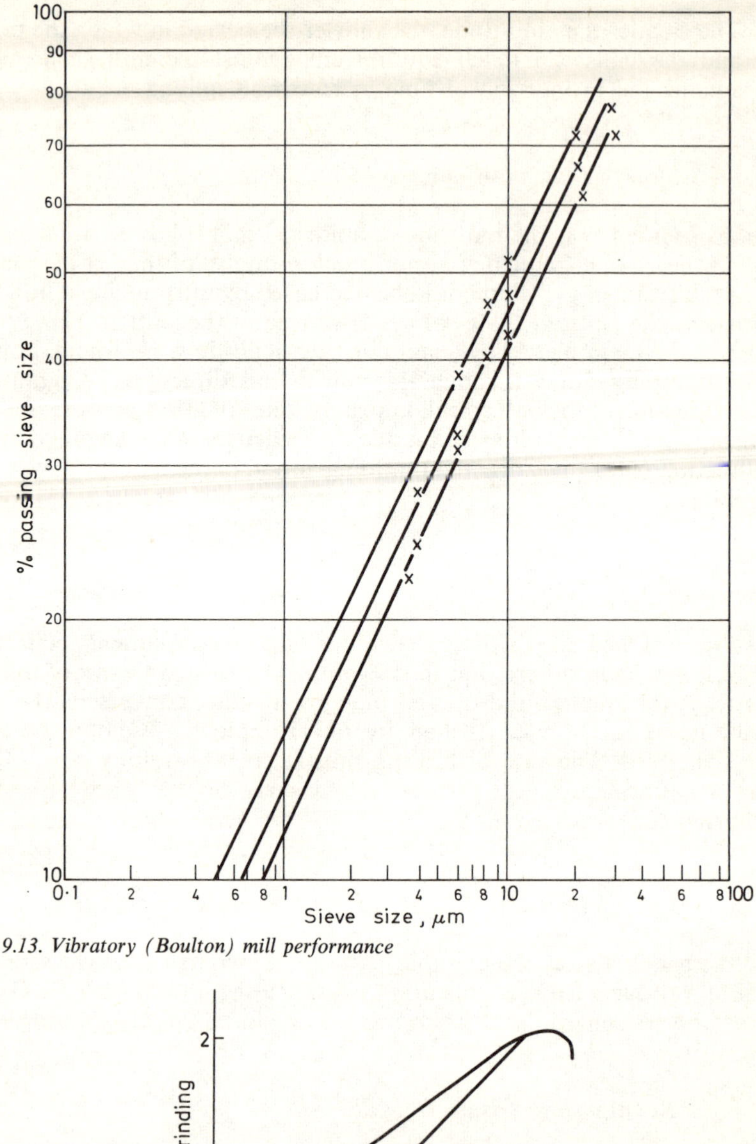

Figure 9.13. Vibratory (Boulton) mill performance

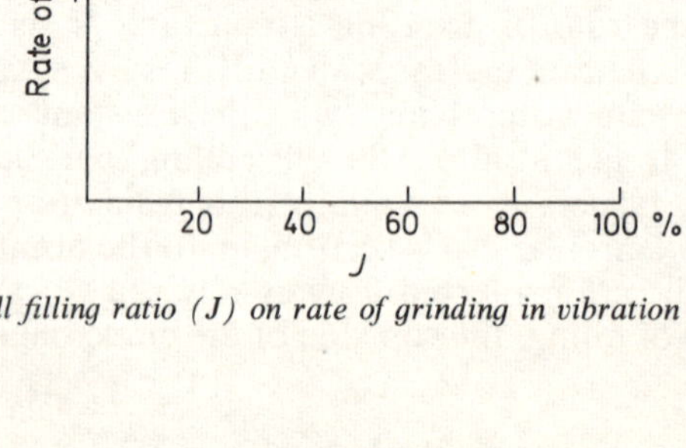

Figure 9.14. Effect of ball filling ratio (J) on rate of grinding in vibration mill. (After H. E. ROSE *and* R. M. E. SULLIVAN[10]*)*

of the balls and projects them downwards. This provides an additional comminuting effect similar to that achieved in the upward projections of the balls, i.e. so that the total effect approaches a doubling in the rate as the filling approaches 100%. As the filling actually approaches 100%, movement of the balls is increasingly hindered—enough, in fact, at 80–90% filling to begin to counteract the effect of the double projection—and there is for additional filling a falling off in the rate of grinding.

Relationship of Balls to Powder

Rate of grinding is increased by increased ratio of ball diameter (d) to particle diameter (b) (in fact it is proportional to $\sqrt{(d/b)}$). Material of greater diameter than 0.1–0.2 times the ball diameter is not ground. Thus this ratio should be larger than 5–10 if the particle diameter d is taken to be the largest particle in the feed which is to be reduced. This could be due to the need for effective nipping. In Chapter 8 (p. 132) the relationship for two balls in contact (case 5) was shown to be b/d less than $(\sec \theta/2) - 1$; for an angle of nip $(\theta) = 30°$, the value of this expression is b/d less than 0.035. The corresponding relationship for vibration mills could be

maximum size (diam.) of feed =

0.035 × ball diam. + a function of amplitude of vibration

Interrelationship of Balls and Mill

The effect of powder filling is to decrease the rate of grinding per unit weight of powder, although filling beyond 50% of the voids between the balls has little effect on this rate. The rate of grinding per unit weight is a maximum at about 10% filling (V) (see *Figure 9.15*). However, it must be borne in mind that the rate of grinding per unit weight is much different from the rate of production of total new surface area, which is, of course, the product of rate of grinding per unit weight and weight of material in the mill.

Small powder charges incur higher ball wear and there is more contamination of the product.

Qualitative Summary of Factors Contributing to High Rates of Grinding

It will be seen from the above discussion that the following factors contribute to high rates of grinding:

(1) The shell should be lined with a hard material and should be circular in cross-section.
(2) The density of balls or rods should be as high as possible commensurate with the effects of ball coating and the cost of increased shell volume.

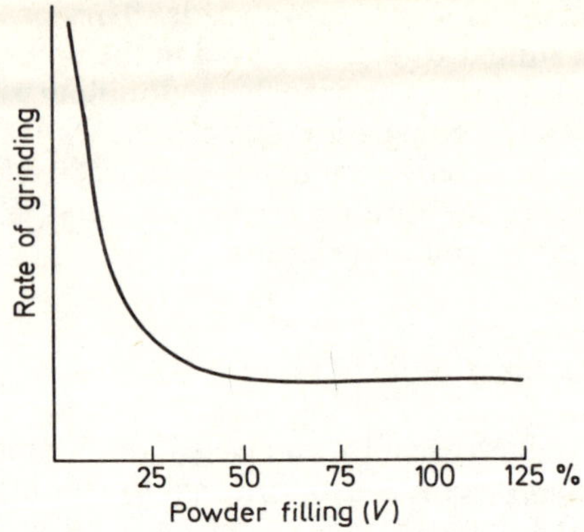

Figure 9.15. Effect of powder filling of vibration mill on rate of grinding. (After H. E. ROSE *and* R. M. E. SULLIVAN[10]*)*

(3) Vibration should be close to resonance conditions with a circular amplitude of co-ordinates about twice the diameter of the largest particle in the feed.
(4) The ball filling should be 80–90% of the volume of the shell.
(5) The ball diameter should be as large as possible and certainly not less than 10 times the diameter of the largest particle in the feed which is being reduced. The diameter of rods should be calculated in the same way: their length should be nearly equal to the length of the mill.
(6) The powder filling should be enough to somewhat overfill the voids between the balls unless rate of grinding per unit weight is more important than rate of production of total new surface. If rate of grinding per unit weight is more important, the filling should not be more than about 10% of the voids between the balls.

Quantitative Expression for Rate of Grinding

ROSE and SULLIVAN[10] used dimensional analysis methods to deduce the rate of production of new surface per unit volume as

$$\frac{\mathrm{d}S}{\mathrm{d}t} = \frac{K\omega^3\alpha^3\rho}{W_\mathrm{i}}\cdot\left(\frac{d}{b}\right)^{\frac{1}{2}}\cdot\mathrm{f}\left(\frac{\omega^2\alpha}{g}\right)\cdot\mathrm{f}(J)\cdot\mathrm{f}(V)$$

in which K is a constant and the other symbols are as defined on p. vii. Now W_i (Bond's index) is approximately proportional to $(b)^{\frac{1}{2}}$ when b is small and $\mathrm{f}(\omega^2\alpha/g)$ does not depart much from unity except when $\omega^2\alpha$ is less than three times g. Thus the equation can be simplified to

$$\frac{\mathrm{d}S}{\mathrm{d}t} = K\omega^3\alpha^3\rho\sqrt{(d)}\cdot\mathrm{f}(J)\cdot\mathrm{f}(V)$$

and, by integration,

$$S = K\omega^3\alpha^3\rho\sqrt{(d)}\cdot f(J)\cdot f(V)\cdot t$$

f(J) can be determined from

J, %	0	20	40	60	80	100
f(J)	1.0	1.0	1.2	1.6	1.9	1.6

and f(V) from

V, %	10	25	50	75	100	125	150
f(V)	5.0	2.0	1.2	1.0	1.0	1.0	1.0

The integrated simplified form of the equation of Rose and Sullivan should mean that, for constant running conditions, the additional surface area per unit volume produced is only dependent upon time and not at all upon the particle size of the material. Results obtained by ROSE[14], depicted in *Figure 9.16*, show approximately straight-line relationships between surface area per unit volume produced and time over a considerable part of the time of operation for five different materials.

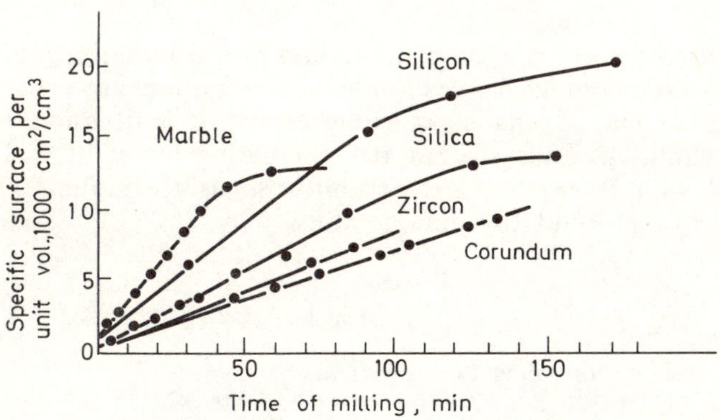

Figure 9.16. Relationship between production of specific surface and time of vibration milling. (After H. E. ROSE[14]*)*

Rate of Production of Total New Surface Area

To obtain the total new surface produced, the rate of increase per unit weight must be multiplied by J and by V, or: total new surface produced = SJV. From this calculation it is found that the greatest rate of production of new surface is obtained when J is about 80% and V is about 100–130%.

Power Requirements

ROSE and SULLIVAN[10] separated the power requirement into two main parts:

(a) that absorbed in the bearings, P_b, and (b) that absorbed by the charge, P_c. The total power requirement is therefore $P_b + P_c$. Rose and Sullivan calculate P_b and P_c as follows:

$$P_b = 2.0 \times 10^{-12} \times (M_t \alpha)^{\frac{3}{2}} \times \omega^4 \times n^{-\frac{1}{2}} \left[\frac{1 - \omega^2/\omega_0^2}{\omega^2/\omega_0^2} \right]^{\frac{3}{2}}$$

$$P_c = 2.6 \times 10^{-11} \times (\alpha \omega^2)^2 M_c \rho$$

in which power is measured in kilowatts and all other measurements are in metric units. α, ω and ρ are as defined in the Table on p. vii; M_t is the total effective mass of the vibrating systems; ω_0 is the natural vibration of the system, so that $\omega_0^2 = K/M_t$; n is the number of bearings supporting out-of-balance weights; K is a constant; and M_c is the mass of the powder charge.

Power absorption per unit weight per unit of new surface produced is at a minimum when (a) the mill is vibrated at resonance frequency ($\omega = \omega_0$); (b) the ball filling is 80% ($J = 80\%$); and (c) the powder filling just fills the voids between the balls ($V = 1.0$).

Ideal Operation from Point of View of Power Absorption and Rate of Grinding

From the discussion of rate of grinding and power requirements it is clear that there is little conflict between optimum conditions for each.

A final point may be made regarding design. It is desirable to arrange banks of cylindrical mills so that the fluctuating forces of each mill are interphased with those of the other mills so as to minimise vibrations transmitted to the foundation for the mills.

REFERENCES

1. MOYES, A. J. and WOODHEAD, P., Private communication
2. GARSIDE, J. and WILDSMITH, J. A., *Trans. Instn Chem. Eng.*, **47** (1969), 285
3. JOISEL, A., *Dechema Zerkleinern Symposion, 1962*, 49
4. CLEWS, F. H., *Heavy Clay Technology*, 1955; Br. Ceram. Res. Assoc.
5. PUFFE, E., *Eng. Min. J.*, **156** (1955), 98
6. FARRANT, J. C., *Crushing and Grinding*, 1958. London; HMSO
7. BATEL, W., *Chemie-Ingr. Tech.*, **32** (1960), No. 7
8. MARSHALL, V. C., *Dechema-Monogr.*, **57** (1967), 729
9. CALDECOTT, W. A., *Trans. IMM*, **19** (1909), 57
10. ROSE, H. E. and SULLIVAN, R. M. E., *Vibration Mills and Vibration Milling*, 1961. London; Constable
11. PODMORE, H. L. and BEASELEY, E. S. G., *Chemy Ind.*, (1967), 1443
12. SMITH, E. A., *Chemy Ind.*, (1967), 1436
13. ALLIS-CHALMERS LTD., Publication 07T 9900-133
14. ROSE, H. E., *Dechema Zerkleinern Symposion, 1962*, 439
15. GRUNDER, W., *Dechema Zerkleinern Symposion, 1962*, 258

10

Tumbling Machines

The distinctive feature of a tumbling mill is the use of loose units occupying not much more than half the volume of a container or shell. The units depend upon the rotation and friction of the shell to be lifted up on the rising side of the shell to a certain height; the units which have reached this height then cascade and cataract down the free surface of the other units in course of being lifted up. The comminuting action takes place between the units in the jostling in the lift up, in the rolling action and the impact in the fall down (*Figure 10.1*).

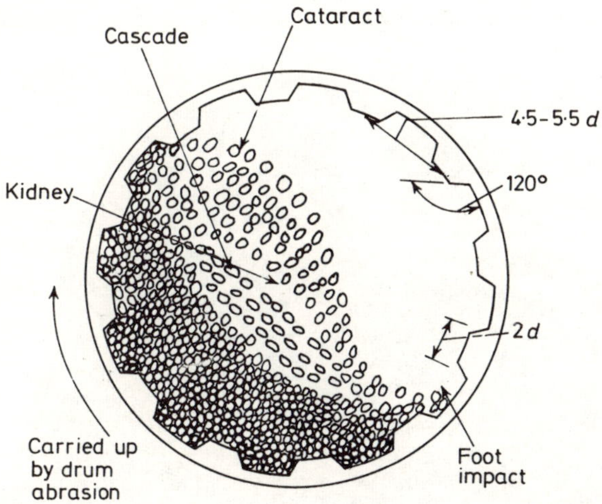

Figure 10.1. Motion of loose units in tumbling mill. Dimensions of lifters in lining (d = maximum diameter of loose unit)

The loose units may be rods, i.e. having one dimension very much larger than the other two; they may be of a shape, usually spherical, in which any one dimension is not greater than twice any other; or they may be the material particles themselves. The two types of tumbling mill using specially added loose units are more clearly distinguished by reference to the loose units they contain, i.e. rods or balls, rather than to the shape of the shell in which they are loaded. They will be classified as rod-loaded and ball-loaded. In some rare types, part of the mill contains rods and part balls. Those tumbling mills which use the material particles themselves as the units will be classified as autogeneous mills.

In general, tumbling mills are simple to operate and can handle a considerable range of size of feed and type of material. They can be operated wet or dry, batchwise or continuously. The equipment is robust and the wear parts are very easily replaced, particularly the loose grinding units, which can usually be added without stopping the mill. On the other hand, tumbling mills are relatively high in power consumption and require expensive foundations.

Batch operation can be used to predict continuous operation behaviour if the time of contact of the material with the comminuting units is used as the connecting factor.

ROD-LOADED MILLS

The method of rods rolling in a cylinder to effect comminution was patented by Zanenko in 1870.

The distinctive feature (see *Figure 10.2*) of the rod-loaded mill is a shell,

Figure 10.2. Rod mill with end peripheral discharge used for wet or dry grinding. (Courtesy Allis-Chalmers Ltd)

usually cylindrical, revolving with axis horizontal (or very nearly so, e.g. at 2°) and from 1.5 to 2.5 times its diameter in length. This length-to-diameter ratio is important because it prevents the rods, which are made within a few centimetres (usually in the range 5–25 cm) of the length of the shell, from being located across the width and yet is not so large for the maximum diameters of shell in use that the rods deform and break. However, it has been reported[1] that a rod mill with a ratio of 0.66 has been used successfully. The cylindrical shells, often called tubes, rest on trunnions and are usually driven through a girth gear-wheel set by means of flanges around the tube shell.

Few present-day machines depart from these basic specifications, although variable slope and even multiplicity of shells have been tried, without advantage.

Mechanism

The mechanism by which the rod mill effects comminution appears to be that of the rods being carried up the shell, by the rotation of the shell to approximately the top of the quadrant and of the rods arranging themselves in the course of this action with a free surface about 45° to the horizontal (see *Figure 10.2*); as the rods reach the top, they are projected either on or near to this slope and either hit the side of the shell or an assembly of rods lying at the bottom of the slope. The rods at the bottom then begin to be lifted again. The object is to have such a quantity of units in the mill that the shell lining is sufficiently covered to prevent its being struck directly or abraded. Units which are projected out and fall without contact are said to cataract. Some of the units do not have sufficient energy to cataract and they slide down the others; they are said to cascade. The distance of the fall, whether as cataract or cascade, is about the same and, owing to the shell's simultaneous movement in the opposite direction, is about two-thirds of the radius of the shell. Individual units make about three cycles per revolution of the mill[2]. In each cycle they exert several kinds of action on suitably located material: shearing, caused by the jostling of the rods in their move upwards; compression between those rods which cascade on their descent; and line impact at the end of the rods' fall, either rod against side of shell or, as quantity of rods and speed of mill are intended to ensure, rod against rod. Until the advent of supercritical speeds, which are discussed below (p. 232), it was generally believed that cataracting was the action to be achieved and that sliding on the mill lining or of the rods on themselves was to be avoided, but there are now those who advocate operation to achieve cataracting, cascading and sliding at the mill wall.

The impact, being on a line, is like the impact of an axe; the radius of curvature is not critical, nor does an angle of nip seem to be critical. The rods set themselves at an angle to one another in the bed in the ascending mass, so that the distance separating them at the feed end of the shell is the size of the feed particles, and at the discharge end the size of the product. The rods are therefore formed into a grand cone which functions as a sieve and permits particles smaller than the particular distance between the rods at their location to pass through the rods until they reach spacings less than their particular size, when they become subject to the action of the rods. Over-large pieces in the feed never penetrate the screen and remain at the feed end until they become numerous enough to affect the setting. This cone of rods has the effect of controlling the size-reduction ratio to a considerable extent. The size range of the product is quite narrow, and a typical Gaudin index (see p. 70) is 0.9.

Advantages of Rod Mill

Advantages of the rod mill are:

(1) The particle size of the product from the hardest to the softest material can be maintained within a narrow range of intermediate size and a closed circuit is unnecessary. Thus the expense of elaborate equipment and high power consumption is avoided.
(2) The range of wetness in the material handled can be wide, but ideally should fall either within 0–4% or be greater than 50% moisture. This contrasts with the ball mill, for which the first range is below 1.5% moisture. Materials of intermediate moisture content to the categories stated for rod mills can be handled but they stick and build up in the mill and output is reduced as moisture is increased.

Some results obtained with coke, which is regularly milled in rod mills, are illustrated in *Figure 10.3*. It can be seen that for a given feed rate the proportion retained on a given sieve size can increase by a factor of about 4 as the moisture of the coke increases from 3 to 18%. Such moisture ranges require that feeding be only at the rate of discharge or serious choking problems result, and the feeding arrangements must be such as to ensure positive delivery into the mill. For example, if a chute is used, it should be at least 3 m above the centre-line of the mill. Additives are available which are stated to assist in the handling of moist or sticky materials. Coal is one such material (see p. 47).

Method of Rating

The usual method of rating is by the two dimensions diameter × length. The diameters are usually taken as the internal ones without the liners in position (although some manufacturers quote diameters with the usual liners in position, which may be confusing because types of liners vary in thickness and shape) and the lengths as including the trunnions to the centre-lines of the bearing.

Feed

The upper limit of the size of feed to rod mills is somewhat critical and sizes

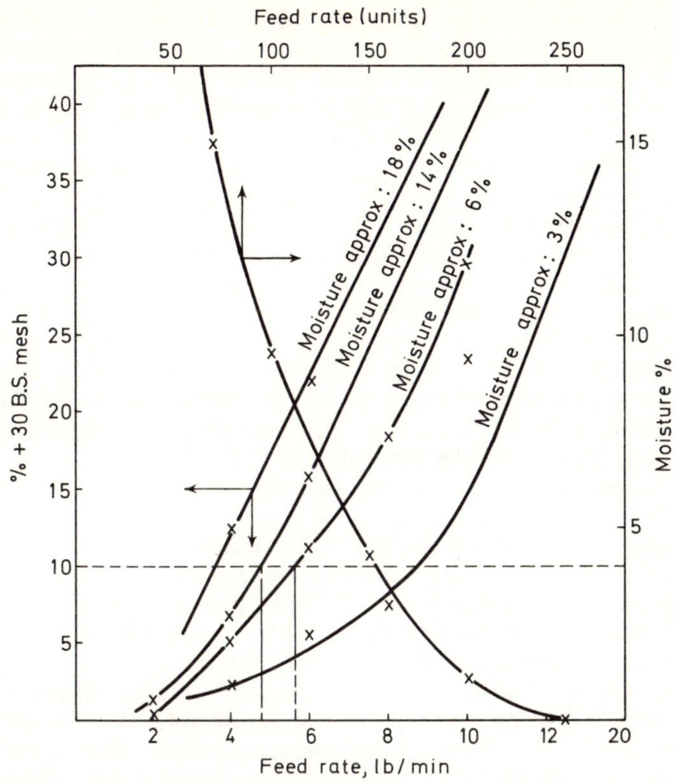

Figure 10.3. Effect of moisture on size distribution (courtesy Newell Dunford Ltd) and on rate of feed accepted by rod mill (courtesy Allis-Chalmers Ltd)

less than 2.5 cm are desirable to prevent build-up. The material may be of any degree of hardness. Its wetness should fall within the limits given above (p. 228). It has been found that optimum rates of grinding are obtained if the volume of material in the mill is about one-third of the total volume of the shell. The feed should not be scalped; fines are necessary to move the coarse material and to set up the cone of rods over the length of the mill. Wholly coarse material tends to choke the mill.

Methods of feeding include the use of chutes (both wet and dry feeds); vibrating conveyor tubes (not for materials of 'intermediate' moisture (see above)); and skips, which are particularly useful for slurries. For materials of intermediate moisture, as coke usually is, screw feeders are a very convenient method.

Product

The range of products which can be obtained from this equipment can be in the sieve size range 80% passing 0.5–5.0 mm. Individual product can be kept within a very close range. Finer products are possible; but size precision is then lost, retention times have to be long and the advantages as compared with ball mills diminish.

Shell

The shells are usually made of mild steel sections welded together. A feature in several designs is the use of conical ends: the rods tend to set with their ends vertically over one another and thus leave a conical space, at each end, bounded by the rod ends and the conical shell ends into which feed and product, respectively, can enter and perform a smoothing effect. There are arguments against this conical form, particularly on the ground that the spaces serve to collect broken rods which obstruct the flow of material and the proper movement of the rods. An efficient compromise appears to be ends with a 3° slope to the vertical.

The bearing trunnions are cast onto the shell. Bearings may be plain or rolling.

Feeding is almost invariably through one hollow trunnion, the opening of which is usually in the diameter range 40–100 cm.

There are three chief methods of discharging: (1) From the centre of the end by overflow, preferably without a retaining grid for the rods. Overflow is usually 5–10 cm lower than the feed opening level to provide a flow gradient. This method is used mainly for wet grinding. (2) From the end of the side of the tube by means of several peripheral holes, the material being directed into a close-fitting circumferential chute. This method is used mainly for dry grinding. (3) By peripheral discharge at the centre of the tube,

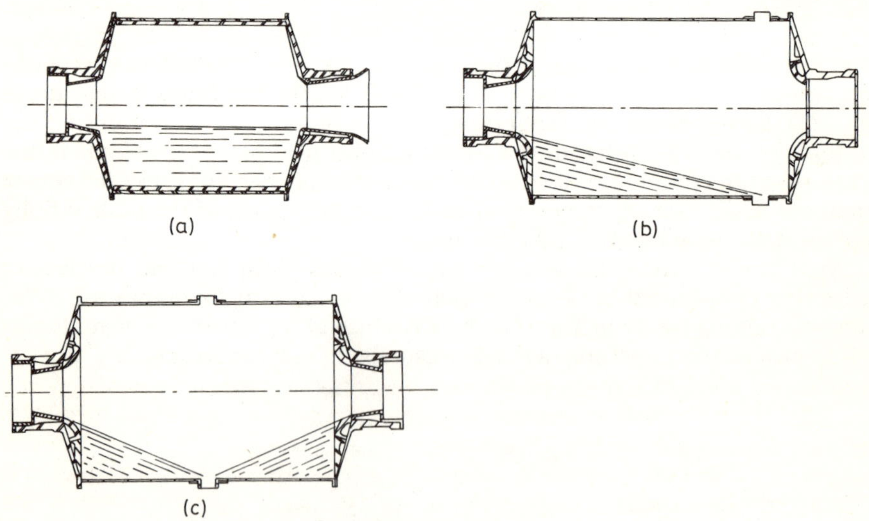

Figure 10.4. Shell discharge arrangements for tumbling mills: (a) overflow, (b) end peripheral, (c) centre peripheral discharge. (Courtesy Allis-Chalmers Ltd)

i.e. feeding from both ends of the tube, which gives a smaller size reduction (somewhat smaller ratio than half of that obtained over the whole length) and is used for both wet and dry grinding. (See *Figure 10.4*.) The method of discharge has an effect on the rate of grinding. The density of the media is decreased according to Archimedes' principle by being immersed in the material. Peripheral discharge means that the buoyancy of some of the media is lower and the effectiveness of the media is increased.

Diaphragms to control product size, and even for use as grids to retain rods, are best avoided because impacting rods tend to peen the openings shut or to break the grid. If diaphragms are essential, their holes should be tapered at 30° with the narrower dimensions on the coarse side.

The shells are lined inside wherever material touches. Materials used are cast steel or Ni-hard cast-iron or cast-iron with about 0.75% chromium. It had been found economical, because cracking of lining is very considerably decreased, to rubber-line the mills and then place the metallic lining on top of the rubber, wedging the lining or bolting it through the rubber onto the shell. Rubber has been used alone in rod mills, though more frequently in ball mills, and methods of rubber-lining are discussed in more detail below (p. 238). Rod mill trunnions are sometimes lined with rubber and even with concrete, worn rods being welded into place as a matrix. The really important quality which rod mill liners must have is freedom from surface irregularities, which would assist the breaking of the rods; the liner surface should, however, be sufficiently rough to prevent sliding.

Trunnion feed end liners are usually flared inwards to the shell body; at the discharge end the liners are flared outwards to assist flow outwards. The heads of the shell (ends) are covered with smooth liners and chrome molybdenum steel has been found economical.

The best main shell lining material is Ni-hard. Rods make line contact and distribute the load, as distinct from balls, which make point contact. The lining design for rod mills does not, therefore, need to be so strong. Disused railway lines welded to the shell and grouted in with concrete are sometimes used; the lining is very effective but it takes about 30 cm off the diameter of the shell.

The main shell liners incorporate lifters, an empirical rule for which is that a lifter is provided for every 15 cm of diameter and a usual protruberance is from 3 to 9 cm. The lifters may be metal or rubber. Recommended dimensions are given in *Figure 10.1*.

Rate of Rotation

For many years, the speed of rotation necessary to achieve most grinding for least power consumption has been considered to be that which generates enough centrifugal force to make the rods lift up on the shell lining to a height of about two-thirds of the diameter of the shell with least sliding and cascading. This speed has been found to be about three-quarters of the critical speed. The critical speed is the minimum speed at which the media are held against the shell by centrifugal force, i.e. the speed at which the centrifugal force is just balanced by the weight of the media (Mg). Since the speed of rotation of the shell is the number of rotations of the shell in unit time (N) multiplied by the circumference of the shell ($2\pi R$), the following relationships apply at the critical speed:

$$\frac{Mv^2}{R} = Mg \qquad \text{or} \qquad v^2 = Rg$$

$$v = N \times 2\pi R = (Rg)^{\frac{1}{2}} \qquad \text{or} \qquad N = \frac{g^{\frac{1}{2}}}{2\pi R^{\frac{1}{2}}}$$

The last expression is equal to $5/R^{\frac{1}{2}}$ if R is in centimetres and N is in revolutions per second or $300/R^{\frac{1}{2}}$ if N is in revolutions per minute and R is in centimetres. Strictly speaking, R is the effective radius of the media and is the mill radius less the radius of the media because the centre of gravity of the media cannot perform a larger radius of gyration than this. The critical speed, then, in terms of revolutions per minute is

$$N_c = \frac{300}{(\text{internal radius of mill} - \text{radius of media})^{\frac{1}{2}}}$$

if the radii are expressed in centimetres.

Typically, this critical speed for rod mills has been modified by the factor 0.65 for dry grinding and 0.70 for wet grinding.

HUKKI[3] has drawn on work published in the early 1900s to bring this whole question of critical speed under review. The above derivation of an expression for critical speed applies to the case of the media being at the highest vertical position in the shell (see *Figure 10.5*), whereas a more critical stage is reached, before the media reach that position, at the most extreme horizontal position at which the rods will carry on round if the frictional force exceeds the gravitational one, i.e. if $\mu Mv^2/R$ is greater than

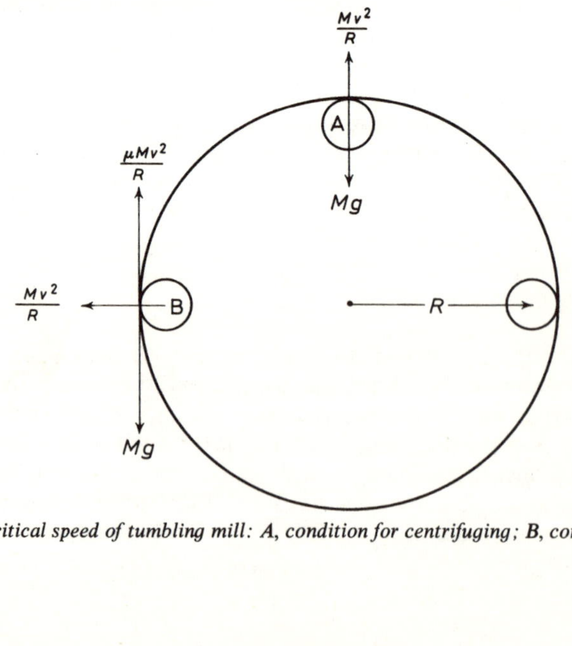

Figure 10.5. Critical speed of tumbling mill: A, condition for centrifuging; B, condition for slipping

Mg or v is greater than $(gR/\mu)^{\frac{1}{2}}$. But $v = 2\pi RN$. Therefore N is greater than

$$\frac{1}{2\pi}\left(\frac{g}{\mu R}\right)^{\frac{1}{2}}$$

Thus $N/N_c = 1/\mu^{\frac{1}{2}}$, and, if μ is less than unity, the media will still slip back at the quadrant position, even if the rate of revolution (N) is greater than the critical speed (N_c).

Hukki states that the coefficient of friction (μ) is decreased by (1) increasing the smoothness of the liner surface; (2) increasing the circularity of the media; (3) increasing the fineness of the material to be ground; (4) decreasing the pulp density; and (5) decreasing the abrasiveness of the material to be ground.

By causing slipping to occur between the shell lining and the media another source of comminution is added in the form of attrition. Speeds up to 20 times the critical speed have given improved grinding results. Slipping of the media on the liners causes wear and a careful assessment is necessary to determine the economics of the faster speed.

HUKKI[3] states that higher effective speeds are also attainable if the mass of the individual grinding unit is increased and the total mass of all the grinding units is decreased. Mill-length rods are heavier than spheres of the same diameter and material and thus it should be possible to run the rods at higher speeds than balls.

Rods

The rods are square, hexagonal or circular in cross-section, and in length are within 1 cm or so of the length of the interior of the tube. The diameter of the rods is not critical for the size of feed or the size of product: rods of diameter 2.5 cm and 4 cm gave similar products from the same size of feed. The density of the rods also is not critical either; rods of density 2.9 gave similar products from the same feed to the products of rods of density 7.3. Rod diameters in use are in the range 25–150 mm. Their selection should be related to the size of product required, a crude relationship being diameter of rod = 14 × diameter of product.

The composition of the rods varies, although 1% carbon steel is commonly used.

The volume of rods found to give optimum rates at sub-critical speeds is 35% of the shell volume when all the rods are new; this reduces to 20–30% with wear and is maintained at this figure by addition of new rods and removal of worn ones. This proportion of rods means that with normal voidage about 45% of the mill volume is occupied. HUKKI[3] states that smaller proportions of rods enable higher supercritical speeds to be attained, but the total effect of decreased volume of rods and increased speeds has not been generally assessed. The diameter at which the rods should be withdrawn should be ascertained by experiment, but BOND[7] has shown that not many rods survive after being worn to 4 cm diameter; the object should be to remove them before they distort or break. Broken rods destroy the conical formation described on p. 227 and the whole control of the particle size. The presence of broken rods may have a much more severe effect on production than regular interruption for their removal.

Wear on the rods varies from 0.01 to 0.1% and on the linings from 0.002 to 0.02% of the material ground, depending upon the fineness, wetness and hardness of the material ground. Wear can also be related to the power used in the grinding, in grams of metal per kilowatt-hour consumed, as follows:

		Wet	*Dry*
Rod mill	rods	75/90	15
	lining	15	3
Ball mill	balls	65	13
	lining	15	3

Capacity

Capacity depends very much on the size reduction required. The following data may be given as an example:

Reduction ratio	5.4	17	25	33	67
Capacity proportion	1	0.8	0.37	0.27	0.13

HUKKI[3] states that capacity is proportional to (speed/critical speed)$^{1.4}$.

Power Consumption

Examples of power consumption for various rocks and size reductions are:

Size reduction ratio	4	5	7	10
kWh/t	3.3	3.3	3.4	3.0
Material	hard ore	calcium silicate	copper ore	bauxite
Dry (D) or wet (W)	D	W	D	W

Remarks under this heading in the section on ball-loaded mills (p. 248) are relevant. HUKKI[3] points out that one of the advantages of supercritical speed is that idling power increases in less than direct proportion to speed and a much higher proportion of useful power is obtained.

BALL-LOADED MILLS

Ball mills have been in use for about 160 years. Much of what is known of the behaviour of these mills is due to the fundamental studies of H. E. ROSE and his co-workers[9] and to the investigations of F. C. BOND and his co-workers[7].

Ball-loaded mills may be classified as ball mills or tube mills. Ball mills have a length-to-diameter ratio not exceeding 1.5 and tube mills have ratios which exceed this figure. Tube mills may be compartmented, so that in effect they are a series of ball mills or, particularly when the size reduction is small and there is no need to separate ball sizes, there may be no division of the tube. Sometimes a further distinction is made between those mills in which the balls are made of iron or steel and those in which the balls are made of other materials: the former are called ball mills and the latter pebble mills. Tube mills may have rods in one part and balls in another part.

Mechanism

It appears that in the upward movement of the side of the shell the balls are interlocked and relatively stationary. At the top of their lift, at about the quadrant, the balls cascade down the free surface or cataract in free fall and hit the toe of the slope. It is the jostling, sliding and impacting of the balls, mainly in their downward path, that causes the comminution of the material in their midst. The material particles appear to take up positions either as anvils at the toe of the slope where they are simply struck by the falling balls or as annuli having a common centre at the point of contact between pairs of balls, so that the particles, ideally, take up positions according to their

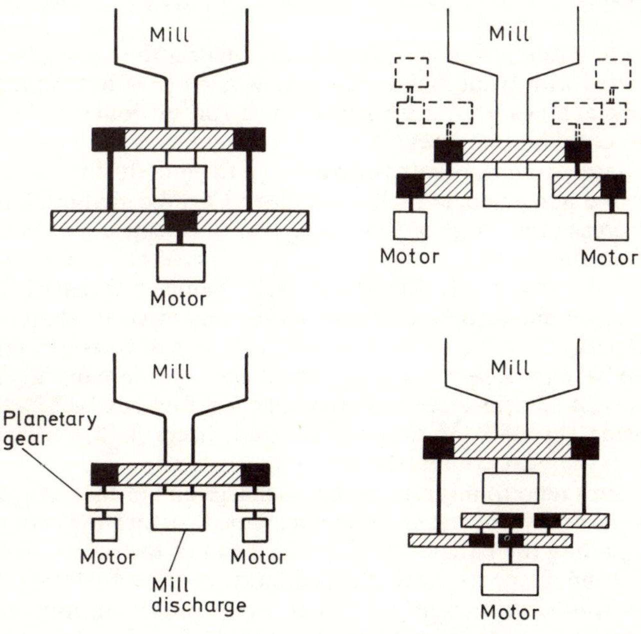

Figure 10.6. Methods of arranging gearing to reduce torque on motor driving tumbling shell. (After A. BELLWINKEL[4]*)*

size, the finest being in the annulus of smallest radius and the coarsest in the annulus of largest radius: in practice, the chance of the finest getting into position would be expected to be small because the coarsest would exclude others, and in fact it is the coarsest sizes that are most attacked (see *Figure 5.11*). The diameter of the balls would therefore be expected to be a very important factor in the determination of the particle size of the product, in contrast to the less critical diameter of the rods in rod mills.

Advantages

The particular advantage of the ball mill over other tumbling mills is that it can, given enough time, produce from material of any hardness particles of almost any degree of fineness subject to the limitations of the material itself (see Chapter 5, p. 87) and that it does this with the minimum of attention and skill on the part of the operator.

Disadvantage

A ball mill is sensitive to moisture and stickiness and will not usually handle material between 1% moisture and a slurry.

Shell

The shell of a ball mill is very similar in construction to that of the rod mill. It is usually fitted with trunnions at each end which run in bearings, although some smaller machines have girth rings which run on rollers. The bearings are subject to considerable stress due to the torque created by the displacement of the charge from the centre of gravity of the mill during rotation. This topic is discussed in more detail below (p. 248). Various methods[4] have been devised to decrease the effect of this torque on the electric motor and these are indicated in *Figure 10.6*.

Tube mills are often up to five times their diameter in length and may consist of three or more compartments, either separated by diaphragms or left open. Diaphragms are used if the size reduction is to be large and the ball sizes therefore have to be kept separate. The diaphragms are usually followed by partitions with central discharge and carrying flights which pick up the material passing through the diaphragms and direct it to the hole at the centre and into the next compartment.

The larger balls tend to migrate to the discharge end of the shell. This tendency can be partially corrected by liners shaped so that they form a connected spiral pulling the balls back to the feed end of the drum. HARDINGE[20] modified the drum shape to combat this migratory effect by fitting a conical section on a cylindrical section about half the normal ball mill length (i.e. about three-quarters of the diameter). The angle of the cone is about 30°. Owing to the centrifugal force generated the balls take up positions (see

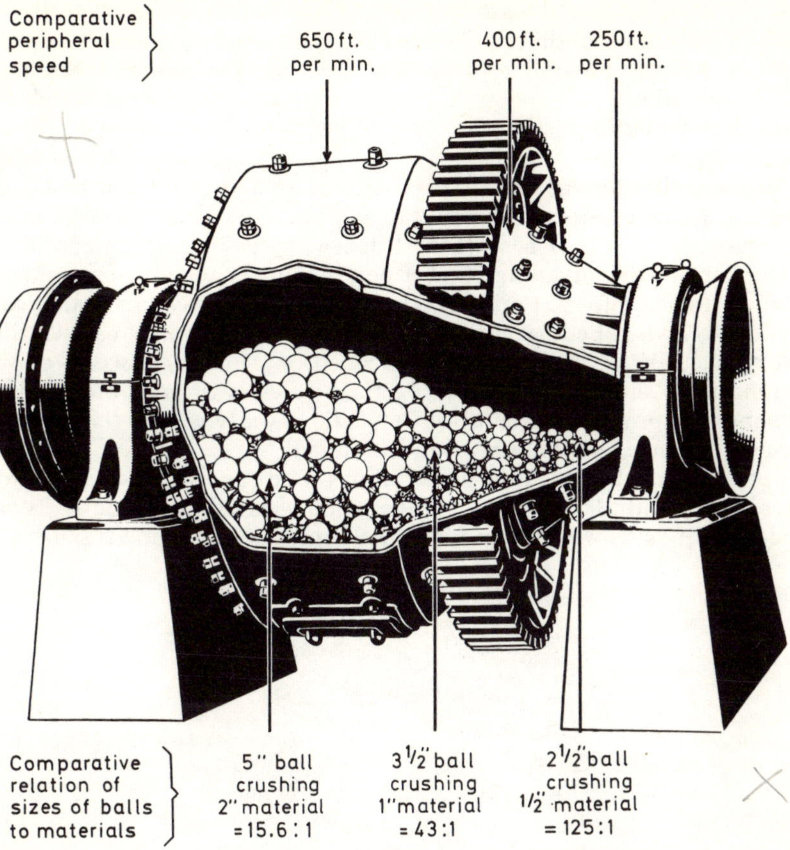

Figure 10.7. The Hardinge mill: cutaway showing how various sizes of balls tend to be located

Figure 10.7) so that the largest are at the feed end of the cone, i.e. the part of largest diameter and greatest centrifugal force, and the smallest are at the discharge end or the part of smallest diameter. By this means a regular gradation of ball size and of size reduction is produced. It is also claimed for the design that it induces flow of material and there is thus less dead space and that there is less wear at the discharge end, where the smallest balls are.

Lining

The lining of the shell of a ball mill is similar to that of the shell of a rod mill, and can be of Ni-hard, high-carbon steel, manganese steel, silica, quartzite or granite (silex).

RUNOLINNA[5] has given the criteria for ball mill lining as:

(1) The material must be most wear-resistant.
(2) The shape of the pieces must be such that no machine work is necessary.

(3) The pieces must be light to handle.
(4) The number of different shapes of liners must be as small as possible.
(5) It must be possible to reline the shell and the ends independently of each other.
(6) It must be impossible for pulp to get behind the lining.

Non-metallic linings have the advantage that they do not add metallic particles to the product; but they can add particles of the material used and they have joints which have to be fitted with cement, which can dislodge.

Rubber has the advantage that it permits higher speeds of rotation to be used, particularly for wet grinding, and decreases ball wear and ball breaking, particularly when pebbles are used. When a construction is used of rings of soft rubber, like so many gaskets, threaded onto the rods (see *Figure 10.8*), eliminating a sheet-metal shell altogether, a lighter drum is obtained which demands less power, decreases noise and allows the tension in the rubber to be carefully adjusted so that abrasion is minimal.

RUNOLINNA[5] suggests that if the linings used are formed of Ni-hard ribs interspersed with dry birch wood, the wood wears slightly and the slots are wedged with balls which then form the lining, and the lining is self-renewed.

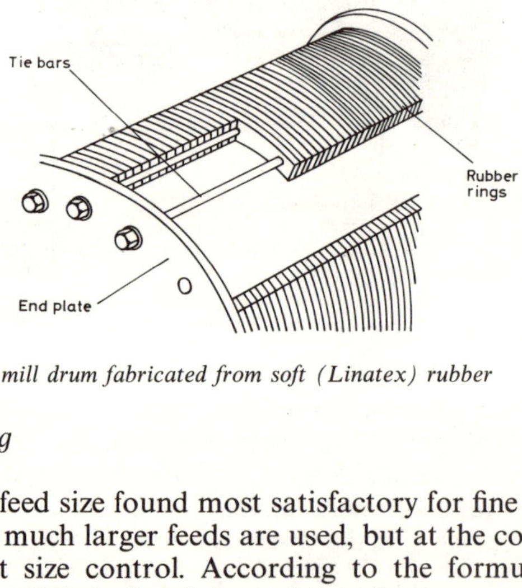

Figure 10.8. Tumbling mill drum fabricated from soft (Linatex) rubber

Feeding

For ball mills the feed size found most satisfactory for fine grinding is about 1 mm. Sometimes much larger feeds are used, but at the cost of much power and poor product size control. According to the formula developed for nipping (case 5, p. 132), balls of about 3 cm diameter would handle 1 mm feed. For tube mills the feed is often more than 2 cm.

Feed may be introduced into the shell through a door port, in which case the process is limited to batchwise operation, or through a hollow trunnion, which permits either batch or continuous operation.

Discharge of Product

The usual product range is 500 μm downwards.

The method of discharging the mill may be from a door port (appropriate to batch operation only) or through a hollow trunnion. If through a hollow trunnion, the discharge may be by overflow, in which case the gradient of the material in the mill is predetermined by the relative levels of the feed and discharge openings (usually a difference of 5–10 cm), or by diaphragm, which allows the gradient in the mill to be determined to some extent by rate of feed or by insertion of a special diaphragm with appropriate levels of perforations. Because the ports would need to be closely gridded to prevent ball discharge, peripheral discharge is hardly ever used on ball mills. If overflow is used, the discharge trunnion is sometimes protected with a cone coaxial with the shell but set so that a space of a few centimetres exists between its sides and the rim of the trunnion; discharge of product can take place but any balls of greater diameter than the setting are held back. This device is called a ball-retaining cone. This purpose is also served by the insertion of spirals in the trunnions of opposite sense to the direction of rotation of the shell.

The fineness of grind is said to be increased as the level of material in the mill is decreased[3], particularly at the discharge end. The reason given is that the buoyancy effect (according to Archimedes' principle) is decreased because the balls are not fully immersed in the material and the full effect of their density can be exerted.

Ball Shape and Composition

Balls may be shaped as cubes, octahedra, cylinders, discs, ellipsoids, spheres and concaved spheres. Nothing has been found better than spheres. Spheres wear irregularly and become concaved, and Allis Chalmers had the idea in the 1930s of making media a unit of which commenced life in the shape of a sphere into which two concaves were recessed of radius equal to that of the sphere, but placed opposite each other and penetrating the sphere only to the extent that the minimum distance between the concaves was one radius. The spherical surface of one ball was intended to fit into the concave of another; voids were fewer and the initial shape of the media was retained throughout their life. To be effective the speed of the mill had to be 65% of the critical. Because faster speeds than this were more efficient, the idea was later abandoned. It is disadvantageous if the balls become polished because the effective maximum angle of nip becomes less. It is also found disadvantageous if the balls are not true spheres, and if, for example, the balls are made in the form of hemispheres, the halves should fit as symmetrically as possible together.

Compositions which have been used include cast-iron, steel flint, alundum, porcelain, steatite and tungsten carbide. Cast-iron is not satisfactory except for small balls below 5 cm in diameter because it produces weak and brittle balls and large differences between the ball sizes cause breakages. Criteria for the choice of composition depend, apart from density and hardness, upon the possibility of tolerating the material in the product. The importance of hardness and density is discussed below (p. 241). Clearly, hollow balls are less dense than solid ones and balls which have been cast should be checked for weight, so that hollowness can be detected.

Ball Diameter

Various formulae have been proposed for the determination of the required diameter of balls (d) to comminute a given size of feed at maximum rate:

$$d = K_1 \cdot (\text{feed size})^{\frac{1}{2}} \qquad \text{COGHILL and DEVANEY}^{6}$$

$$d = K_2 \cdot (\text{feed size}) \qquad \text{ALLIS CHALMERS}^{7}$$

$$d = K_3 \, (\text{feed size})^{\frac{2}{3}} \qquad \text{BOMBLED}^{8}$$

in which K_1, K_2 and K_3 are constants. K_1 for chert is said[6] to be 37.4 and for dolomite 29.8.

Other investigations[6] suggest that above a certain ratio of feed size to ball diameter no breaking occurs, and that below this threshold value the breaking rate increases to a maximum and then falls off. These studies support the theory that breaking occurs by particles being nipped by balls in contact with one another or with the walls of the lining. Using the arguments given in Chapter 8 (p. 129) and applying them to two balls in contact (case 5), we see that, if the angle of nip is taken to be 30°, no breaking is possible with a particle exceeding 0.035 × ball diameter. On the basis of the results of these investigations, which were on chert and dolomite, the values of the ratio of particle diameter to ball diameter are shown in *Table 10.1*. Within the error

Table 10.1. SHOWING NO BREAKING OCCURS IF RATIO PARTICLE SIZE (b) TO BALL DIAMETER (d) EXCEEDS 0.035 AND MAXIMUM BREAKING RATE OCCURS AT A RATIO CLOSE TO THIS FIGURE. VALUES OF BALL DIAMETER AT MAXIMUM BREAKING RATE ARE COMPARED WITH FORMULA $d = K_1\sqrt{b}$

Particle size, mm	*Zero breaking*		*Maximum breaking rate*					
	Particle size / Ball diameter		*Particle size / Ball diameter*		*Value of ball diameter*		*Value of ball diameter according to formula*	
	Chert	*Dolomite*	*Chert*	*Dolomite*	*Chert*	*Dolomite*	*Chert*	*Dolomite*
0.36	0.02	—	0.019	0.023	19	15	22	18
0.62	0.038	0.080	0.024	0.031	26	20	29	23
1.00	—	0.108	—	0.030	—	33	—	30
1.24	0.080	—	0.032	—	38	—	42	—
1.41	—	0.108	—	0.042	—	34	—	35
2.00	0.120	0.160	0.030	0.039	64	52	53	42

of the measurements from the graphs, the ratios are all seen to be about 0.035 and only two results exceed this value. The ratios at which no comminution occurs are seen to be mostly in excess of 0.035. Since large particles can be broken in a ball mill, e.g. 5 cm particles can be broken below 5 mm with 12 cm balls, it appears that two distinct steps are necessary: (1) particles must be broken by impact below a size corresponding to 0.035 times maximum ball diameter before the particles can be broken further by compression between the balls; and (2) if the feed is just above this maximum size, it is only slowly broken.

The formulae given at the beginning of the section, though not in accord with the above reasoning, can nevertheless give correct answers if the feed diameters are within certain ranges, while all the results reported in the papers supporting these formulae[6,8] can be shown to be in accord with the above reasoning.

That other processes besides nipping occur in a ball mill is also clear from the lower limit of the particle size range obtainable in a ball mill. Few balls are worn less than 1 cm before accompanying the discharging product from the mill. A pair of 1 cm balls would nip a particle not larger than 0.35 mm and it could be argued that, although enough particles of this size are present to be nipped, entry to smaller particles will be blocked, whereas products containing large fractions considerably smaller than this size are commonly obtained. Again, impact and, additionally, attrition can be expected to have contributed.

Ball Density

ROSE and SULLIVAN[9] deduce that the ball density (ρ) should be related to the density of the material (σ) for maximum breaking rate by the formula

$$\rho = (0.016\sigma^2 + 20\sigma)^{\frac{1}{2}} - 0.4\sigma$$

The ball density should be as great as possible compared with the material density, because the operation of Archimedes' principle decreases the effectiveness of the balls by giving them buoyancy which is greatest where the depth of material is greatest in proportion to the diameter of the balls and least in such places as the point of peripheral discharge, where the depth of material is least.

Ball Hardness and Wear

Balls wear into irregular shapes with a tendency to form concaves, so that they appear multi-sided. They discharge from a mill when they are about 1 cm in diameter and utilisation of the ball therefore is the greater the larger the original diameter. Balls are usually cheapest in the 5–7.5 cm range.

Wear on a set of balls of graduated diameter and in equilibrium with the system appears to be a uniform decrease in diameter over the whole range; in other words, wear appears to be related to the volume (or weight) of the ball per unit surface.

Speed of grinding appears to be greatest when ball hardness is less than that of the material (see *Figure 10.9*), while the wear on the balls is greatest when the balls and material are of equal hardness (see *Figure 10.10*). Therefore the worst circumstance from the point of view of both rate of grinding and wear is to have equal hardnesses of balls and materials. In one

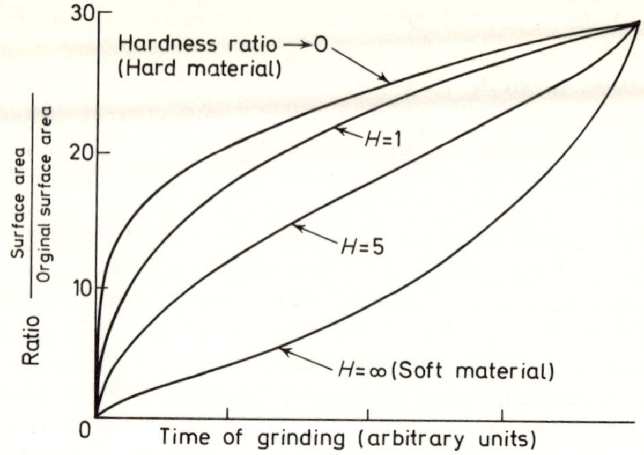

Figure 10.9. Rate of production of new surface for various ratios (H) of ball to material hardness. (After H. E. ROSE *and* R. M. E. SULLIVAN[9]*)*

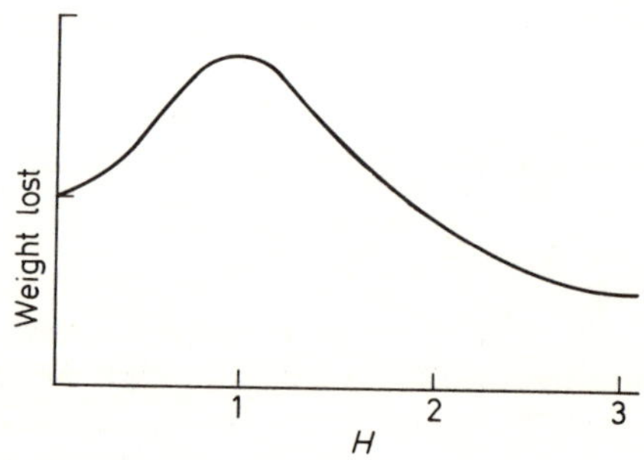

Figure 10.10. Ball wear for various ratios (H) of ball to material hardness. (After H. E. ROSE *and* R. M. E. SULLIVAN[9]*)*

such instance of pebbles being used of about the same hardness as the material being ground, the pebbles wore 30 times as much as steel balls.

Wear also appears to depend upon the filling ratio (V):

$$V = \frac{\text{bulk volume of powder}}{\text{voids between balls}}$$

and is largest when V is smallest.

HUKKI[3] says that ball wear is dependent upon speed of rotation of the mill and gives the formulae

$$\text{ball wear} \propto d^{\alpha}\left(\frac{N}{N_c}\right)^{\beta} \cdot D^{\gamma}$$

$$\text{lining wear} \propto \left(\frac{N}{N_c}\right)^{\beta} \cdot D^{\gamma}$$

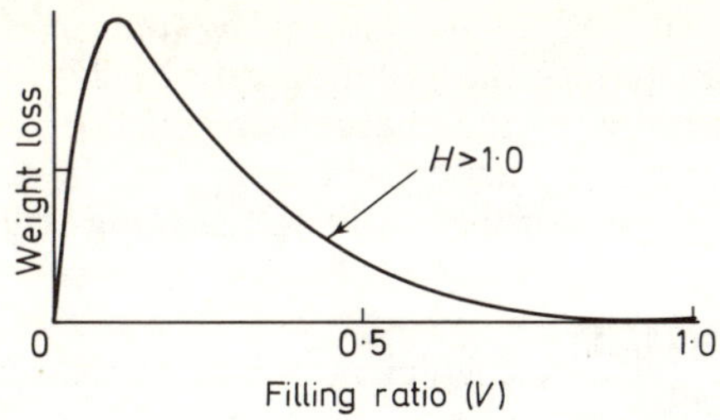

Figure 10.11. Ball wear for a fixed hardness ratio H = 1 for various ratios of V = bulk volume of material/voids between balls. (After H. E. ROSE *and* R. M. E. SULLIVAN[9]*)*

in which α is said to have a linear relationship to N/N_c and β and γ are constants.

Wear is a minimum when: (1) the filling ratio (V) is about unity (wear is maximum when V is about 0.1; see *Figure 10.11*); (2) the hardness ratio (H) is either very high or very low (wear is maximum when $H = 1$); and (3) the ratio of speed of rotation to the critical speed is least. According to DAVIS[10], if the diameter of the balls fed to the mill is d_1 and the diameter of the balls removed from the mill because they are too small is d_2, then the weight of the ball fraction between sizes d_a and d_b in the mill is

$$\frac{d_a^3 - d_b^3}{d_1^3 - d_2^3}$$

Ball Loading

The loading of the balls in the mill to effect maximum rate of production of

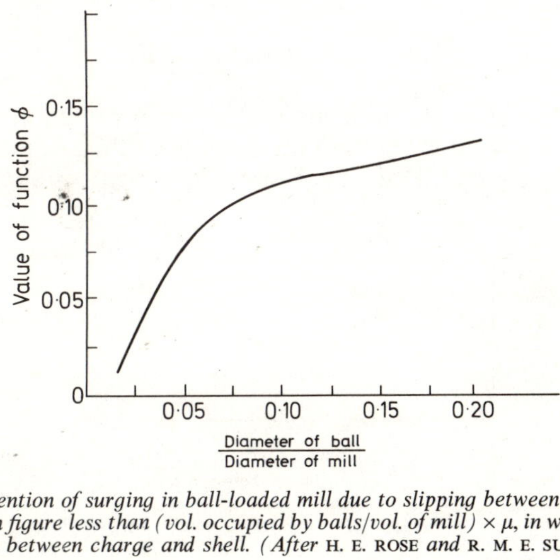

Figure 10.12. Prevention of surging in ball-loaded mill due to slipping between charge and shell by making ϕ read from figure less than (vol. occupied by balls/vol. of mill) $\times\ \mu$, in which μ = coefficient of dynamic friction between charge and shell. (After H. E. ROSE *and* R. M. E. SULLIVAN[9]*)*

new surface is about 50% of the internal volume of the mill; a drop to a loading of 40% can reduce the grinding rate by 25%.

Surging or oscillation of the charge (see p. 248), which can have serious effects on the gearing[9,11], arises when slipping is occurring between the charge and the shell. According to ROSE[9], surging will not occur from this cause if

$$\frac{\text{volume occupied by balls}}{\text{volume of mill}} \times \text{coefficient of dynamic friction between charge and shell} > f\left(\frac{\text{diameter of balls}}{\text{diameter of mill}}\right)$$

as read from *Figure 10.12.* The coefficient of dynamic friction between the charge and shell is usually about 0.2.

Sufficient loading of balls will therefore prevent such surging as well as give maximum rate of production of new surface.

Speed of Rotation of the Shell

It has long been argued, and supported by much practical experience, that the optimum speed of rotation is that at which balls riding on top of the highest material in the mill are thrown to land at the toe of the material, where the largest pieces tend to accumulate. This optimum speed has been found to be between 50 and 75% of the critical speed, which has been defined above (p. 231).

HUKKI's study of earlier work on supercritical speeds[3], discussed above (p. 232), applies to all tumbling mills. Judgement must be reserved until the economies have been fully demonstrated.

Loading or Charge of Material in Mill

The optimum filling of the mill for maximum production of new surface appears to be somewhat more than the complete filling of the voids between the balls (see *Figure 10.13*), although some investigators suggest a figure as low as 60%.

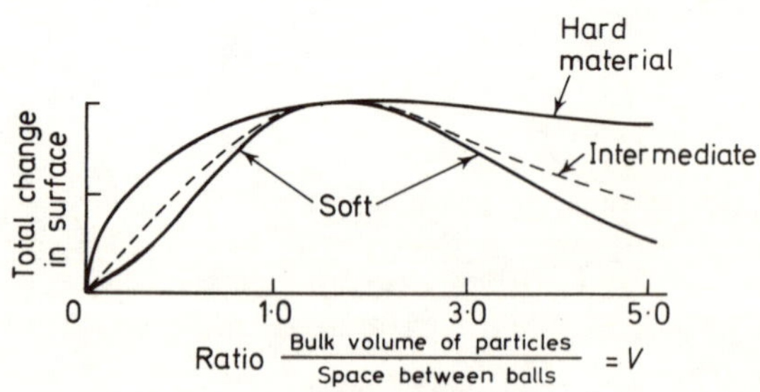

Figure 10.13. Change of total surface with ratio of material charge filling. (After H. E. ROSE *and* R. M. E. SULLIVAN[9]*)*

Fineness, as measured by specific surface, i.e. surface per unit volume, appears to depend, to some extent, on the material loading of the mill. The finest material is obtained when the loading corresponds to a function of hardness of the powder; this loading is 10% of the voids between the balls for very hard materials and rises to 100% for very soft materials.

Range of Ball Sizes

A factor of far greater influence on product-size distribution is that to which reference has already been made in the section on ball diameter (p. 240). The important practical consideration is that there be present in the mill a sufficient number of balls of the size appropriate to the sizes which are required to be reduced. The largest nip will determine, to some extent, the particles to be broken, because, once the nip has been filled, the entrance of smaller particles is blocked, i.e. they are excluded from attack or size reduction. The finest material is therefore ground with the smallest balls and the size distribution is seen to be made up of the largest sizes which can be nipped by the particular sizes of ball present, the very smallest pieces of material remaining mostly unchanged; the absolute size of these smallest pieces depends upon the composition of the feed, the size of the balls and the degree of impact and attrition which has taken place.

The advantages of a tube ball mill become apparent; the length of the mill, particularly when compartmented, allows a succession of ball size ranges to be presented to the material and for the gradation of ball sizes to be controlled and maintained.

PAPADAKIS[12] states that the ball-size distribution is obtained from the formula

$$d^m = \frac{d_0^m}{\dfrac{at_s l}{L} + 1}$$

where d_0 is the original ball size corresponding with the feed; d is the ball size at any position of distance l along the total tube length L; and a, m and t_s are factors which are defined below:

$$a = \frac{1.65K\delta}{\alpha \times D^{\frac{1}{2}}}$$

in which K is the proportionality factor connecting the volume of material and the cube of the particle diameter (if the particle were a true sphere, K would be equal to $\pi/6$); α is the ratio of weight of material to weight of balls in the mill; D is the internal diameter of the mill; and δ is the effect of the wall of the mill on the balls, negligible when d/D is small.

$$t_s = \frac{S}{ak} - \frac{1}{a}$$

in which S is the additional specific surface required to be generated and k has the value

$$k = \frac{2900B\gamma D}{K(bd_0^m + a)}$$

in which B is the surface produced per unit of energy; γ is the coefficient of transfer of energy linked to the average volume of crushing in a single ball impact; and b and m are the coefficients in Beke's formula for agglomeration (see p. 88).

Papadakis deduces these factors by determining curves relating specific surface and time for three different ball sizes, all other factors remaining constant.

It is also possible by ball-size control to exclude particles of a particular size from attack.

Rate of Grinding

In the above discussion the rate of grinding, i.e. the rate of production of new surface, has been stated to depend upon: (1) ratio of particle size to ball diameter; (2) ratio of particle density to ball density; (3) ratio of particle hardness to ball hardness; (4) ball loading; (5) speed of rotation of mill; and (6) volume of charge in mill.

CAREY and STAIRMAND[11] showed that in a particular mill (0.66 m diameter and 0.60 m long) the rate of grinding is initially very fast; that it rapidly drops from this initial rate to less than half this rate, rises to a peak of about three-quarters of the initial rate and then drops away to about one-fifth. While not giving any explanation for the peak Carey and Stairmand associate the decay in the rate with the cushioning effect of the fines. SLEGTEN[13] showed a peak in the power consumed compared with particle size. NIJMAN[14] shows a peak in the breakage rate corresponding with certain ball sizes. It is suggested that each of these results can be explained, at least to some extent, by the ratio of particle to ball diameter, as discussed above (p. 240).

Nijman associates the rate of grinding with the rate at which energy is transferred to the balls as they are projected from the top of their rise against the mill wall to the toe of the ball heap, and gives the equations:

$$\text{breakage rate between balls} = c\rho D^{0.5}bd^{-2}$$

and

$$\text{breakage rate between ball and wall} = c\rho D^{0.5}bd^{-1}$$

in which ρ and d are, respectively, the density and the diameter of the ball; b is the diameter of the particle; and D is the diameter of the mill.

PAPADAKIS[12] gives the rate as $cLD^2\rho/t_s$, in which the value of t_s is given by the equation on p. 245. JOISEL[15] considers the area of impact between two balls, taking account of the friction between the material and the ball, and shows that on average this is of the order of 0.083 μ^2 per unit surface of the ball, where μ is the coefficient of friction; assuming all the material to be equally distributed over all the balls, when the balls collide the proportion of material in the mill concerned in one impact is 0.166 μ^2. A ball, says Joisel, circulates on average 1.25 times as often as the mill (this compares with 3 times as often for rods). If the mill is turning at, say, 0.75 times the critical rate, if D is the diameter of the mill in metres, and if t is the residence time of the material in the mill, then in t minutes the number of impacts submitted to the material is

$$0.75 \times \frac{32}{D^{\frac{1}{2}}} \times t \times 1.25 \times 0.166\, \mu^2 \text{ or nearly } \frac{5t\mu^2}{D^{\frac{1}{2}}}$$

A typical value for μ is 0.2; thus the number of impacts is $0.2t \times D^{-0.5}$. Now number of impacts multiplied by mass of material in the mill is proportional to the amount of grinding done in time t. But rate of grinding is amount of grinding per unit time, which is proportional to

$$\frac{0.2t\,D^{-0.5} \times LD^2JV\sigma}{t} = 0.2LD^{1.5}\cdot JV\sigma$$

The meanings of the symbols L, J, V and σ are defined below.

ROSE and SULLIVAN[9], using methods of dimensional analysis, suggest that the rate of grinding per unit volume of mill can be expressed as

$$SJV = cD^{0.5}\rho tB\cdot \mathrm{f}\left(\frac{d}{D}\right)\cdot \mathrm{f}\left(\frac{b}{D}\right)\cdot \mathrm{f}\left(\frac{N}{N_c}\right)\cdot \mathrm{f}\left(\frac{\rho}{\sigma}\right)\cdot \mathrm{f}(J)\cdot \mathrm{f}(V)\cdot \mathrm{f}\left(\frac{u}{D}\right)$$

in which the symbols used before retain their meaning and the additional ones have the following meaning:

B is change of surface area per unit of energy;
c is a constant;
J is $\dfrac{\text{volume of balls} + \text{space (voids) between balls}}{\text{volume of mill}}$;
L is length of mill;
N is speed of rotation of shell;
N_c is critical speed of rotation of shell;
S is surface area per unit volume of material;
u is diameter of exit of the mill;
V is $\dfrac{\text{volume of material in mill}}{\text{voids between balls}}$; and
σ is density of material.

Rose and Sullivan show that for practical purposes this reduces to

$$D^{0.5}\rho tB\frac{N}{N_c}\mathrm{f}(J)\mathrm{f}(V)$$

Now

$$J = \frac{\text{volume of balls} + \text{space (voids) between balls}}{\text{volume of mill}}$$

which is a function of d^3/LD^2, and

$$V = \frac{\text{volume of material}}{\text{volume of voids between balls}}$$

which is a function of b^3/d^3, and Rose and Sullivan show that the value of f(J) varies with J something like

J, %	5	10	20	30	40	50	60	70	80	90	100
f(J)	0.93	1.0	0.97	0.93	0.79	0.71	0.58	0.50	0.39	0.29	0

whereas f(V) varies with V depending upon whether the material is hard or soft:

V		0.1	0.2	0.3	0.4	0.5	0.6	0.7
f(V)	hard ore	0.97	0.67	0.50	0.34	0.27	0.20	0.17
	soft ore	0.10	0.53	0.50	0.34	0.27	0.20	0.17

V		0.8	0.9	1.0	1.1	1.2	1.3	1.4
f(V)	hard ore	0.13	0.12	0.11	0.10	0.06	0.06	0.06
	soft ore	0.13	0.12	0.11	0.10	0.06	0.06	0.06

Power Consumption of Ball Mill

CAREY and STAIRMAND[11] draw attention to the fact that a considerable consumer of power in the ball mill is the torque created by the displacement of the charge from the centre of gravity of the mill during rotation. They quantify this by the equation

$$WNy \times 0.32 = P$$

in which W is the total weight of material + grinding media; N is the rate of rotation of the mill in revolutions per minute; y is the horizontal displacement from the centre of gravity in feet; and P is the power input to maintain displaced charge in kilowatts. *Figure 10.14*, taken from their paper, relates percentage of critical speed at various percentages of filling with this horizontal displacement (effective eccentricity). Full lines denote non-slip conditions, i.e. for mills with optimum number of lifters, and dotted lines denote smooth mill conditions.

As with the rate of grinding, various investigators have tried to deduce equations for the power consumption of ball mills.

JOISEL[15] relates the energy consumed by the mill to $d^3\rho D$ and suggests that mills are energetically similar if the cube of the diameter of the balls is inversely proportional to the diameter of the mill. NIJMAN[14] relates energy to $\rho D^{2.5}$. Rose has developed an equation for power consumption based upon the principle of dimensional similarity:

$$P = D^5N^3\rho\left(1+\frac{0.4\sigma}{\rho}\right)\frac{L}{D}\,\mathrm{f}\left(\frac{N_c}{N}\right)\mathrm{f}(J)\,\mathrm{f}\left(\frac{D}{d}\right)\mathrm{f}(n)\,\mathrm{f}\left(\frac{h}{D}\right)\mathrm{f}\left(\frac{D}{b}\right)\mathrm{f}(U)$$

stable, London, 1958)

	0.8	1.0
	0.2	0
	10 15 >20	10 15 >20
	0.97 1.0 1.0	0.99 1.0 1.0
7	2 4 6 2 4 6 2 4 6 >7	
1.0 1 1 1	1.26 0.99 0.92 1.33 1.04 0.97 1.4 1.1 1.02 1.0	
0.10 0.20	0.02 0.10 0.20 0.02 0.10 0.20 0.02 0.10 0.20	0.02 0.10 0.20 0.02 0.10 0.20 0.02 0.10 0.20
1.00 0.98	0.90 0.95 0.80 0.93 0.97 0.89 0.96 1.00 0.98	0.90 0.95 0.80 0.93 0.97 0.89 0.96 1.00 0.98
	1.00	1.00
	1.00	1.00
	0.8	1.0
		>400
		1.00

n in the table in that part appropriate to the value of J read off the value of $\phi(J)$; (ii) select the value of D/d, n, ropriate to their value (interpolating where necessary) J, and read off $\phi(D/d)$, $\phi(n)$, $\phi(h/D)$ and $\phi(U)$ as is appro- cularly in the case of $\phi(U)$ the need to relate to the

of $\phi(D/b)$ appropriate to the value of D/b.
lues of $(D^5N^3\rho)$, $(1+0.4\sigma)/\rho$ and L/D.
functions so determined in accordance with the formula le, simply omitting (or, what is the same thing, taking as irrelevant functions.
determined is the relative power consumption of the mill

) BY A BALL MILL (BASED ON GRAPHS IN H. E. ROSE AND R. M. E. SULLIVAN, *Ball Tube and Rod Mills*, Co

$$P = (D^5N^3\rho)\left(1+\frac{0.4\sigma}{\rho}\right)\left(\frac{L}{D}\right).\phi\left(\frac{N_c}{N}\right).\phi(J).\phi\left(\frac{D}{d}\right).\phi(n).\phi\left(\frac{h}{D}\right).\phi\left(\frac{D}{b}\right).\phi(U)$$

	0.4	0.6
	1.0	0.6
	10 15 >20	10 15 >20
	0.92 0.99 1.0	0.95 1.0 1.0
6 2 4 6 >7	2 4 6 2 4 6 2 4 6 >7	2 4 6 2 4 6 2 4 6 >
81 0.91 0.95 0.85 0.96 1 1	0.85 0.89 0.90 0.90 0.94 0.95 0.95 0.99 1 1	0.94 0.91 0.90 0.99 0.96 0.95 1.04
02 0.10 0.20 0.02 0.10 0.20	0.02 0.10 0.20 0.02 0.10 0.20 0.02 0.10 0.20	0.02 0.10 0.20 0.02 0.10 0.20 0.02
93 0.97 0.89 0.96 1.00 0.98	0.90 0.93 0.80 0.93 0.96 0.89 0.96 1.00 0.97	0.90 0.95 0.80 0.93 0.97 0.89 0.96
	0.95	1.00
	1.00	1.00
	0.4	0.6
		200
		0.95

f using table
ler to determine the relative power consumed in driving the mill, s stated hereunder. To convert relative power into absolute power ow the same procedure for a ball mill of known power consumption ose as possible to the same operating conditions) and thus deter- relationship between the units as determined by this procedure ıte units of power.
ate the values in consistent units of all of the relevant symbols in st for the mill in question.
r whether the interior of the mill shell has lifters or is smooth for attention the appropriate part of the table.
te the value of $\phi(N_c/N)$ according to the appropriate formula

(5) Select the colun determined and (i) h/D and $\phi(U)$ app and to the value of priate, noting part value of ρ.
(6) Select the value
(7) Calculate the va
(8) Multiply all the at the top of the tab equal to unity) any
(9) The product so in unit time, P.

Table 10.2. POWER CONSUME

	Smooth					Lifters
$\phi(N_c/N)$	$3(N_c/N)^{2.15}$					$0.04(N_c/N)^{2.0}$
J ↓	0.2	0.4	0.6	0.8	1.0	0.2
$\phi(J)$	0.6	1.0	0.6	0.2	0	0.6
D/d	10 20 >30	10 20 >30	10 20 >30	10 20 >30	10 20 >30	10 15 >20
$\phi(D/d)$	0.3 0.9 1.0	0.65 0.92 1.0	0.7 0.95 1.0	0.92 0.97 1.0	1.0 1.0 1.0	0.86 0.97 1.0
n						2 4 6 2 4
$\phi(n)$						0.80 0.87 0.90 0.
h/D						0.02 0.10 0.20 0.
$\phi(h/D)$						0.90 0.95 0.80 0.
$\phi(U)$ when $\rho = 2.6$	1.03	0.95	1.00	1.00	1.00	1.03
$\phi(U)$ when $\rho = 7.8$	1.17	1.00	1.00	1.00	1.00	1.17
J ↑	0.2	0.4	0.6	0.8	1.00	0.2
D/b	0		200		>400	0
$\phi(D/b)$	0.80		0.95		1.00	0.80

NOTES

Meaning of symbols

L = length of mill
D = diameter of mill shell
N = revolutions in unit time of shell
N_c = critical rate of revolutions of shell
ρ = density of balls
σ = absolute density of powder

$$J = \frac{\text{volume of balls} + \text{volume of voids between balls}}{\text{volume of mill}}$$

d = diameter of ball
n = number of lifters in mill shell interior
h = height of lifter
b = diameter of powder particle
$\phi(U)$ = torque factor
P = power consumed in unit time to drive mill

Method o
(1) In or
proceed a
units, foll
(and as c
mine the
and abso
(2) Calcu
the abo
(3) Con
and selec
(4) Calc
in that p

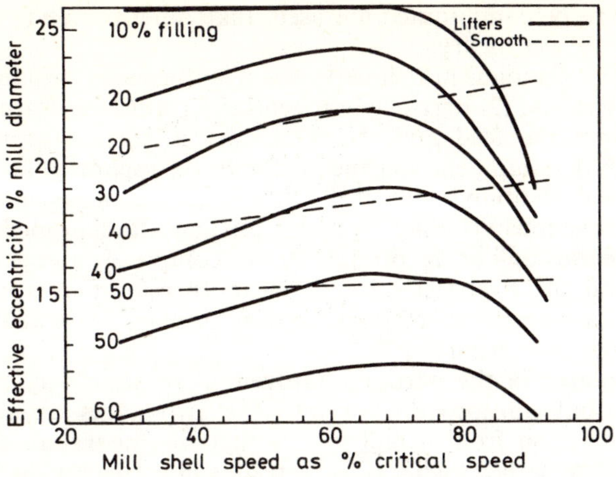

Figure 10.14. Effective eccentricity of charge in ball-loaded mill for % filling of shell and % critical speed. Broken lines for smooth shell interior; continuous lines for lifters. (After W. F. CAREY *and* C. J. STAIRMAND[11]*)*

A method of solving this equation for particular cases is presented in *Table 10.2.* From the figures given and from $N = K/D^{\frac{1}{2}}$ (see p. 232), it is seen that, for practical purposes,

$$P = D^{2.5}\rho L\left(1+\frac{0.4\sigma}{\rho}\right)\cdot f\left(\frac{N_c}{N}\right)\cdot f(J)\cdot f(U)$$

If power to drive the ball mill is considered to be essentially the energy required to lift the balls to a certain position in the shell cross-section, which is related to the diameter of the mill (D), it can be seen that the work done is proportional to

weight of balls × diameter of mill × revolutions of mill

Now the weight of the balls is related through their density (ρ) to the volume of the balls and the volume of balls is related to the volume of the mill (LD^2). The revolutions of the mill in unit time are proportional to the critical speed, which is related inversely to the square root of the diameter of the mill ($D^{-0.5}$). Thus work done in unit time is proportional to

$$\rho LD^2\cdot D\cdot D^{-0.5} = \rho LD^{2.5}$$

AUTOGENOUS MILLS

The principle of the autogenous mill was established as soon as primitive man aimed a large piece of stone to break another piece, but the earliest report of a tumbling mill being worked autogenously is by HARDINGE in 1908[16]. The autogenous mill uses tumbling to effect comminution, but instead of special milling bodies such as rods and balls being added, lumps of the material to be comminuted are used. Like other tumbling methods, the process may be wet or dry.

HARDINGE[17] distinguishes three types of autogenous grinding: (1) completely autogenous; (2) partially autogenous, in which the larger pieces are separated from the feed and the remainder is partially comminuted in another form of equipment and then fed with the separated larger pieces to a conventional tumbling mill shell, the lumps acting as media; and (3) modified autogenous grinding, in which auxiliary foreign media are added to the autogenous mill or the product from a completely autogenous process is scalped and the oversizes reduced in another form of equipment. The three types are sometimes distinguished as primary, secondary and intermediate autogenous grinding[18].

The advantages of the partially autogenous or secondary process compared with a ball mill process are not clear; two stages are necessary and the only distinction from a ball mill is that the media are the lumps of material itself. Whether such a process results in an over-all saving must be left to investigation of particular cases. The modified or intermediate process must depend for its advantage on the degree of additional foreign media necessary.

There are three broad categories of mill: (1) drum, which is simply a ball mill without any foreign media; (2) cascade, which is a drum with a very large diameter-to-length ratio of, say, 3–4; (3) flighted or Aerofall type, which is like a cascade except that flights are set around the periphery or near the circumference of the sides of the drum. Examples of the flighted type are shown in *Figure 10.15*.

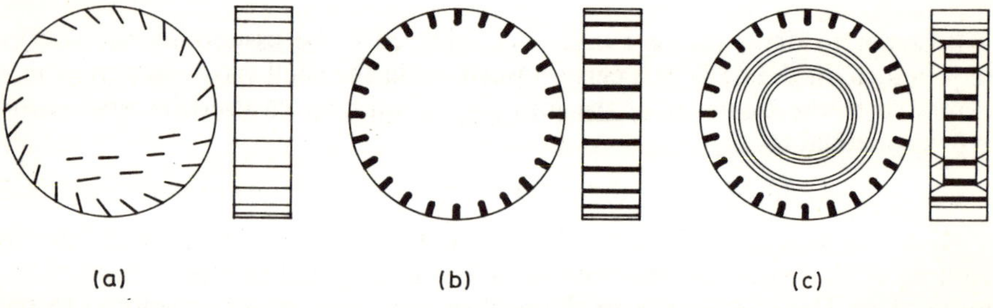

Figure 10.15. Types of flighting to autogenous mill shells: (a) Hardinge–Hadsel, with impact plates; (b) Western Hadsel; (c) Aerofall

The cascade and Aerofall types may or may not contain breaker plates. These plates may be arranged in echelon formation across a horizontal chord at about the lowest third of the diameter of the drum. They hang from a beam which is cantilevered through the feed trunnion and is thus independent of the drum and of its rotation. The breaker plates sometimes have small upstands round their edges so that fine material collects on them and protects them from abrasion.

In some earlier types of flighted mill the drums were set as pairs on the same shaft with a flywheel. Sometimes the periphery is interspersed with gaps. The drums run in troughs so that their lower third is submerged in water and, as the drum rotates, the finer material is washed from the drum into the troughs; the contents of the troughs are agitated with air and overflow to the next stage of the process.

All autogenous mills, like any other tumbling mill, can be air-swept and the air can be heated so that *in situ* drying can take place.

Advantages

The advantages of the autogenous mill are:

(1) Size reduction from 25 cm down to 0.1 mm is achievable in one piece of equipment.
(2) It is possible for ground material to be free of contamination since the lining and the media can both be of the material being ground.
(3) Materials tend particularly to break in these mills along grain sides and separation takes place over lines of different composition; thus separation of mineral constituents takes place in a fairly clean way, which can assist subsequent mineral dressing.
(4) The capital cost is usually less than the equivalent of the cost of a ball- or rod-loaded tumbling mill; the ratio is said to be about 0.7 to 1.
(5) Because most minerals are less dense than steel balls, both the effectiveness of the milling and the energy required to lift the material in the mill are less than for ball-loaded mills. The over-all power advantage is claimed to be in favour of autogenous milling.
(6) Although lining wear is greater in autogenous than in other tumbling mills, wear on the media is of course entirely absent; over-all, the wear costs are about 55–60% less than with a steel-ball-loaded mill[18].
(7) The power demands of the autogenous mill are more sensitive to the load of material than other tumbling mills and therefore it is more amenable to automatic control of feed based upon power demand.

Disadvantage

The disadvantage of the process is that not all materials respond to it. Those which are of a friable, grainy nature are best, but, as indeed for all comminution processes, careful trials with truly representative samples of the actual material to be processed are essential before the process is adopted. Materials which have responded well include silica rock, asbestos ore, basic slag, bauxite, dolomite, ferrosilicon, limestone, taconite and cement clinker, but variations within a quarry can be critical to the success of the process.

Mechanism

The mechanism of autogenous milling does not differ in principle from that of any other tumbling milling. It relies upon large lumps of the material being in the feed and upon reduced material being capable of being broken further by the lumps. If pieces of intermediate size are produced which are relatively greater resisters to breaking than the largest lumps, then further reduction will not proceed well and the intermediate size will accumulate in the mill. In the modified form of grinding, such intermediate sizes are removed and separately treated; and in the partial form, steel balls are added. These superposed operations may, however, minimise the advantages of the process.

The sizes of the largest lumps are usually between 25 and 40 cm and steps must be taken to ensure that an adequate number of the minimum size, found by experiment, are always present.

Because most of the comminuting effect is due to the impact of the larger pieces on the smaller of equal density, the bed of material in the mill has to be kept shallow to avoid cushioning and buoyancy, and the total load in the mill is usually below 30% of the effective mill volume.

The fundamental action in the mill may be considered to be the imparting of the potential energy of the fall of the large lump to itself and to smaller pieces so as to cause rupture. It may be postulated that the condition for rupture consists in the supply of sufficient strain energy to the particle concerned, i.e.

$$Mgh = \frac{\sigma^2}{2E} V$$

in which M is the mass of the large lump; h is the distance through which it falls; σ is the ultimate compressive strength of the material (see p. 16); E is Young's modulus; and V is the volume of the particle broken. If the lump has an average diameter of b and density ρ, then

$$Mgh = \rho b^3 \cdot gh = \frac{\sigma^2 \cdot b^3}{2E} \quad \text{and} \quad h = \frac{\sigma^2}{2Eg\rho}$$

For a material such as limestone $\sigma = 14$ kgf/mm², $\rho = 3 \times 10^{-6}$ kg/mm³ and $E = 2000$ kgf/mm², and, because the g-value is present in both σ and E,

$$h = \frac{(14)^2}{4 \times 10^3 \times 3 \times 10^{-6}} = 16.3 \times 10^3 \text{ mm} = 16.3 \text{ m}$$

which is about three times larger than that needed in practice.

Another approach is that of JOISEL[15], who suggests that the action of breaking is the severance of the lump across a diameter when the lump reaches the floor of the mill, the floor acting as an anvil. Joisel suggests that the force necessary to perform this breaking action acts through a distance equivalent to that of about half the diameter and over an area A equal to the cross-sectional area across a diameter, thus:

$$Mgh = \rho b^3 gh = \frac{\sigma A b}{2} = \frac{\sigma b^3}{2}$$

Thus

$$\rho gh = \frac{\sigma}{2} \quad \text{and} \quad h = \frac{\sigma}{2\rho g}$$

Joisel claims that the value of σ that should be taken is the ultimate tensile strength, which for limestone is about 0.4 kgf/mm^2. Thus

$$h = 0.4/2 \times 3 \times 10^{-6} = 67\,000 \text{ mm} = 67 \text{ m}$$

This figure is about ten times larger than is required in practice: 6 or 7 m is usually large enough for most stones.

Besides the objection to both of these derived figures on grounds of their size, there is also the objection that they both make the size of the lump irrelevant to the breaking process. In fact the size of feed is very relevant to the comminution. The feed must include large lumps, the minimum size being best found by experiment. The size is rarely found to be less than 250 mm. To ensure proper control of this process, facilities should be provided for separating lumps larger than this experimentally determined minimum and for feeding them into the plant with the remainder of the feed in at least the minimum proportion found to be necessary by experiment. If the feed as received does not contain this minimum proportion of the minimum size of lumps, then steps should be taken to ensure that the feed is appropriately changed or that autogenous grinding is abandoned. The proportion of the large lumps necessary is usually found to be about 10%.

Size of Product

A product of 5 mm is quite typical and smaller sizes are often possible, though sometimes at the expense of disproportionately increased power consumption.

Shell

The end plates to the shell sometimes carry scroll-shaped lifters which are intended to combat size segregation.

Wear on the linings is greater in autogenous mills than in other tumbling mills and various attempts have been made to trap material on the periphery of the mill so that wear is material on material. HUKKI[19] has suggested Ni-hard strips, e.g. 50 mm × 10 mm in section, placed on edge with wooden spacers, say 50 mm in thickness and somewhat less in height, so that the Ni-hard strips are slightly proud of them. Material then collects in the slots and forms a lining.

Speed of Rotation

The economic value of speed of rotation above that generally specified for tumbling mills, i.e. about 75% of the critical, has not been fully substantiated. Claims are made on the basis of experiments for both subcritical[17] and supercritical[19] speeds; but it seems generally agreed that speeds of the order of 80–85% of the critical speed are optimal.

Capacity and Power Consumption

There does not appear to be any reason why the formulae for capacity and power consumption which have been deduced for ball mills should not apply to autogenous mills.

An indication of the production of limestone from autogenous mills reducing from 250 mm to 80% less than 300 μm is as follows:

Diameter, m	3·6	5.1	6.9
Width (length), m	0.9	1.2	1.6
Production, ton/h	18	69	200
Power consumption, kWh/ton			14.5

REFERENCES

1. COGHILL, W. H. and DEVANEY, F. D., *Trans. Am. Inst. Chem. Eng.*, **34** (1938), 113
2. BOND, F. C., *Eng. Min. J.*, March (1960)
3. HUKKI, R. T., *Trans. Am. Inst. Mech. Eng.*, May (1958), 581
4. BELLWINKEL, A., *Dechema Zerkleinern Symposion, 1962*, 373
5. RUNOLINNA, U., *Dechema-Monogr.*, **57** (1967), 385
6. COGHILL, W. H. and DEVANEY, F. D., *U.S. Bur. Mines Tech. Paper* 581 (1937)
7. ALLIS-CHALMERS LTD, Publication 07T 9900-133
8. BOMBLED, J. P., *Dechema-Monogr.*, **57** (1967), 633
9. ROSE, H. E. and SULLIVAN, R. M. E., *Ball, Tube and Rod Mills*, 1958. London; Constable
10. DAVIS, E. W., *Trans. Am. Inst. Min. Metall. Eng.*, **61** (1919), 250
11. CAREY, W. F. and STAIRMAND, C. J., *IMM Symposium on Mineral Dressing, 1952*
12. PAPADAKIS, M., *Dechema Zerkleinern Symposion, 1962*, 341
13. SLEGTEN, J. A., in discussion of Reference 12
14. NIJMAN, J., *Dechema Zerkleinern Symposion, 1962*, 249
15. JOISEL, A., *Dechema Zerkleinern Symposion, 1962*, 49
16. HARDINGE, H. W., *Trans. Am. Inst. Mech. Eng.*, October (1908)
17. HARDINGE, H., *Mining Congress J.*, October (1958)
18. ROWLAND, C. A., *Trans. Am. Inst. Min. Metall. Petrol. Eng.*, February (1964)
19. HUKKI, R. T., *Dechema Zerkleinern Symposion, 1962*, 394
20. ANON., *Crushing and Grinding*, **1**, July–August (1931), 47

11

Cutting Machines

Cutting is the most widely known method of comminution. The distinctive feature of the equipment is the application of a force to an area of material of which at least one dimension is very narrow. The equipment used to do this can be classified as:

Knives, which apply such a force slowly through a single edge onto material supported on a table or quickly onto material held still by its inertia.

Shears, which apply such a force through a pair of edges on opposite sides of the material, but so that the projections of the areas on which they act are very nearly contiguous.

Saws, which apply such a force through the points of teeth over the length of the area on material supported by a table.

Spikes, which apply such a force to an area, both the dimensions of which are very small, on material supported by a table.

The cutting pieces may be stationary, rotating or oscillating, and are invariably made of material which is hard and will sharpen to and retain a fine edge.

The chief advantages of the method are that it allows of considerable control on the location of the rupture and consequently the shape of the product. The method is particularly useful for the size reduction of plastic, elastic and fibrous materials which are otherwise difficult to comminute.

The chief disadvantages of the method are that its rate is slow and the minimum size of product is large. Cutting machines also pose very considerable safety problems.

MECHANISM OF CUTTING

The mechanism of cutting would appear to be penetration of the finest edge of the cutter as a result of plastic deformation of the material being cut. First the material is displaced much as a furrow is made in the soil. If the cutter is moved, abrasion also is caused by the saw-like edge of the knife, which high magnification will show to be present on the sharpest, smoothest blade, digging into the material and colliding with similar hillocks on the exposed surface of the material and sweeping them away. The relative influence of the plastic deformation and the abrasion depends upon the nature of the knife and the material: clearly, the action of a sharp knife through butter is almost wholly one of plastic deformation, whereas the action of a diamond saw through a ceramic is almost wholly one of abrasion. The application of knives and shears is mainly to plastically deformable materials, whereas saws are applied to more brittle materials. Spikes can be used for all kinds of material.

Materials which are bouncy, such as rubber and plastics, or fibrous, such as asbestos, are most easily comminuted by cutting.

MATERIALS OF CONSTRUCTION

The cutting tools for this class of equipment are usually made from steel containing between 0.8 and 1.5% carbon. To produce steel most suitable for sharpening, the cementite constituent is converted into fine nodules by raising the temperature of the steel above about 720°C and quenching it, with subsequent tempering processes to form martensite, which provides the good wearing properties.

The hardness can be increased by adding manganese (0–2%), chromium (0–2%) or vanadium (0–0.2%). An alloy with very superior cutting power is produced with tungsten (3–6%) and for working at high speed on hard materials (14–18%), at which composition the alloy will retain its cutting properties even when it is red hot. Toughness may be added with 5–12% cobalt when an alloy capable of cutting very hard gritty materials can be produced. Molybdenum steels if prepared with due metallurgical care are also very good cutters. Another method of adding hardness is to splutter the surface of the steel with a thin layer of chromium; the grinding of the knife is carried out so that the spluttered chromium forms the peak of the edge. Razor blades and the shoes of paper beaters are so treated. The process is called fescalising.

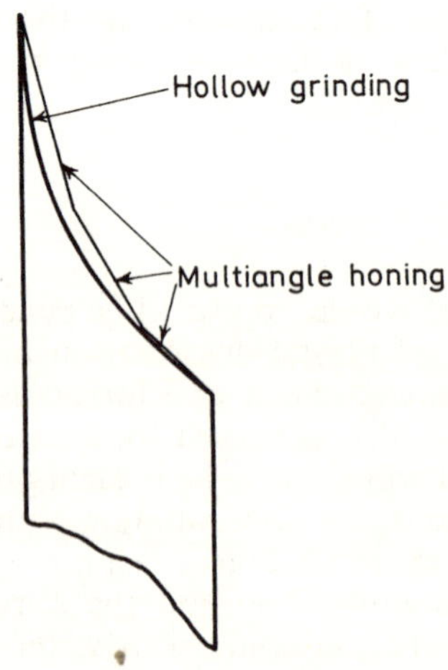

Figure 11.1. Knife edge design

The application of the high-carbon steels can be roughly classified as follows, increasing carbon content corresponding with decreasing toughness and increased hardness:

Carbon content, %	*Application*
0.8 toughness	large knives, chisels
1.0	milling cutter, mining drills
1.10	saws
1.20	twist drills
1.30	razors
1.50 hardness	roll corrugators

Cutting tools are also made by sintering or setting materials such as stellite (35–80% cobalt, 10–40% chromium and smaller quantities of tungsten and carbon), tungsten carbide or diamond into matrix. Such tools are used particularly in cutting ceramics or in borehole drilling.

TYPES OF CUTTING TOOL

Knives

Knives are in common industrial and agricultural use: examples range from the plough and the scythe to the beet cassettes which macerate the beet into slices for sugar extraction.

The forming of the knife edge requires design expertise, the object being to have the sharpest angle at the extreme edge commensurate with the relative compressive strengths of the material of the knife and the material to be cut while maintaining sufficient lateral strength to prevent buckling. This compromise is usually achieved by restricting the sharpest angle to close to the tip and buttressing that angle with successive steps of larger angle, or alternatively, by curving the surface away from the tip (see *Figure 11.1*), the one being achieved by multiangular honing and the other by hollow grinding.

Shears

By far the major application of cutting is through the use of shears. KAISER[1] distinguishes two forms of shears: (1) those which require a continuous sheet or cord of material to be fed to them which they cut, usually with precision, into regular rectangular pieces; and (2) those which will cut pieces of assorted sizes into random shapes and sizes. *Figures 11.2* and *11.3*, taken from Kaiser's papers, illustrate the two types. They are both used extensively for plastics, rubber, leather, man-made fibres, asbestos, cellulose pulps, rags and non-ferrous metals.

An application of shears on a large scale is in the steel scrap business. The first manufacture of hydraulically moved shears was in 1959 for

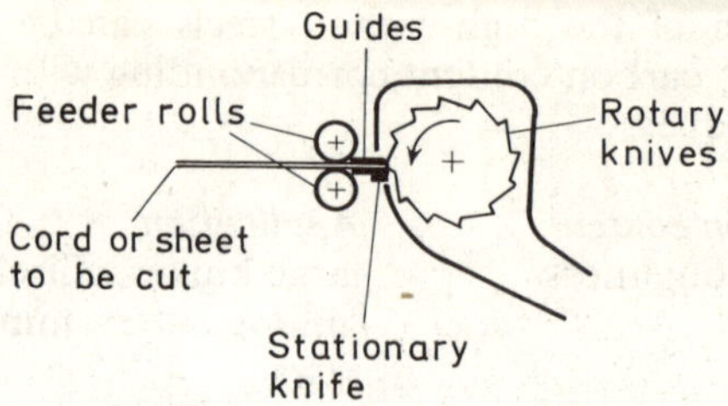

Figure 11.2. Sheet or cord cutter. The feeder rolls are turned synchronously with the knife rotor. (After F. KAISER[1]*)*

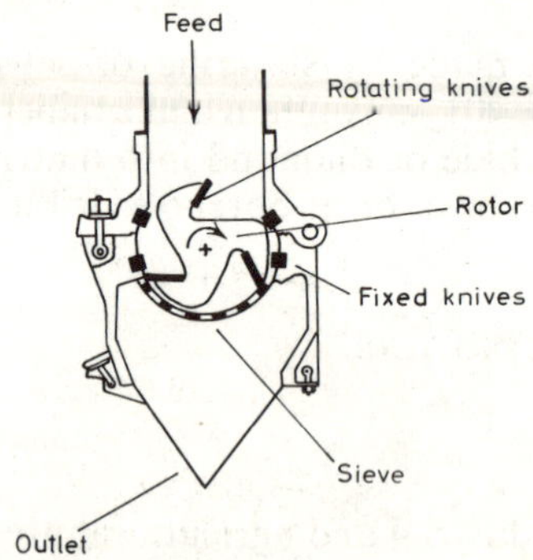

Figure 11.3. Cutting mill. (After F. KAISER[1]*)*

comminuting ship-breaking scrap. The shear force of the machine was 8200 kN, with a blade width of 2050 mm. Machines are now being made with forces of over 20 000 kN (see *Figure 11.4*). The scrap is fed into a chamber and compacted into a prism and then fed at set lengths over a fixed blade and under a moving blade, the pair of blades being set so that their nearer sides come within a very short distance of each other. The size of the shear force required depends upon the width of the cut and the thickness of the individual pieces of scrap composing the feed, but a crude relationship can be obtained from *Table 11.1*[2].

Table 11.1

Type of scrap	*Shear force*, t
Motor cars, frames and axles: light construction	250–350
Heavy vehicle frames	400–550
Anything up to 10 cm thickness	750–1 200
15 cm thickness	2 000

The lower fixed blade is adjusted to the upper moving blade with shims or sometimes by means of a movable holder. The blades are made of Ni-chrome alloy and bolted into place. Blades have to be kept sharp. Up to four sharpenings can be taken before a blade is finally discarded. Feeding is by

Figure 11.4. The 1128 ton Fielding Richards Shear—the 'Iron Shark'. (Courtesy the Bird Group of Companies)

means of a hydraulic pusher or sometimes by means of a vibrating conveyor.

A device also used in the recovery of steel scrap has a strong resemblance to a hammer mill, but its action is a cutting one on the edges of the hammer bars. The rotating bars push through the metal and tear it into pieces, which allows the pieces to fall through a grid. This type of machine has many forms depending upon its use. The rotor can be made up of a series of discs which intermesh with fixed shredding teeth or with discs which are either free to rotate or are driven. These forms are used in refuse disposal. Also into this class must be put the intermeshing discs and 'doctor knife' used for breaking up filter-cake.

A smaller scale of subdivision, still working on the shear principle, is provided by the paper beater (see *Figure 11.5*), which is like a grand lawn

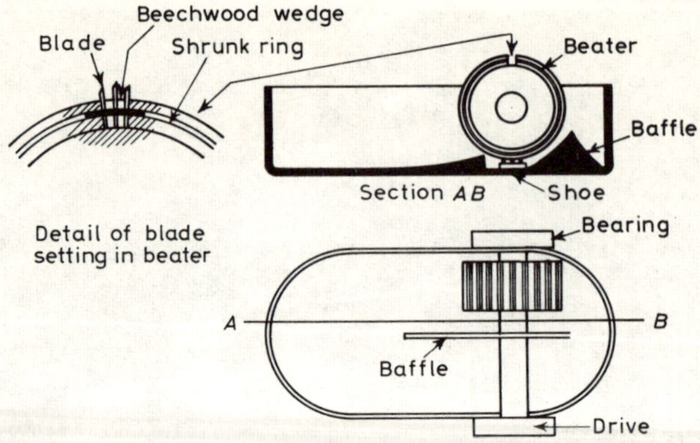

Figure 11.5. Comminution by cutting: paper beater

mower except that all the blades are parallel to the axis of the rotor. Similar machines are used on a smaller scale for reducing waste plastics sheet.

Guillotines are in common use for trimming of sheet metal, paper and leather.

Saws

Saws have certainly been used in prehistoric times; there is some slight evidence that animal teeth embedded in jaw bones formed the first saws. The builders of the Egyptian pyramids used bronze saws, in which were set jewels, to cut the granite. Saws have been designed to cut nearly every material in use, ranging from those for Carrara marble for statuary, which is cut into blocks at Carrara with a band wire, to the surgeon's bone saws.

The idea of the circular saw is attributed to BRUNEL[3] and dates from about 1790.

Spikes

Examples of the use of spikes are harrows and tines for breaking the soil. Spikes are an essential part of some roll crushers (see p. 160).

REFERENCES

1. KAISER, F., *Dechema-Monogr.*, **57** (1967), 741
2. BIRD, W. T. and BIRD, A. P., *Iron Steel Inst./Instn Mech. Eng. Meeting on Scrap Handling, October, 1971*
3. HOTCHKISS, G. W., *Encyclopaedia Britannica,* 9th edn., Vol. 21 (1886), 344

12

Attriting Machines

The distinguishing feature of attriting machines is that size reduction is effected by particles breaking each other having acquired energy to do so from a solid or fluid impeller. Included in this class are buhrstones, colloid mills, fluid-energy mills and sand grinders.

Advantages of the class are:

(1) Material can be produced in very small sizes, of the order of less than 10 μm.
(2) The product can be controlled in a narrow size range.

A disadvantage is that equipment is usually only large enough to handle small quantities of material.

The mechanism of this class of mill changes with type: in the buhrstone the action is abrasion; in the colloid mill the discs mainly impart energy and rely upon collision between particles; in the so-called petit pulveriser enormous energy is imparted to the individual particles and spin induced in them produces enough centrifugal force to cause disintegration; in the fluid-energy mill the high energy and other properties of fluids such as steam, compressed air or other gas are used to assume flow patterns in carefully designed vessels and thereby induce solid particles to move into similar flow patterns and interact upon one another; in the sand grinder particles are brought into contact with sand particles.

Thus there are the sub-classes: (1) *rotating discs*: buhrstones, colloid mills, petit pulverisers; (2) *patterned fluid*: fluid-energy mills; (3) *abrading sand*: sand grinder.

BUHRSTONE

The buhrstone is as old as the quern in which prehistoric people ground their flour. Dr. Livingstone thus describes a mill he saw in use in Central Africa.

'The mill consists of a block of granite, syenite or even mica schist, 15 or 18 inches square and 5 or 6 inches thick, with a piece of quartz or other hard rock, about the size of half a brick, one side of which has a convex surface and fits into a concave hollow in the large and stationary stone. The work woman kneeling, grasps this upper millstone with both hands and works it backwards and forwards in the hollow of the lower millstone. . . . This is perhaps the most primitive form of mill and anterior to that in oriental countries, where two women grind at one mill, and may have been that used by Sarah (wife of Abraham, Genesis 18:6) when she entertained the angels.'

In Deuteronomy 24:6 the importance of both parts of the mill is emphasised. 'No man shall take the nether or upper millstone to pledge for he taketh a man's life to pledge.'

Buhrstones were extensively used in the old flour mills, the stones being up to 2 m in diameter and 30 cm or so thick. The buhrstone is usually a siliceous stone, the best of which, it was considered, came from the Seine and Maine district of France. The stones were made in segments and bound together with plaster of Paris in an iron hoop. Grooves were cut in the stones, usually parallel to one radius of the segment, but so that the grooves in the top and the bottom stones matched exactly and made scissor action when they revolved against one another. The effect of the grooves, when grain is fed to the middle of the upper stone, is to cause the grain gradually to move out to the circumference of the stone.

Various mills of this same type have been developed with different stones for different materials. Emery has been used for pigments and steel for rubber.

COLLOID MILL

Colloid mills consist of two discs set between 0.02 and 1 mm apart and revolving at peripheral speeds between 10 and 50 m/s. According to the peripheral speed, so is the particle size produced, of which the following may be taken as examples:

Peripheral speed, m/s	10	30	40
Average particle size, μm	1	0.1	0.01

Feed less than 200 μm is sent in the form of a slurry either through the axis of one disc or from the top. The discs can be plain or grooved and can be flat or male or female cones (see *Figure 12.1*). The discs are made of steel, alumina (corundum) or stone.

Colloid mills are also used to disperse agglomerations of particles produced by another comminution process.

PETIT PULVERISER

The petit pulveriser is an interesting form of attrition mill[1] which consists of a single disc set in a casing with clearance from the casing of about 0.5 mm. The disc is about 15 cm in diameter and about 2.5 cm thick. The disc revolves at about 40 000 rev/min (peripheral speed, 300 m/s). Feed is entrained in a gas and delivered to one side of the disc. Outlet is from the other side (see *Figure 12.2*). It is suggested that such are the turbulence and micro-eddies created that the particles are made to spin at 10–100 million rev/min and that

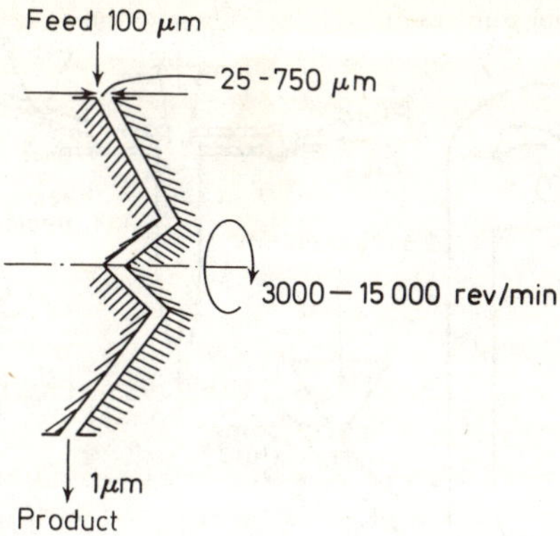

Figure 12.1. Colloid mill

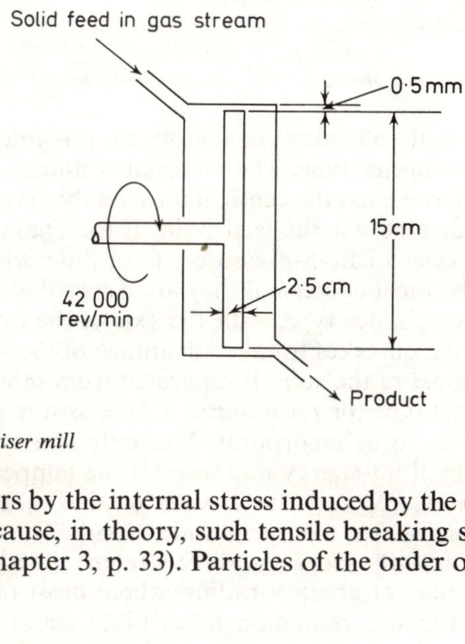

Figure 12.2. Petit pulveriser mill

disintegration occurs by the internal stress induced by the centrifugal forces. It is interesting because, in theory, such tensile breaking should require the least energy (see Chapter 3, p. 33). Particles of the order of 1 μm are said to be produced.

FLUID-ENERGY MILL

The fluid-energy mill in its original form and as patented by Willoughby was of two jets of material fluidised by air or steam, fed diametrically opposite one another so as to collide. Subsequent fluid-energy mills take advantage of the varying velocity across a fluid vortex to provide particles with differential speeds and therefore with opportunities to collide. The method of creating the vortex is either to feed a fluid into the periphery of a cylinder, short relative to its diameter (a type usually called a Micromiser, or sometimes pancake type); or into the side of a torus, long relative to its diameter

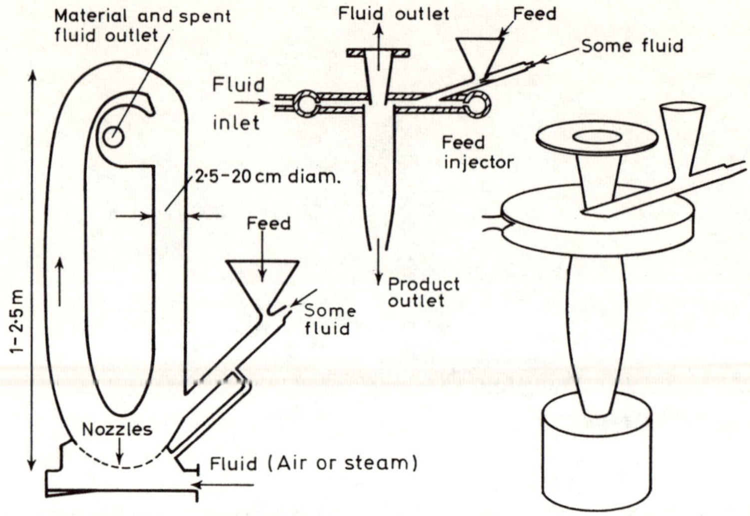

Figure 12.3. Fluid-energy mills showing torus and cylinder types

(see *Figure 12.3*), with off-takes as remote as possible from the intakes, usually called the Wheeler type. The material is fluidised and injected into the chamber, and removed at the centre if the chamber is a cylinder, or at near the complete distance round the feed point if the chamber is a torus. The material meets a series of high-pressure jets of fluid which in the cylinder type are found to be most effective if they are directed at 45° to the radius of the cylinder. In the cylinder type, with the axis of the cylinder vertical, it is possible to design the off-takes to take advantage of the vortex and the force of gravity so that most of the solid is separated from most of the fluid at the off-take. In the torus type (or race-course shape, as it is sometimes called) a separating device has to be incorporated, usually of a cyclone design.

Advantages of the fluid-energy mill are: (1) the temperature is not raised during comminution; (2) very little wear of the equipment occurs and therefore the product is free from contamination.

It appears that, broadly speaking, three streams may be considered to be set up in the chamber: of greatest radius, where most of the collisions and therefore most of the size reduction takes place (MORI[2] considers that all comminution takes place in the vicinity of the jets); of intermediate radius, where classification of heavier and lighter particles occurs, the heavier being returned by centrifugal action to the former stream; of smallest radius, the stream in which withdrawal of finished product is effected, only sufficiently small particles being able to be entrained in this slowest moving stream.

TEMPERLEY and BLYTHE[3] show that there is a considerable momentum change in the solid particle from its relatively stationary state to its meeting the very fast gas flow and that the energy transferred to the solid is a maximum when the mass of solid equals the mass of fluid. Thus the fluid intake capacity is a measure of the feed capacity of the machine under optimum conditions for transfer of energy. (They point out that the optimum conditions for transfer of energy could be different from the optimum conditions for size reduction.) Feed control must therefore be very strict. Fineness of product depends very much on the height of the pressure of the fluid at the inlet.

Method of Rating

The cylinder type of machine is usually rated according to the inside diameter of the mill. *Table 12.1* shows typical relationships.

Table 12.1

Mill diameter, cm	5	10	20	30	40	50	60	80
Air flow, m^3/min	0.6	1.2	2.8	6.3	10	15.5	28	42
Steam flow, kg/h	—	—	110	270	410	680	1 200	1 800

A diameter of 80 cm is considered to be the maximum economic size.

Materials of Construction

Abrasion is not high if the jets are correctly set and mild steel is satisfactory provided there are no corrosion problems. Stainless steel and rubber-lined mild steel are both used.

Fluid

Fluids which have been used in this type of mill include steam, air and other gases. Steam has the advantage that at the outlet it can be condensed and all the entrained solid can be removed. This is of particular value when the material is toxic, pyrophoric or radioactive. Superheated steam is used at from 0.7 to 3.5 N/mm^2 pressure. Steam is usually cheaper than compressed air. In spite of theory, steam usages run as high as 80 kg of steam per kilogram of product for a very fine material.

Feed

Feed up to 10 mm can be used in some equipment, but minus 150 μm is the more usual size. HENDRY[4] found that a finer product is obtained with a closely sized feed even though it is two or three times the size of the average of a wider-sized feed. He also found that the selection function tended to reach an optimum at about the same feed size for a wide range of feed rates.

Fibrous and waxy materials and some metallic materials cannot usually be treated in the fluid-energy mill. Typical materials which react very well in the mill are talc, sulphur, cryolite, barytes, pyrolucite, pyrethrum and, contrary to what was said about fibrous materials, cotton.

When materials are required to be ground which can burn in air and there is some possibility of air entering the grinding system, this type of equipment lends itself to the incorporation of explosion discs as a precaution.

Capacity

MORI[2] found that capacity increased with the power of the compressor raised to the power of 2.5, or power multiplied by 10 increases capacity by a factor of 300. RAMANUJAM and VENKATESWARLU[5] found the relationship

$$\frac{\text{optimum feed weight rate}}{(\text{weight rate of fluid})^{2.5}} = \text{a constant}$$

which confirms the finding of Mori.

SAND GRINDER

In the sand grinder the material to be ground is brought into contact with a foreign attriting material in the form of sand. Ottawa sand of size range 400 μm to 600 μm is quoted as being a suitable attriting material. The material to be ground, in the form of a slurry, is pushed up through a series of rotors set on a single shaft (see *Figure 12.4*) and during its passage upwards meets the downcoming sand. Finally, at the top of the multi-stage

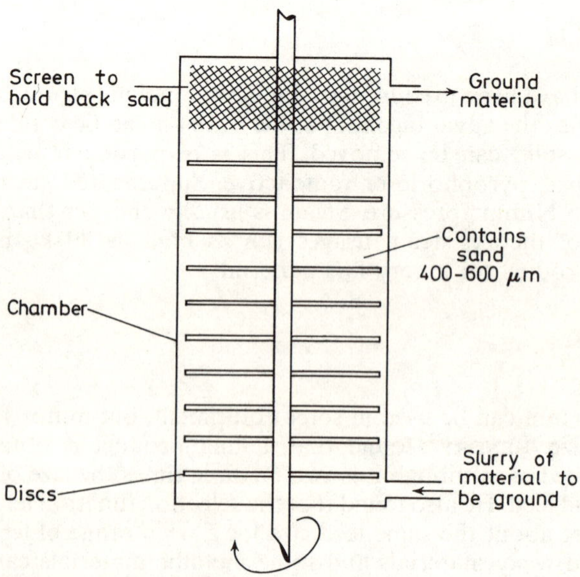

Figure 12.4. Sand grinder

compartment a screen, of smaller mesh than the lowest mesh of sand, separates the sand from the ground material, which then leaves the system.

The volume of sand resident in the chamber should equal the volume of slurry in the chamber at any time; differences from this ratio either stifle grinding or prevent its taking place. The slurry should be heated to 50–65°C to obtain optimum viscosity, i.e. neither skidding over the discs nor incurring intense wear on the discs, and giving adequate mobility to the sand.

The feed should be below 70 μm.

The product can be below 5 μm.

Materials of Construction

The discs should have a hardness of not less than 40 Rockwell C and the conditions of operation should be controlled, as stated above, to avoid intense wear.

ULTRA-ROTOR

The ultra-rotor (see *Figure 12.5*) was developed by the Altenburger Maschinen GmbH of Hamm, Germany, to mill the chemicals they made

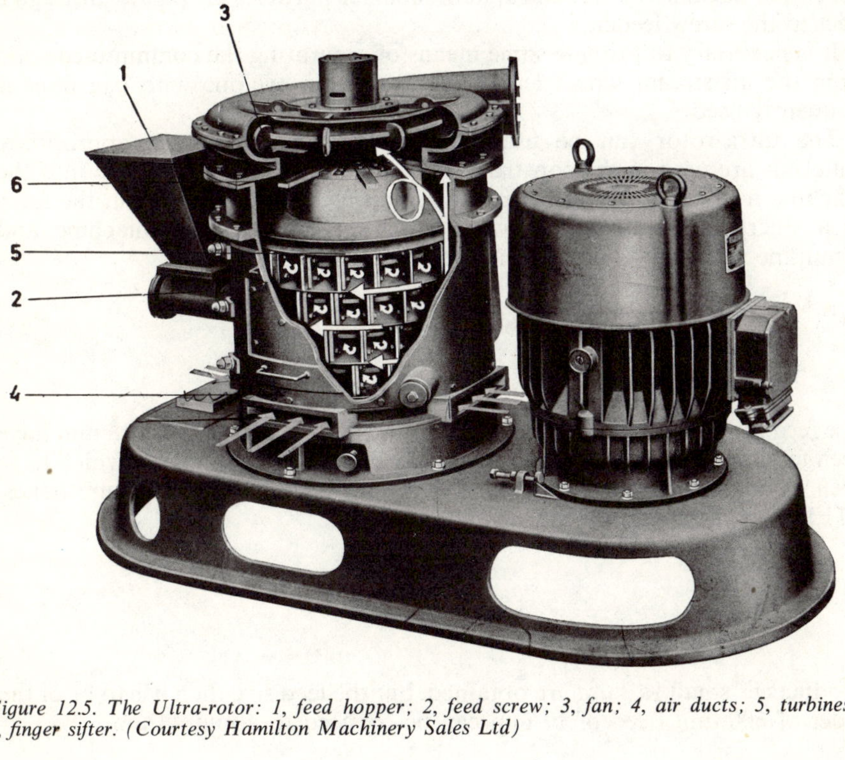

Figure 12.5. The Ultra-rotor: 1, feed hopper; 2, feed screw; 3, fan; 4, air ducts; 5, turbines; 6, finger sifter. (Courtesy Hamilton Machinery Sales Ltd)

from flour. The principle of the mill is related to that of fluid-energy mills. A series of impellers or turbines are all set on one central shaft. Each impeller holds radial vanes which are adjustable. An impeller looks somewhat like the paddle of a water mill. The shaft with its impellers is set coaxially with a cylindrical housing. The clearance between the housing and the circumference of the vanes has a maximum of about 15 mm and can be reduced to 2 mm by adjusting the vanes. The impellers are about 740 mm in diameter and they are rotated at 2200 rev/min (36.6 Hz).

Material is fed into a hopper at the side of the machine and induced into the mill from the hopper by a screw feeder. A fan at the top of the machine, also set on the central shaft, induces air into the machine through an opening in the ducting set round the bottom of the cylinder. The air entrains the material and carries it in a complex spiral path up through the annulus between the housing and the impellers. So great is the velocity of the air in the annulus that the air is turbulent and the particles abrade each other. The air and the entrained particles pass up through the vanes of a device called a finger sifter, which controls particle size by obstructing the exit of the particles with rotating vanes or fingers, the number of which can be readily altered by adding fingers or subtracting them from the hub. The particle-and-air mixture which has passed through the fingers approaches the involuted casing of the fan, the entry to which is monitored by a baffle whose distance from the axis of the shaft is adjustable and which controls further the size of particle to leave the system: coarser particles drop back through a duct to the screw feeder.

It is necessary to provide some means of separating the comminuted solid from the air-stream which leaves the system: a cyclone with bag filter is frequently used.

The ultra-rotor can be used for simultaneous drying if appropriate materials are used in the construction of the mill. Gases can be fed into the machine from a burner at temperatures up to 300°C. Material in the form of a slurry containing 98% water has been fed into the machine and simultaneously dried and comminuted.

Feed

The feed may be up to 10 mm in size; some materials as large as 18 mm have been fed, but with some deterioration in performance. Materials which have been comminuted in the machine have included rubber, polyurethanes, PTFE, soya flour, sugars and animal glue.

Product

Products as small as 3 μm are obtained, but the feed size then has to be of the order of 600 μm. Feed of 10 mm can be reduced to about 44 μm.

Maintenance

The machine is most suitable for handling soft sticky materials, when the maintenance is very small indeed.

Power Consumption

For limestone, the actual Bond index (W_e) is about 26.

Capacity

The capacity of the machine described is 1 tonne per hour of limestone reduced from 10 mm to 44 µm. PTFE has been reduced from 500 µm to 30 µm at the rate of 0.24 t/h.

REFERENCES

1. TANNER, H. G., U.S. Patent 2 552 603 (1951)
2. MORI, Y., *Dechema Zerkleinern Symposion, 1962*, 515
3. TEMPERLEY, H. N. V. and BLYTHE, G. E. K., *Nature, Lond.*, **219** (1968), 1218
4. HENDRY, R., *Dechema-Monogr.*, **57** (1967), 695
5. RAMANUJAM, M. and VENKATESWARLU, D., *Powder Technol.*, **3/4** (1970), 213

13

Other Methods of Comminution

Many methods of comminution have been proposed which cannot conveniently be included in any of the five classes of equipment described in previous chapters. Some of these methods are used on a large industrial scale, though they are usually restricted to a particular material; others have not been developed fully and the economics of using them are doubtful or unknown. Most of the methods described below are listed in a review by IOHN[1].

WEATHERING

Weathering is much used in the clay industry to break down the somewhat moist lumps dug from the pits. The clay is placed in squat heaps and left to the effects of the weather. The agents of weathering may be listed as rain, frost, sunshine and wind; their physical results as wetting, dissolving, drying, heating, cooling, entraining, deposition; their chemical effects as hydration and hydrolysis, oxidation, reduction and carbonation. There are also biological results. Weathering can therefore cause many changes other than comminution, but the effects causing comminution include differential thermal expansion, stresses created by water freezing in cracks, swelling due to water absorption, and abrasion caused by dust in the wind.

THERMAL SIZE REDUCTION

Many rocks split when subjected to heat, owing to unequal expansion of their parts, development of steam pressure from moisture or decomposed water of crystallisation or allotropic modifications resulting in volume changes. The process is sometimes described as decrepitation, particularly when it is accompanied by the production of sound. The method is particularly attractive when the resulting comminution discriminates between the composition of the materials broken and of those remaining unbroken. Thus barytes and zinc blendes have densities which are very close to each other (4.48 and 4.06), but on roasting this mixture for 10 min on a 2 m square hearth and raking onto the floor, the size distribution of the product was such that the greater than 1 mm fraction contained 75% barytes; that between 0.64 and 1.0 mm, 20%; and the fraction less than 0.64 mm, 5%. In a series of experiments[2] there was some evidence that the more moist the barytes, the finer the product. Examples of other minerals which decrepitate are celestine ($SrSO_4$), crocoite ($PbCrO_4$), salt (NaCl) and lead nitrate.

The property of materials rupturing owing to thermal shock is called their 'spallability' and is found to be a function of Young's modulus, the coefficient of linear expansion, the grain size and the inverse of both the thermal conductivity and unity minus the Poisson ratio. A material has a critical temperature at which it becomes sufficiently plastic not to spall.

Fluorspar is a mineral which readily splits because of the unequal expansion of its parts and forms a powder, leaving contaminants largely unchanged in lump form and easily separated from the powder by sieving. Sylvinite can be separated from sodium chloride and other salts from potash deposits in a similar way, but it is the sylvinite which remains in lumps.

The effect of slaking quicklime lumps to form powdered calcium hydroxide is well known.

Vermiculite, which is a form of mica but having water of crystallisation between its micaceous layers, defoliates between 700 and 900°C as a result of the conversion of the water of crystallisation to steam, the resulting pressure between the micaceous laminations causing them to separate into leaves.

Magnesite containing quartz can be broken down by taking advantage of the volume change in the allotropic transformation of α to β quartz, the temperature for which is 575°C and which can be made to take place with shattering of the quartz by imposing a high temperature (600–800°C) suddenly on the mixture. The powdered quartz can then be separated from the magnesite by air-separation or sieving.

Processes have been devised which use burners playing on a steel conveyor belt and carrying the material to be comminuted. The burners are followed by water sprinklers and devices are included for combing out lumps which have not been sufficiently reduced in size.

PRESSURE ALTERATION

Methods in this category depend either upon different materials having different resistances to pressure or upon the binding at grain or fibre boundaries being weakened by alternating pressures. The process is usually expensive and justified only for high-priced products; it has been used for separating diamonds from other minerals.

Asbestos fibres can be unbound in a stirred tank subjected to a pulsating pressure between the extremes of, say, 10 and 400 kg/cm^2. Coal can be shattered by compression in an autoclave by superheated steam or compressed air followed by sudden release of the pressure.

ULTRASONICS

Impulse waves are imposed upon material through a liquid medium from a magnetic or quartz vibrator producing waves of the appropriate ultrasonic length. The main application of the process is for relatively small size ranges and for comparatively low energy inputs since the vibrating system is vulnerable at higher energies.

A sketch of an apparatus used for recovering mica from waste is shown in *Figure 13.1.* A is an ultrasonic vibrator producing waves with a frequency

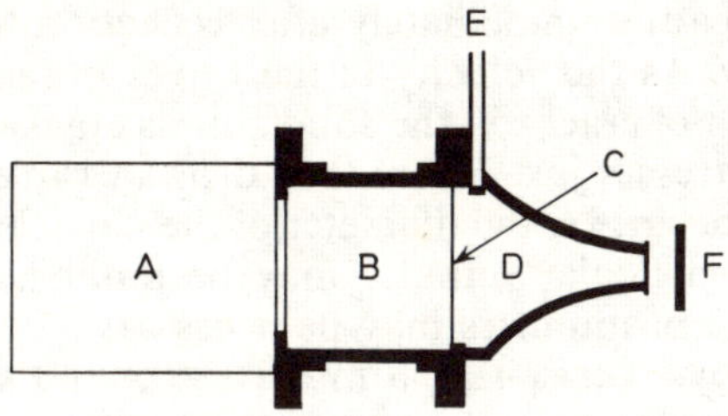

Figure 13.1. Equipment used by SNADOW, PARRIS *and* BAKER[1] *for ultrasonic size reduction: A, ultrasonic vibrator; B, water compartment; C, plastics membrane; D, nozzle; E, feed point; F, impact plate*

of 500 kHz. The vibrator subjoins a water compartment B with a plastics membrane C. Material is introduced through E and the wave impulses eject the material through the nozzle D onto an impact plate F.

In addition to the use of liquid shock waves as in the example above, gas shock waves have been used at ultrasonic velocities, gas being introduced at pressures between 20 and 40 atm through carefully designed nozzles into specially constructed vessels.

ELECTRICAL

Electrical methods depend upon the production of either thermal effects (covered in the section on thermal size reduction) or hydraulic effects. Electrical heating methods involve the use of high-frequency electromagnetic waves to heat minerals on which they impinge. Essentially material is passed through a loop through which a high-frequency electrical current is passing. Another method is to use the heating effect of direct current flow by first making the surface of the mineral conducting by spraying it with an electrolyte of, say, salt or copper sulphate solution and then bringing the lumps into contact with electrode brushes.

Electrohydraulic methods depend upon the effect of the sudden discharge of a quantity of electricity between electrodes in a fluid. An electrical condenser is used to build up the required quantity of electricity. The discharge in the fluid causes the fracture of solids placed between the electrodes.

The pressure developed depends inversely on the time of discharge and the disrupting stresses depend upon the pressure developed. A stress wave, similar to that which is generated by an explosion but of much smaller wavelength and duration, produces an expanding sphere of compression with tensile hoop stresses on the circumference of the wave-front; the wave is reflected back into the solid at the solid–fluid interface in proportion to the impedance to the wave velocity imposed by the fluid, and a solid–air interface is thus more effective than a solid–water interface. After reflection of the wave, the nature of the stress exerted changes from being compressive to being tensile.

It is the tensile waves which cause the disruption. The stress created by these waves rises steadily immediately after reflection to a peak stress and then drops abruptly. As the velocity of the waves is a small multiple of the speed of propagation of cracks in the solid, unless cracks are already present to concentrate the stresses just before they drop in value, considerable fracture is not likely to occur, i.e. the diameter of the particle which it is possible to fracture with a particular impulse may be limited, and for a range of particle sizes a range of impulses may be necessary.

Equipment used by CARLEY-MACAULY, HITCHON and MAROUDAS[5] entrains the feed particles in water, delivers them to a polythene-lined spark chamber and lifts them onto a separator (such as a centrifuge) which takes out the specified small particles and returns the oversize to the spark gap.

The electrohydraulic process was found by YIGIT, JOHNSTON and MAROUDAS[6] to show some selectivity for rocks which were either the more conducting or the smoother rocks in a mixture provided the size range of the mixture was about 25 mm. Thus a mixture containing 50–60% of a constituent distinguished from the other constituents by smoothness or conductivity was comminuted so that the smaller fractions contained more than 98% of that constituent (see *Table 13.1*). The selectivity disappeared when the size of the feed was less than 1 mm.

Table 13.1. SELECTION OF SMOOTHER OR MORE CONDUCTING CONSTITUENT BY ELECTROHYDRAULIC COMMINUTION

Mixture	*Percentage by weight* Feed (−37 to 31 mm)	Product (−6 mm)
Pyrrhotine (more conducting) and	60	98.3
Flint	40	1.7
Flint (smoother) and	50	98.7
Granite	50	1.3

Experiments on a small scale[5] indicate that over-all energy requirements are 20 times as great as for mechanical methods; but theory suggests that such efficiencies could be improved.

CHEMICAL

Hydrogen peroxide when heated decomposes violently; if materials are present, the violent evolution of gas may cause disintegration. Asbestos fibres and mica sheets are both unlocked by the presence of hydrochloric acid in solution. Asbestos when heated to about 90°C under a pressure of about 5 atm in the presence of 0.5% sulphuric acid, and then passed to a chamber at 250°C and about 30 atm, disintegrates. Sometimes such materials are first soaked in a sodium carbonate solution in an autoclave at a high temperature and subsequently, when treated with dilute acid, are split up by the carbon dioxide which is formed. Another method is to soak the material in a liquid which will ferment, e.g. a solution of molasses; the carbon dioxide produced will then break up the fibres or leaves.

EXPLOSION

Explosives are commonly used to remove minerals from their natural beds, particularly in surface quarries, and much of the rock so produced is used without further comminution. Quarrying is a highly specialised operation and it will be discussed only in the briefest outline.

No alternative is known to explosives for removing highly resistant rock in large quantities from its bed. Explosives fall into two classes: those which produce their effect by rapid expansion of gases and the production of high pressures, called low explosives; and those which produce shock waves, in addition to these effects, and are called high explosives. Explosives may consist of capsules of highly compressed gases, gunpowders, dynamite (kieselguhr soaked with nitroglycerine), gelignite (nitrocellulose soaked with nitroglycerine), ammonium nitrate mixed with fuel oil and sometimes with finely powdered metals, slurries of ammonium nitrate, or nitro-organic compounds. More recently nuclear explosives have been used. High explosives require a primer to ensure that the reaction proceeds sufficiently quickly. This primer can be guncotton or tetryl in which is embedded a detonator.

The effect of the detonation of an explosive in a hole in the rock, provided the explosive is closely confined, is to send out a high-pressure pulsation which initially crushes and fractures the rock; when the stress wave meets the free face, it is reflected back through the rock, causing more crushing and fracturing.

The first objective in a quarrying operation is to make a large cavity so that an open face as nearly vertical as the rock will allow is exposed. A profile similar to that shown in *Figure 13.2* is then produced. The bench height (see figure) is determined by the design of the shovel to be used in loading the rock, i.e. so that the heap produced by blasting is not so high from the floor of the operation that when the heap is scooped up with the shovel it leaves overhang (which could collapse and be dangerous) and yet not so low as to create unnecessary shovel movement.

The system of holes to take the explosive is designed, commensurate always with safety, according to the type of operation planned and the type of rock. For conveyor and rail haulage the benches are long horizontally and the material is required to be placed parallel to the pit walls in relatively short depth; for lorry haulage the benches are short and the material is

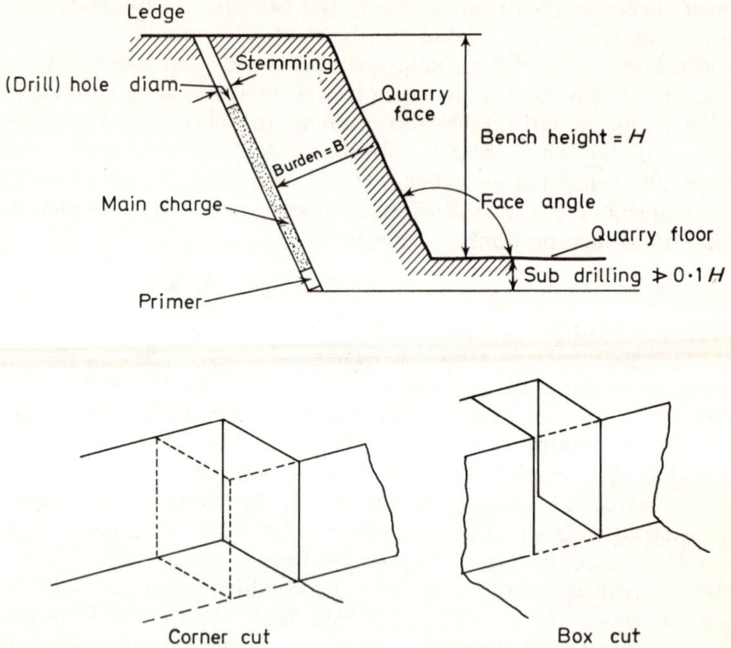

Figure 13.2. Quarry profile and form of cuts

required to be placed deep enough for the shovel to be able to feed on two sides; for drag-line scrapers the material should be placed long and deep.

The burden is the thickness of material between the first row of holes and the quarry face (see *Figure 13.2*). The single hole is designed according to the bench height (H). The maximum burden, to suit the explosives, should be not more than about one-third of the bench height. The relationship between the burden (B) and the weight of explosive (W) is $W = KB^3$ or $W = KH^3/27$. The object is to shatter all the rock in the prism from the charge hole to the quarry face down to the quarry floor and about $B/2$ either side of the hole and to displace the rock so that it is accessible to the shovel but does not spread too far out for economic working.

The diameter of the hole is determined by the requirements that the explosive charge (a) have a diameter above the minimum to ensure reaction propagation in the explosive; (b) be of a diameter large enough to permit the required volume of explosive and the required primer and stemming (to seal the hole) to be placed in a hole extending downwards from the ledge in a direction parallel with the quarry face but not to a greater depth than that corresponding to a vertical distance below the level of the quarry floor of more than the equivalent distance of 10% of the bench height (see *Figure 13.2*); and (c) be of a standard charge diameter which fits into the hole without much space (decoupling) around it, because decoupling as expressed by the ratio of diameter of hole to diameter of charge, causes the amplitude of the effect of the charge to fall off inversely as an exponent of the decoupling of about 1.5.

The layout of the holes relative to one another and the quarry ledge is influenced by (a) the method of operation to be employed, the object being a minimum of all kinds of useless movement, and (b) the nature of the rock, e.g. position of jointing or stratification.

The material removed by one round of charges is called a cut. The initial cut into a face is called a box cut (see *Figure 13.2*), while subsequent cuts into that face which are continuous with the initial box cut are called corner cuts.

Materials have different densities when located in the rock from what they have when they are separated from the rock face. The usual measurement for this difference is called the swell factor (S_f), which is given by

$$S_f = \frac{\text{density in free condition}}{\text{density in the rock}}$$

This factor varies with the rock, but may be in the range 0.7–1.0.

The spread of fractured material on the floor from the original face, assuming simple cataract unprojected fall, is given approximately by the expression[7]

$$c = (\tfrac{1}{2}S_f^x)\cdot 2y\cdot(1-S_f^x)+H$$

in which c is the forward distance in metres of the heap from the original face; S_f is the swell factor of the material; y is the distance of the farthest row of charges from the open face; H is the bench height; and x is an exponent equal to $\frac{1}{2}$ for box cuts and $\frac{1}{3}$ for corner cuts. From this expression is determined the depth to which the charges should be placed to give the spread acceptable to the shovel.

If the firing procedure is to be such as to allow sufficient time between adjacent charges individually and successively to complete their fullest possible effect (delayed or relay firing), then the spacing between charges should equal the burden both parallel and perpendicular to the face; moreover the arrangement should be square or chequer form. If simultaneous firing is to be used, the spacing between charges parallel to the face should be approximately double the burden and the spacing between the adjacent parallel rows should equal the burden but the charges should be placed in diamond form (see *Figure 13.3*).

For every quarrying operation there is a maximum size of rock which the machinery can handle, and all rock must be blasted down to this size. This maximum size increases as the quarry throughput increases, because the maximum feed size of a primary crusher is a function of its throughput, i.e. a large-capacity machine has a larger feed aperture. This means that there is a tendency for the smaller quarry to employ more secondary blasting, because a crusher large enough to handle the largest products of primary blasting would be idle most of its time.

Stemming is the name given both to the operation of sealing the hole after the explosives have been put in place and to the material used to effect the seal. The material is usually composed of powdered rock and is tamped into the hole. Its depth should be such that it offers about the same resistance to the explosive as the burden; in practice the depth of stemming is about equal to the burden.

The method of hole drilling or rock penetration to be used is determined by the kind of rock. A broad classification of rock penetration methods is given in *Table 13.2.* The principal methods employ one of the two main

Table 13.2. METHODS AND APPLICATION OF ROCK PENETRATION

Method of penetration	*Application: Rock type (Mohs No.)* 0–2	3–4	5–6	7
Rotary drag drilling (blade, saw)	×	×		
Rotary roller drilling	×	×	×	×
Percussion		×	×	×
Thermal piercing (jet piercing)			×	×
Fluid				
Sonic				
Chemical dissolution	According to chemical reaction with reagent			
Electric arc				
Light laser	×	×	×	×
Nuclear methods	×	×	×	×

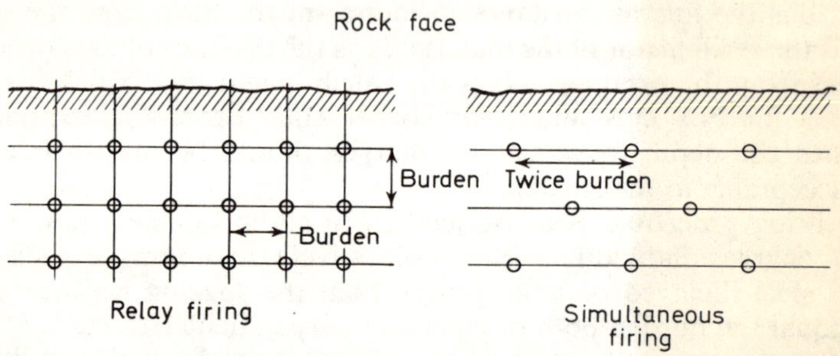

Figure 13.3. Arrangement of holes for charges

types of rotary drag drill (blade and roller), the percussion drill or the jet-piercing drill, which burns fuel oil and oxygen. The drag drill carries a bit in which is set removable cutting teeth of tungsten carbide or a bit with two or more rolling cutters set at an angle to the axis of the drill. The percussion drill is pulsated by compressed air and carries a bit which may be studded or cruciform chisel-ended. Jet piercing, which becomes the most economical method for very hard rocks, depends upon the tendency of rock to fracture because of thermal shock or the spallability of the rock (see p. 272). A typical jet-piercing burner consists of a mixing chamber into which the fuel oil and oxygen are fed (see *Figure 13.4*); from the chamber issue two flame nozzles, one of which is directed downwards and the other to the side; the burner rotates and is water-cooled. The temperature attained by the burner is about 2500°C, and the flame velocities are of the order of 1500 m/s.

EDISON[3] thought that the expense of the size reduction of rocks by the use

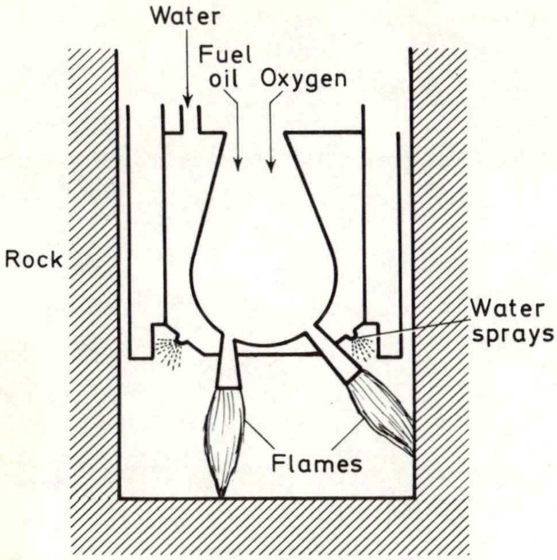

Figure 13.4. Jet piercing burner

of explosives could be decreased by the development of large crushers; but explosives have been suggested even for secondary size reduction. One method[4] proposes to have a series of steel cylinders arranged in the form of a column, by connecting them with flexible pipes loosely bolted to the flanges of the cylinders. To various positions in the inside of the column are fed explosives. Ore to be crushed is air-lifted with water up the cylinders; the explosives are periodically fired electrically. Crushed material is said to overflow the tops of the cylinders.

REFERENCES

1. IOHN, P., *Aufbereitungstechnik*, **1** (1967), 12
2. LOWRY, T. M. and MCHATTON, L. P., *Trans. Faraday Soc.*, **17** (1922), 82
3. MCGREW, B., *Rock Prod.*, June (1950) onwards
4. U.S. Patent 3 207 447
5. CARLEY-MACAULY, K. W., HITCHON, J. W. and MAROUDAS, N. G., *Trans. Instn Chem. Eng.*, **44** (1966), T395
6. YIGIT, E., JOHNSTON, H. A. and MAROUDAS, N. G., *Trans. Instn Chem. Eng.*, **47** (1969), T332
7. ASH, R. L., in *Surface Mining* (ed. E. P. PFLEIDER), 1968. New York; Am. Inst. Min., Metall. Petrol. Eng.

Index

Author index

Adams, J. T., 66
Agrawal, P. K., 99
Aljawdin, 92
Allis-Chalmers Ltd, 224, 240, 254
Anderson, J., 98
Andreasen, A. H. M., 4, 12, 66, 70, 98, 118, 128
Arbiter, N., 88, 92, 99
Ash, R. L., 279
Austin, L. G., 10, 13, 77, 98
Axelson, J., 50, 66

Badger, G., 190
Batel, W., 224
Batra, N. K., 91, 99
Baumann, V. A., 140, 190
Beaseley, E. S. G., 224
Beilby, G., 39, 47
Beke, B., 99, 246
Bellwinkel, A., 235, 254
Bendicks, C., 47
Bennett, J. G., 66, 73, 98
Bhrany, U. N., 88, 92, 99
Bird, A. P., 260
Bird, W. T., 260
Biswas, A. K., 91, 99
Blanc, F. C., 79, 98
Blythe, G. E. K., 264, 269
Bombled, J. P., 240, 254
Bond, F. C., 47, 52, 61, 66, 85, 98, 233, 235, 254
Bonwetsch, A., 144, 190
Born, M., 66, 81, 98
Bostock, W., 13
Bowden, F. P., 41, 47, 88, 91, 99
Bowdish, F. W., 92, 99
Bowes, E. A., 13, 50, 66, 69, 81, 98
Bredig, J., 99
Broadbent, S. R., 75, 95, 98
Brown, R. L., 66, 73, 95, 98, 99
Brunauer, S., 5
Brunel, M. I., 260
Burns, J. H., 99

Caldecott, W. A., 224
Callcott, T. G., 75, 95, 98
Carey, W. F., 57, 66, 159, 246, 248, 254
Carley-Macauly, K. W., 274, 279
Clarke, F. J. P., 30, 47
Clelland, D. W., 99
Clews, F. H., 224
Coghill, W. H., 99, 128, 240, 254
Cooper, B. S., 82, 98
Coulson, J. M., 184, 190
Cristelow, J. W., 13
Croft, P. E., 48
Crone, H. G., 73, 98
Cross, A. H. B., 174, 176, 190

Davidge, R. W., 47
Davis, C. W., 13
Davis, E. W., 243, 254
Del Fa, C., 102, 113
Devaney, F. D., 99, 128, 240, 254
Djingheusian, L. E., 33, 47
Dobie, W. B., 66

Edison, T. A., 157, 278
Elam, C. F., 27
Emett, P. H., 5
Engelhardt, W. von., 114
English, J. E., 145, 190
Epstein, B., 75, 98
Evans, I., 98
Ewald, P. P., 27
Exner, F., 98

Fahrenwald, A. W., 35, 47
Fairs, G. L., 51, 66, 79, 98

Farrant, J. C., 66, 128, 204, 224
Ferrara, G., 102, 113
Forty, A. J., 45, 48
Forwood, C. T., 45, 48
Friederich, E., 38, 47
Frisch, B., 99

Gardner, R. G., 10, 13, 77, 98
Garside, J., 78, 98, 198, 224
Gaudin, A. M., 70, 80, 98, 99
Gauldie, K., 141, 142, 153, 155, 190
Ghigi, G., 109, 114
Gieskieng, D. H., 140, 190
Gilbert, L. A., 109, 114
Gilliland, E. K., 66
Gilvarry, J. J., 73, 98
Goldsmith, J. R., 99
Gottschalk, V. H., 13
Greer, M. R., 48
Gregg, S. J., 5, 13, 85, 90, 99
Griffith, A. A., 24, 47
Gross, J., 66
Grunder, W., 224

Hardinge, H., 236, 250, 254
Hardinge, H. W., 249, 254
Harris, C. C., 92, 99
Hauge, T., 99
Heaver, A. A., 90, 99
Heddle, M. F., 36, 47
Hendry, R., 78, 95, 98, 265, 269
Herkenhopf, E., 35, 47
Hersam, E. A., 141, 190
Heywood, H., 13, 35, 47, 61, 66, 68, 71, 83, 98
Hickman, J., 85
Higi, G. G., 13
Hitchon, J. W., 274, 279
Holmes, J. A., 54, 66, 101, 102, 113
Hotchkiss, G. W., 260
Hughes, T. H., 109, 114
Hukki, R. T., 54, 66, 232–234, 242, 244, 253, 254
Hurst, W., 7, 13

Inglis, C. E., 24, 47
Iohn, P., 271, 279

Jamieson, J. C., 99
Jenson, I. H., 66
Joffe, A., 31, 32, 47
Johnson, J. F., 13, 66
Johnston, H. A., 274, 279
Joisel, A., 22, 47, 117, 128, 145, 157, 190, 197, 224, 246, 248, 252, 254

Kaiser, F., 257, 260
Kanowitz, S. B., 128
Kapur, P. T., 99
Kelleher, J., 128
Kempis, E. R., 99
Kick, F., 51, 66
Kihlstedt, P. G., 8, 13
King, R. F., 31, 47
Koehler, W. A., 190
Kurth, A., 38, 47
Kusnezow, W. D., 66, 81, 98
Kwong, J. N. S., 50, 66

Larin, A. P., 190
Lea, F. M., 5, 13
Lewis, W. K., 66
Lockwood, H. C., 12
Lowry, T. M., 279

Machlin, E. S., 31, 47
Mackenzie, R., 99
Makarov, V. I., 190
Mallikarjunan, R., 114
Manning, A. B., 98
Maroudas, N. G., 274, 279
Marsh, D. W., 38, 47
Marshall, D. F., 174, 176, 190
Marshall, V. C., 208, 224
Martin, G., 5, 13, 50, 66, 69, 81, 98
McAdams, W. H., 66
McCance, A., 38, 47
McGrath, L., 90, 99
McGrew, B., 139, 155, 190, 279
McHatton, L. P., 279
McKenzie, E. C., 190
Messenger, J. F., 145, 190
Miller, W. T. W., 190
Milne, A. A., 99
Mori, Y., 264, 266, 269
Mott, N. F., 30
Moyes, A. J., 196, 202, 203, 224
Mullin, J. W., 4, 12
Murray, G. T., 31, 47

Nadai, A., 34
Naeser, G., 99
Newton, J., 35, 47
Nijman, J., 246, 254
North, R., 105, 113
Nurse, R. W., 5, 13

Orowan, E., 48

Papadakis, M., 245, 246, 254

Patching, S. W. F., 101, 102, 113
Piret, E. L., 38, 47, 66
Podmore, H. L., 224
Prescott, T. W. L., 74, 98
Preston, F. W., 47
Price, J. D., 48
Pryor, E. J., 83, 98
Puffe, E., 224

Rabinowicz, E., 39, 48
Rabottino, L., 13, 109, 114
Ramanujam, M., 266, 269
Rammler, E., 71, 92, 98
Randall, J. T., 82, 98
Rehbinder, P. A., 108, 113
Richardson, J. F., 184, 190
Richie, P. D., 99
Rittinger, P. R. von, 50, 66
Rooksby, H. P., 82, 98
Rose, H. E., 99, 145, 190, 211, 217, 218, 220, 222–224, 235, 241–244, 247, 248, 254
Rosin, P., 71, 92, 98
Rowland, C. A., 254
Rumf, H., 1, 2, 31, 47, 106, 113

Sargant, R. J., 174, 176, 190
Schellinger, A. K., 38, 47
Scholz, W., 99
Schreiner, L. A., 108, 113
Schulz, N. F., 33, 47
Schuhman, R., 70, 98
Selim, A. A., 114
Siebel, J., 114
Sitwell, G., 157
Slegten, J. A., 246, 254
Smalley, I. J., 90, 99
Smeaton, J., 157
Smekal, A., 82, 88, 89, 98
Smith, E. A., 99, 218, 219, 224
Smith, R. W., 99
Speciale, P., 157
Spence, G., 7, 13
Stadtler, H., 57, 66, 69, 98
Stairmand, C. J., 57, 66, 246, 248, 254
Steinherz, A. R., 85, 87, 98
Stern, O., 66, 81, 98
Sullivan, R. M. E., 217, 218, 220, 222–224, 241–243, 247, 254
Szantho, E. von, 114

Tabor, D., 31, 41, 47, 48, 88, 91, 98
Taggart, A. F., 113, 138, 157, 169, 190
Talbot, J. H., 99
Tanaka, T., 40, 47, 101, 102, 113
Tanner, H. G., 269
Taplin, T. J., 51, 66
Tappin, G., 47
Taylor, J., 74, 98
Teller, E., 5
Temperley, H. N. V., 264, 269
Thomson, W. (Lord Kelvin), 98
Trzebiatowski, W., 88, 89
Turner, F. B., 50, 66, 69, 98
Tylor, E. B., 1, 2

Vegard, L.,
Venkateswarlu, D., 266, 269
Verney, E. J. W., 66

Walker, W. H., 66
Wang, J. T., 61, 66
Webb, F. C., 74, 98
Wildsmith, J. A., 78, 98, 198, 224
Williams, J. G., 31, 47
Woodhead, P., 196, 202, 203, 224

Yagit, E., 274, 279
Yancey, H. F., 48

Zeisel, H. G., 114
Zhigach, K. F., 108, 113
Zimmerley, S. R., 66
Zwicky, F., 47, 82, 89, 98

Subject index

Abrasion, 41
  methods of measuring, 42
Abrasive grains, separation of, 29
Additives, to aid grinding, 108
Adsorption method of measuring surface area, 5
Aluminium oxide, 91
Apron feeder, 125
Asbestos, 257, 272, 274
Attriting machines, 261ff
Autogenous tumbling mills, 249ff

Babcock and Wilcox ring ball mill, 186
Ball loaded mills, 234ff
  advantages and disadvantages of, 236
  ball density, 241
  ball diameter, 240
  ball hardness and wear, 241
  ball loading in, 243
  ball shape and composition, 239
  discharge of product, 238
  feeding, 238

Ball loaded mills *continued*
lining, 237
shell, 236
Barytes, 271
Basalt, 53
Bauxite, 53
Bearings, for rotary impactors, 203
Beilby layer, 39
BET method of measuring surface area, 5
Blake type jaw crusher, 133
Blakedene rock, 112
Bond method of measuring abrasion, 44
Bond methods of measuring grindability, 58
Bond work indices, table of, 53
Bond's energy equation, 52
relationship with Kihlstedt's equation, 56
Bond's particle shape, 85
Breakage and selection functions, 75
Breaker plates, in roll crushers, 162
Breaking, conditions for, 23
Brinell hardness, 36
Brittle body, definition of, 18
Buhrstone, 261

Calcite, 39, 89, 91
Capacity, of gyratory crushers, 155
of jaw crushers, 141
of ring roll and ring ball mills, 188
of rod loaded mills, 234
of roll crushers, 167
Carbon black, 110
Celestine, 272
Cement clinker, 53, 85, 109, 110
Chemical methods of comminution, 274
Chemical reaction during grinding, 90
Chilean mill, 173
Chocolate, 4
Choke feeding, effect on energy consumption, 65
Classification of comminution equipment, 115
Clay, 271
Closed-circuit calculations, 94
Closed-circuit milling, 93
Coal, 53, 78, 272
Coke, 53, 228
Colloid mill, 262
Compression, mechanism of, 32
Concaves, for gyratory crushers, 150
Concentration of stress, 24
Cone crushers, 148
Continued comminution, effect of, 87
Control of mills, 126
Corundum, 53
Cracks, condition for self-propagation of, 30
healing of, 31
propagation in amorphous and crystalline materials, 30
Cracks, speed of propagation of, 30
Critical speed of rotation, of tumbling mills, 231
Crocoite, 272
Cutting machines, 255ff
Cylindrical vibration mills, 216ff

Diamond, 26, 35, 272
Dimensional analysis, use of, in predicting mill performance, 118
Disc crushers, 170ff
Dissolution method of measuring surface area, 5
Dolomite, 53, 102, 110, 174
Ductility, 44

Eccentricity, in gyratory crushers, 151
Electrical methods, of comminution, 273
Electrohydraulic methods, of comminution, 273
Elutriation method, of measuring particle size, 5
Energy, in comminution, 49, 60, 65
Equipment, classification of, 115
Explosion, during comminution, 113
Explosive methods, of comminution, 275

Feed, size of, for roll crushers, 164
Feeding equipment, 124
Feldspar, 53
Fineness, degree of, 79
Flint, 53, 274
Fluid-energy mill, 263ff
Fluorspar, 53, 112, 272
Foodstuffs, 33
Force, effect of rate of application of, on breaking, 34
Friction, internal, 21

Gaudin's size distribution law, 70
Geometry of broken particles, 81
Giants Causeway, Northern Ireland, 84
Glass, 39, 53, 85
strengthening of, by superficial ion exchange, 31
Glue, 268
Grain size, natural, 79
Granite, 53, 274
Granular forms, behaviour of, 29
Graphite, 81, 85, 89, 109, 110
Grindability, 57
of components in mixture, 102
Grinding aids, 108
Grinding path on rotary hammer mill, 195
Gritstone, 202, 205

Grizzly screen, 126
Gypsum, 53, 109, 110, 188
Gyratory crushers, 146

Hammer mills, 193ff
Hardgrove test for grindability, 58
Hardness, 35
  relative effect of, on grinding of mixtures, 39
Hardness ratio of components in mixtures, effect of, on relative rate of grinding, 102
Hardness vectors, for single crystals, 82
Hematite, 53
Herzian fracture, 41
Heywood–Pryor particle shape, 85
Hooke's Law, 16

Ice, 31
Impacting machines, 191ff

Jaw angle, 140
Jaw crushers, 133ff
  comparison of single and double-toggle machines, 136

Kaolin, 5, 81, 89
Kick's energy law, 51
Kinetics of grinding, 91
Knives, 257

Lea and Nurse method of measuring surface area, 5
Lead nitrate, 272
Lead ore, 53
Lead sulphide, 31
Leather, 257
Limestone, 39, 53, 78, 84, 101, 109, 110
Linear expansion and strain compared, 28
Loesche mill, 185
Lopulco mill, 185

Macerator, for waste, 195
Magnesite, 53, 110, 272
Magnesium hydroxide, 89
Magnesium oxide, 25
Magnetic coercivity method of measuring surface area, 6
Magnetite, 53
Maintenance, 128
Malartic, 53
Marble, 53, 89, 109, 260
Mechanics of comminution equipment, 117
Mica, 90, 110, 273, 274
Microscopy method, of measuring particle size, 5
Mixtures, 101
Mohr's circle, 20
Mohs' scale of hardness, 36
Molybdenum sulphide, 89
Morgan Marshall test for abrasion, 42

Nip angle factor, for jaw crusher capacity, 141
Nipping machines, 129ff
Nylon, 33

Open-circuit milling, 93

Packing density, 75
Pan mills, 173ff
  ploughs and scrapers, 178
  rollers, 177
Paper, 259
Particle size, 4, 10
Particles, broken geometry of, 81
Pastiness, in wet grinding, 103
Pebble mills, 235
Permeability method, of measuring surface area, 5
Petit pulveriser, 262
Phosphate rock, 53, 190
Pin mills, 205ff
Plant layout, 123
Plastics, 257
Poisson's ratio, 20
Polyethylene, 31
Polyurethanes, 268
Pores, effect of, on Young's modulus, 25
Potash, 53, 272
Potassium dichromate, 102
Power consumption
  of ball loaded mills, 248
  of jaw crushers, 144
  of rod loaded mills, 234
  of roll crushers, 167
Pressure alteration, as method of comminution, 272
Product size and rotor speed in impact mills, 198
Product size from roll crushers, 165
Proton beam energy loss method, of measuring surface area, 6
PTFE, 268
Pumice, 110
Pyrites, 39, 53
Pyrrhotine, 274

Quartz, 22, 39, 53, 58, 85, 90, 91, 101, 109, 110, 205

Rate of application of force, effect of, 34
Rate of grinding, in ball loaded mills, 246
Raymond mill, 183
Reduction ratio, 10
Ring ball mills, 181ff
Ring roll mills, 181ff
Rittinger's energy law, 51
Rockwell hardness, 36
Rod loaded mills, 226ff
  advantages of, 228
  capacity of, 234
  effect of moisture on performance of, 228
  feed for, 228
  method of rating, 228
  power consumption in, 234
  product from, 229
  rate of rotation, 231
  rolls, 160, 162, 166
  shell, 230
Rods, for rod loaded tumbling mills, 233
Roll crushers, 157ff
  advantages and disadvantages of, 159
Rolls, in roll crushers, force between, 167
Rosin and Rammler's size distribution law, 71
Ross feeder, 125
Rotary hammer machines, 193ff
Rotation, speed of, for ball loaded mills, 244
Rubber, 257, 268
Rutile ore, 53

Salt, 31, 39, 272
Sand grinder, 266
Sandstone, 53
Saws, 260
Scalping, 92
Scoop feeder, 127
Sedimentation method, of measuring particle size, 4
Selection functions, 75
Shape, definition of, 85
Shears, 257
Shore hardness, 37
Sieving method, of measuring particle size, 4
Silica, 53, 102
Silica grains, separation of, 29
Silicon carbide, 53
Silicone rubber, 34
Single crystals, behaviour of, 25
Size, 4, 67ff, 75, 87
Size control, during milling, 92
Size distribution, 71, 102
  quantification of, 69
  representation of, 10
Slag, 53
Slate, 53
Soya flour, 268
Spallability, 272
Stamp mills, 209ff
Steinherz particle shape, 85
Strain and linear expansion compared, 28
Stress concentration, 24
Sugar, 268
Sulphur, 110
Superplasticity, 46
Surface energy and hardness, relationship between, 38
Surging, in ball loaded mills, 244
Sylvinite, 272

Talc, 46
Telsmith–Wheeling jaw crusher, 134
Temperature, effect of, on breaking, 33
Tetrakaidecahedron, as stable solid shape, 83
Thermal size reduction, 271
Torus vibration mill, 213ff
Toughness, 44
Tumbling machines, 225ff
  critical speed in, 231

Ultra-rotor, 267ff
Ultrasonics, 272
Universal jaw crusher, 133

Vermiculite, 110, 272
Vibration mills, 211ff
  power requirements of, 223
  rate of grinding of, 222
Vickers hardness, 37

Walking mechanisms, for rotary impactors, 203
Wear, in tumbling mills, 234, 241
Weathering, 271
Wedge gauge method, of measuring particle size, 4
Wet comminution, 103

Yancy, Geer and Price method, for measuring abrasion, 43
Yield stress and hardness, relationship between, 38
Young's modulus, 16
  effect on, of pores, 25

Zinc blende, 109, 271
Zinc ore, 53
Zircon, 53